"十三五"江苏省高等学校重点教材(编号：2020-2-077)

21世纪应用型高等院校示范性实验教材

大学物理实验

（双语）

主　编　戴　俊　王　颖　仲志强

副主编　李永壮　高国军　厉淑贞　荆庆丽

扫码加入学习圈　轻松解决重难点

南京大学出版社

图书在版编目(CIP)数据

大学物理实验 ：汉文、英文 / 戴俊，王颖，仲志强主编. -- 南京 ：南京大学出版社，2021.8(2024.1 重印)
ISBN 978-7-305-24307-3

Ⅰ. ①大… Ⅱ. ①戴… ②王… ③仲… Ⅲ. ①物理学-实验-双语教学-高等学校-教材 Ⅳ. ①O4-33

中国版本图书馆 CIP 数据核字(2021)第 051130 号

出版发行 南京大学出版社
社　　址 南京市汉口路 22 号　　　邮　编 210093

书　　名 大学物理实验(双语)
DAXUE WULI SHIYAN (SHUANGYU)
主　　编 戴 俊 王 颖 仲志强
责任编辑 吴 华　　　编辑热线 025-83596997

照　　排 南京南琳图文制作有限公司
印　　刷 常州市武进第三印刷有限公司
开　　本 787×1092 1/16 印张 24 字数 569 千
版　　次 2021 年 8 月第 1 版 2024 年 1 月第 4 次印刷
ISBN 978-7-305-24307-3
定　　价 59.80 元

网址：http://www.njupco.com
官方微博：http://weibo.com/njupco
微信公众号：njupress
销售咨询热线：(025) 83594756

☞ 扫码学生可免费观看本书部分实验视频

Preface

University Physics Experiments is an independent course taken as subsidiary to the lecture portion of *University Physics*. This course aims to train undergraduates of non-physics majors to master important laws of physics and key physical instrument operation techniques in the field of Engineering Technology, emphasizing cross discipline applications between physics experiments and Engineering topics. This textbook can be used by Universities or Colleges for overseas undergraduate students in physics experiment course teaching or domestic undergraduate general physics experiment course with bilingual teaching.

目 录 / Contents

第1章 绪 论

1.1 物理实验的地位和作用

物理学是人类认识自然界的基础，它揭示和阐述了物质世界基本构成及其运动和相互作用的基本规律。物理学也是一门建立在实验基础上的科学，在物理学的发展过程中，实验起着决定性的作用。

首先，任何物理概念的确立、物理规律的发现、物理理论的建立都必须是以严格的科学实验为基础。科学实验是人们认识自然和改造客观世界的基本手段，科学技术越进步，科学实验就显得越重要。任何一种新技术、新材料、新工艺、新产品都必须通过实验才能获得。由实验观察到的现象和测得的数据，加以总结和抽象，找出其内在的联系和规律就得到理论，实验是理论的源泉。牛顿总结了伽利略、开普勒等前人的实验研究成果，创建了经典力学理论；麦克斯韦总结了库仑、奥斯特、安培和法拉第等人的实验研究成果，创建了电磁场理论。

其次，理论一旦提出，则必须借助实验加以检验理论的正确性，是否具有普遍意义。麦克斯韦提出的电磁场理论(预言电磁波的存在)是在赫兹成功完成电火花实验后才被人们公认；近代物理的发展是在某些实验的基础上提出的假设。例如，普朗克根据黑体辐射提出“能量子”假设，再经过大量的实验证实，“能量子”假设才成为科学理论；爱因斯坦的狭义相对论是以“光速不变”假设为基础的，在经过长期大量的实验后，才普遍被人们接受。

物理学上的任何理论都必须由实验验证，正确的理论得到发展，错误的理论就被摒弃。当旧理论不能解释新的物理现象时，就会诞生新理论。在物理学发展的进程中，物理实验和物理理论始终是相互促进，相互制约，相得益彰的。没有理论指导的实验是盲目的，实验必须经过总结抽象上升为理论，才有其存在的价值，而理论又需要依赖于实验来检验，这种需要又促进实验的发展。

物理实验不仅对于物理学的研究极其重要，而且对于物理学在其他学科中的应用十分重要。20世纪以来，物理学的发展以及在各行各业中的应用使我们的世界发生了巨大变化，由物理学研究带来的新技术、新产品层出不穷，极大地改变了人们的生活方式和生产方式，如无线电广播、电视、电话、手机、汽车、飞机、火箭、卫星、计算机、因特网、微波炉、电磁灶、摄像机、DVD、液晶电视等等。物理学的发展带动了化学、天文、生物、医学、能源、材料、信息等学科的发展，为其他学科提供了强有力的研究方法和探测手段。如X射线应用于生物大分子的结构研究,发现了DNA的双螺旋结构;CT、核磁共振、超声、微波、激光、光纤成像被广泛应用于医学和生物学研究;同步辐射强光源应用于研究材料的结构和性能、化学反应过程、生物细胞活动;快中子束和离子束、微光机电器件、扫描显微镜应用于纳米材料的研究。这些应用无不和物理实验有关,实验正是连接物理学发现和其他学科的桥梁。物理实验是研究物理测量方法与实验方法的科学,是其他一切实验的基础,其他学科的复杂实验经过分解,其中绝大部分是常见的物理实验。在工程技术领域中,产品的研制、生产、加工、运输等都普遍涉及物理量的测量及物理运动状态的控制,这正是成熟的物理实验的推广和应用。现代高科技发展,设计思想、方法和技术也来源于物理实验。因此,物理实验是自然科学、工程技术和高科技发展的基础,科学技术的发展离不开物理实验。

物理实验在其发展过程中,形成了一套特有的理论、方法和技术,它们是各类科学实验的基础。掌握好这套理论、方法和技术是很不容易的,需要由浅入深、由简到繁逐步学习、训练和提高。物理实验具有丰富的实验思想、方法、手段,同时能提供综合性很强的基本实验技能训练,体现了大多数科学实验的共性——科学实验的基础。因此,大学物理实验是理工科类院校学生进行实验训练的一门独立的必修基础课程,也是理工科学生进入大学后接受系统的实验方法和实验技能训练的开始。通过学习大学物理实验,学生初步了解科学实验的主要过程和基本方法。大学物理实验培养学生通过实验手段去观察、发现、分析和研究问题,最终解决问题的能力方面起着重要的作用,也为学生能够独立进行科学实验研究,设计实验方案,选择、使用仪器设备和提出新的实验课题,提升能力,为进一步学习后续的实验课程打下良好的基础。

1.2 物理实验课的目的和任务

一、物理实验课的目的

(1) 学生通过对实验现象的观察、分析和对物理量的测量,学习物理实验知识和设计思想,理解和掌握相关物理理论知识;

(2) 培养和提高学生的实验能力,包括查阅实验教材和资料,借助教材或仪器说明书,正确使用常用仪器,运用物理理论知识对实验现象进行初步分析判断,正确记录和处理实验数据,绘制实验曲线,说明实验结果,撰写合格的实验报告;

(3) 提高学生实验素养,包括严肃认真的工作态度、理论联系实际和实事求是的科学作风,主动研究的探索精神以及遵守纪律、团结协作和爱护公物的优良品德。

二、物理实验课的任务

通过对实验现象的观察、分析和对物理量的测量,掌握物理实验的基本知识、基本方法和基本技能,通过对物理原理的运用、物理实验方法的训练,加深对物理实验知识物理学原理的理解,培养学生的科学思维能力、创新能力和创新精神,物理实验课的任务具体可分为以下几个方面:

(1) 学习和掌握运用实验原理和方法研究某些物理现象,通过具体实验,得出某些结论。

(2) 初步培养学生的科学实验能力,即如何从实验目的出发,依据哪些原理,采用什么方法,选用哪种合适的仪器与设备,确定合理的实验程序获取准确的实验结果。

(3) 进行实验技能的基本训练,熟悉常用仪器的基本原理、结构、性能、调整操作、观测分析和排除故障等。

(4) 学习和掌握处理实验数据的方法,以及分析试验方法、测量仪器、周围环境、测量次数和操作技能对实验结果的影响。

(5) 通过实验,尤其是设计性实验,有意识地培养学生的创新意识和创新能力。

(6) 通过实验培养学生严肃认真、细致踏实、一丝不苟、实事求是的科学工作态度,克服困难、坚韧不拔、勇于进取的探索精神,以及团结协作的优良品质,为后续课程的学习打下良好的基础。

1.3 物理实验的基本程序

物理实验是学生在教师的指导下独立进行和完成的。每次实验学生必须主动、努力、自觉获取相关实验知识和实验技能,不仅仅是测出一些实验数据。如果还能进一步领悟实验中的物理思想和方法,那将会获得更大的受益。物理实验课程在教学环节上分为实验预习、实验操作和课后实验报告3个部分。

一、实验预习

预习是上好实验课的基础和前提,在规定的时间内高质量地完成实验任务,必须在实验前做好充分的预习,只有这样,才能正确分析实验中各种现象、掌握物理现象的本质,充分发挥主观能动性,自觉地、创造性地获取实验知识和锻炼实验技能。

预习时,应仔细阅读实验教材,明确实验目的,了解实验原理和方法。弄清实验的内容是什么,实验采用什么方法,依据什么原理,实验用到哪些实验仪器,实验的关键环节是什么。通过预习,应对实验有大致了解,在此基础上撰写预习报告。预习报告的主要内容包括:实验名称、实验目的、简要的实验原理、实验步骤、实验数据记录表格、作答预习思考题或给出不清楚的问题。其中实验目的、实验原理部分应自己组织语言,写出实验所依据的主要原理、公式,以及公式中各物理量的含义,画出原理图、电路图或光路图。实验数据记录表格中,直接测量量和间接测量量应在表中分开,避免混淆。

二、实验操作

进入实验室前,必须详细了解实验室的各项规章制度。这些规章制度是为保护人身和仪器设备的安全,维护正常的教学秩序而制订的,进入实验室应严格遵守。

实验操作包括仪器仪表的布局、线路的连接、仪器仪表的安装调试、实验现象的观察、故障分析和排除、数据测量和记录。做实验时,要做到大胆心细、严肃认真,多动手、勤动脑,遇到问题多思考。这样才能在实验中发现问题、分析问题并解决问题，提高动手实践能力和思维判断能力。实验操作一般按照如下程序进行：

实验前，教师会简要讲解实验内容，分析实验的难点及强调操作中的注意事项。学生应结合预习内容，认真听讲，并且对照实验仪器，了解仪器的使用方法。

在实验台上合理安排仪器仪表布局，连接线路，调试实验装置。在实验测量前，仪器和元件必须处于最合理的工作状态。在使用任何仪器前，必须了解和清楚注意事项或说明书；实验装置调节符合要求后方可进行实验操作、测试数据；在电学实验中，接好的线路必须检查无误才可接通电源；调节仪器时，应先粗调再微调；在读数时，应先取大量程后取小量程。

根据实验要求记录实验数据和相关实验现象，实验记录应该按照科学、实事求是的原则进行，应全面反映实验的全过程。除了记录直接的测量数据外，还应记录实验条件(如环境温度、湿度等)、仪器型号规格、仪器的准确度等级、仪器误差、实验中发现的问题。不仅要记录与预想一致的数据和现象，更要记录与预想不一致的数据和现象。记录应尽量详尽、清晰、条理清楚。实验过程中如发生反常现象或仪器故障，应冷静地分析可能出现的原因，恰当处理，排除故障，必要时可请实验老师帮助解决。

实验完毕数据交给老师审阅通过后，整理还原好实验仪器后，再离开实验室。

三、实验报告

实验后要及时进行数据处理，并写出完整的实验报告。实验报告是实验工作的总结，要求字体工整、文理通顺、数据齐全、图表规范、结论明确、布局整洁。一份完整的实验报告应包含以下内容：

(1) 实验名称、姓名、学号、实验日期。

(2) 实验目的。要求自己组织语言书写，简明扼要。

(3) 实验原理。写出简要的原理和实验依据的主要公式，以及公式中各物理量的含义，画出原理图、电路图或光路图。要求自己组织语言书写，简明扼要。

(4) 实验仪器。包括仪器的型号及准确度等级。

(5) 实验内容。包括实验内容和主要步骤。依据实验进行的具体情况，简要写出实验内容和步骤。

(6) 数据处理。该部分是实验报告的核心。首先将实验所测得的原始数据整理列表，再按照实验要求计算测量结果。如需作图则应使用坐标纸作图。计算时应有数据代入的简单过程，并按照不确定度理论评估测量结果。最后写出实验结果表达式。鼓励使用 Excel、Origin、Matlab 等计算机软件来进行处理、分析、作图、拟合及统计。

(7) 实验结论。根据测量结果给出实验结论、误差分析，以及对实验的建议和体会等。

(8) 思考题。

(9) 原始数据。实验过程中记录的实验数据。

以上各个部分中的(1)、(2)、(3)在预习阶段完成，(4)、(9)在实验操作阶段完成，其余的(5)、(6)、(7)、(8)在实验后完成。完成实验报告后，在规定的时间内交由教师批阅。

Chapter Ⅰ Introduction

1.1 The Motivation and Function of Physics Experiments

Physics is the basis of human understanding of nature. It reveals and expounds the basic constitution of the material world as well as the basic laws of its movement and interaction. Meanwhile, Physics is also a science based on experiments and experiments play a decisive role in the development of physics. First, the establishment of any physical concept, the discovery of physical laws and the establishment of physical theory must be based on strict scientific experiments. Scientific experiments are the basic means for people to understand nature and transform the objective world. The more science and technology progresses, the more important scientific experiments become. Any new technology, new material, new process and new product can only be obtained through experiments. Through summarizing and abstracting the phenomena and data observed in the experiment and uncovering internal relations and laws, we can acquire the theory and the experiment is the source of the theory. Newton summarized the experimental research results of Galileo, Kepler and others, establishing the classical mechanics theory; Maxwell summarized the experimental research results of Coulomb, Oersted, Ampere and Faraday, establishing the electromagnetic field theory. Secondly, once the theory is put forward, its accuracy must be tested with experiments. Maxwell's electromagnetic field theory (which predicted the existence of the electromagnetic wave) was recognized by people only after Hertz successfully completed the electric spark experiment; the development of modern physics is to suggest hypotheses on the basis of some experiments, such as when Planck put forward the "quantum hypothesis" based on black-body radiation theory, and after numerous experiments, plank's quantum hypothesis became a scientific theory. Notably, Einstein's special theory of relativity is based on the assumption that the speed of light is constant. It was only after a long period of experiments that this was generally accepted by people. Any theory in physics must be verified by experiments, if it is right, it will be developed and if it is wrong, it will be abandoned. When the old theory cannot explain a new physical phenomenon, a new theory will be born. In the process of the development of physics, physics experiments and physics theory always promote, restrict and complement each other. Specifically, experiments without theoretical guidance are blind. Experiments must be summarized and abstracted to become theories before they have their existence value, and it is necessary for theories to be tested by experiments, which promotes the development of experiments.

Physics experiments are very important not only for the research of physics, but also for the application of physics in other disciplines. Since the 20th century, significant changes have taken place in our world with the development of physics and its application in all walks of life. New technologies and new products created by physics research have emerged in an endless stream, greatly altering people's way of life and production. This can be seen in products such as radio broadcasting, televisions, telephones, mobile phones, automobiles, airplanes, rockets, satellites, computers, the internet, microwave ovens, electromagnetic range, cameras, DVDs, LCD TVs, etc. The development of physics promotes the development of chemistry, astronomy, biology, medicine, energy, materials, information as well as other disciplines, and provides powerful research methods and detection means for other disciplines. For example, X-ray is applied to the structural research of biological macromolecules, and the double helix structure of DNA has been discovered. Meanwhile, CT, NMR, ultrasound, microwave, laser, and optical fiber imaging are widely applied in medical and biological research. In other fields, strong synchrotron radiation light source is used to explore the structure and properties of materials, chemical reaction processes, and biological cell activity. Meanwhile, fast neutron beam and ion beam, micro-optical electromechanical devices, scanning electron microscopes and the application of microscopes are all used in the research of nano materials. These applications are all related to physics experiments, which act as the bridge between the discovery of physics and other subjects. Physical experiments are the science of studying physics measurement methods and experimental methods—it is the basis of all other experiments. The complex experiments of other disciplines are decomposed, and the majorities are common physics experiments. In the engineering technology field, the development, production, processing and transportation of products are generally related to the measurement of physical quantities and the control of the physical motion state, which is the promotion and application of mature physics experiments. Modern high-tech development, design ideas, methods and techniques also come from physics experiments. Consequently, physics experiments are the basis of the development of natural science, engineering technology and high technology, and the development of science and technology is inseparable from physics experiments.

In the process of its development, physics experiments have formed a set of its own theories, methods and techniques, which act as the basis of various scientific experiments. However, it is not easy to master this set of theories, methods and techniques and we need to learn, train and improve them step by step. Physics experiments are rich in experimental ideas, methods and means, and can provide comprehensive training of basic experimental skills. They embody the common characteristics and are the basis of most scientific experiments. Therefore, University physics experiments are an independent compulsory basic course for science and engineering college students to carry out experimental training. It is also the beginning of systematic experimental methods and experimental skills training for science and engineering college students after they enter university. It also enables students to initially understand the primary process and basic methods of scientific experiments. It plays an essential role in cultivating students' ability to observe, discover, analyze and study problems, and finally resolve problems through experimental means. It also lays a good foundation for students to independently implement scientific experimental research, design experimental schemes, select and use instruments and equipments, and put forward new experimental courses.

1.2 The Purpose and Task of Physics Experiments

1.2.1 The Purpose of Physics Experiments

1) Enable students to learn physics experiments knowledge and design ideas, understand and master relevant physics theories through observation and the analysis of experimental phenomena and measurement of physical quantities.

2) Cultivate and improve students' experimental ability. This includes consulting experimental teaching materials, using teaching materials or instrument instructions, correctly using common instruments, utilizing physics theory to preliminarily analyze and judge experimental phenomena, correctly recording and processing experimental data, drawing experimental curves, explaining experimental results, and writing qualified experimental reports.

3) Improve students' experimental literacy. This involves a serious work attitude, a scientific style of integrating theory with practice and seeking truth from facts, an exploratory spirit of active research, and a good moral character of abiding by discipline, unity and cooperation, and taking good care of public property.

1.2.2 The Task of Physics Experiments

Through the observation and analysis of experimental phenomena and the measurement of physical quantities, we can master the basic knowledge, basic methods as well as the basic skills of physics experiments. Through the application of physical principles and the training of physical experimental methods, we can deepen the understanding of physical principles of physical experiments, and cultivate students' scientific thinking ability, innovation ability and innovation spirit. It can be divided into the following aspects:

1) Learn. Master using experimental principles and methods to examine some physics phenomena and draw some conclusions through specific tests.

2) Cultivate students' ability of scientific experiment. Specifically, this involves knowing how to begin from the requirements of experimental purpose, according to which principle, through what method, choose the appropriate instruments and equipments, and determine reasonable experimental procedures to obtain accurate experimental results.

3) Basic training of experimental skills, familiar with the basic principle, structure, performance, adjustment operation, observation analysis and troubleshooting of common instruments.

4) Learn and master the methods of processing experimental data, and analyze the influence of experimental methods, measuring instruments, surrounding environment, measurement times and operation skills on experimental results.

5) Through experiments such as designed experiments, we can consciously cultivate innovative consciousness and ability.

6) Through the experiment, we can cultivate a serious, meticulous, down-to-earth, practical and realistic scientific working attitude. We can also overcome difficulties, practice perseverance, an enterprising spirit of exploration, and the excellent quality of unity and cooperation. This lays a good foundation for follow-up courses.

1.3 Basic Procedures for Physics Experiments

Physics experiments are independently implemented and completed by students under the guidance of teachers. Each experimental student must take the initiative in acquiring relevant experimental knowledge and skills, this not only involves measuring some experimental data. Notably, it will be more beneficial if one can further understand the physical methods of the experiment. The physics experiment course is divided into three parts: experimental preview, experimental operation and after-class experiment report.

1.3.1 Experimental Preview

Previewing is the basis and premise of a good experiment class. To complete the experiment task with high quality in the specified time, we must thoroughly prepare before the experiment. Only in this way can we accurately analyze all kinds of phenomena in the experiment and master the essence of physical phenomena. This is also in addition to giving full play to subjective initiative while consciously and creatively acquiring experimental knowledge and exercising experimental skills.

We should read the experimental materials carefully, clarify the purpose of the experiment, and understand the principle and method of the experiment. This specifically includes the contents of the experiment, the methods that are used in the experiment, according to what principle, the experimental instruments that are used in the experiment, and the key links of the experiment. Through previewing, we should obtain a general understanding of the experiment that is based on the preparation report. The main contents of the preview report include: the name of the experiment, the purpose of the experiment, the brief principle of the experiment, the steps of the experiment, the record form of the experimental data, the answer to the preview questions, or some unclear questions. Among them, the purpose and principle of the experiment should organize the language with the students' own words, detail the main principles, formulas as well as the meaning of each physical quantity in the formula while drawing a schematic diagram, circuit diagram or optical path diagram, etc. In the experimental data record table, the direct and indirect measurements are clearly separated from the table to avoid confusion.

1.3.2 Experimental Operation

Before entering the laboratory, we must obtain a detailed understanding of the rules and regulations of the laboratory. These rules and regulations are formulated to protect the safety of people, instruments and equipments and to maintain normal teaching anecdotes. Entry into the laboratory should be strictly observed.

The experimental operation includes the layout of the instrument, the connection of the line as well as the installation and debugging of the instrument. It also includes the observation of the experimental phenomenon, fault analysis and troubleshooting, the data measurement as well as the recording. When conducting experiments, we should be bold, careful, serious, more hands-on, and hard to think, encounter problems and think more. Only through such methods are we able to find, analyze and solve problems in experiments, and improve the ability of hands-on practice and thinking judgment. The experimental operation is generally implemented according to the following procedures:

1) Before the experiment, the teacher will make a brief introduction, analyze the

difficulties of the experiment and emphasize the points of attention in the operation. Students should listen carefully and understand the use of the instrument against the experimental instrument.

2) Arrange the instrument layout, connect the instruments and reasonably debug the experimental device on the test table. Before the experimental measurement, the instruments and components must be in the most reasonable working state. Before using any instrument, the precautions or instructions must be observed—such as whether the experimental adjustment of the experimental device meets the requirements, whether the experimental operation and test data can be carried out. In the electrical experiment, the connected line must be correctly reviewed before connecting to the power supply. When adjusting the instrument, it should be adjusted first and then fine-tuned. When reading, a large range should be used first and then a small range.

3) According to the requirements of the experiment, the experimental data and related experimental phenomena should be recorded, and the experimental records should be implemented according to the principle of science and seeking truth from facts, and the entire process of the experiment should be fully reflected. In addition to recording direct measurement data, experimental conditions (such as ambient temperature, humidity, etc.), instrument model specifications, accuracy grade of the instrument, instrument errors, problems identified in the experiment should also be recorded. The data and phenomena that are consistent with the expectation should be recorded along with the data and phenomena that are inconsistent with the expectations. Records should be as detailed, clear and organized as possible. In the event of abnormal phenomenon or instrument failure in the course of the experiment, we should calmly analyze the possible causes, properly deal with it, troubleshoot and ask the teacher to assist in solving them if necessary.

4) After the experiment, the data was reviewed by the teacher for approval, and the experimental instruments were sorted and restored.

1.3.3 Experimental Report

After the experiment, the data should be processed in a timely manner, and the complete experimental report should be written. The experimental report is a summary of the experimental work, which requires neat font, smooth arts and science, complete data, standard chart, a clear conclusion and an organized layout. A full experimental report should contain the following:

1) The name, reporter name, number, date of the experiment.

2) The purpose of the experiment. Here writing should be organized concisely and to the point.

3) Experimental principles. Write out the brief principle and the main formula

based on the experiment, and the meaning of each physical quantity in the formula, draw the schematic diagram, circuit diagram or the optical path diagram, etc. Organize the language writing in a concise manner.

4) Experimental instruments. Include the type and accuracy of the instrument.

5) The content of the experiment. The specific conditions of the experiment should be briefly written out. The language writing should be organized in a concise manner.

6) Data processing. This part is the core of the experimental report. The original data measured by the experiment should be arranged and the measurement results should be calculated according to the experimental requirements. If you need to draw, use the coordinate paper to do so. The calculation should involve a simple process of data substitution, and the measurement results should be evaluated according to the uncertainty theory. Finally, the expression of experimental results is written. The use of Excel, Origin, Matlab and other computer software is encouraged for processing, analysis, mapping, fitting and statistics.

7) Experimental conclusions. According to the measurement results, the experimental conclusions, error analysis, suggestions for the experiment experience are given.

8) Think about the experimental questions.

9) Raw data. Experimental data recorded during the experiment.

1), 2), 3) of the above steps are completed in the preview stage, 4) and 9) are completed in the experimental operation stage, and the remaining 5), 6), 7), 8) are completed after the experiment. After the completion of the experiment report, the teacher will review it within the specified time.

第 2 章 测量误差、不确定度及数据处理

2.1 测量的定义与基础知识

一、测量

1. 测量的定义

测量就是将待测物理量与选作计量标准的同类已知物理量进行比较，得出其倍数的过程。倍数值称为待测物理量的数值，选作计量标准的已知量称为单位。因此，一个物理量的测量值应由数值和单位两部分组成，两者缺一不可。

2. 测量结果

由测量所得到的赋予被测量的值。这里的“赋予”二字，指明测量结果不是“真值”（被测量的真实值），而只是真值的一个估计结果。对于直接测量来说，如果只做了单次测量，则此测量值可作为测量结果；如果对同一物理量做了多次重复性测量，得到多个测量值，则取其多次测量值的算术平均值作为测量结果。

二、物理量的单位

按照中华人民共和国计量法规定，物理量单位均是以国际单位制（SI）为基础的，其中长度、质量、时间、电流强度、热力学温度、物质的量和发光强度是基本单位，其他物理量的单位可由这些基本单位导出，故称为导出单位。

三、测量的分类

测量的分类标准很多，本教材只介绍按测量值的获取方法进行的分类：直接测量和间接测量。

1. 直接测量

无需对被测量与其他量的量值进行函数关系的辅助计算，而直接得到被测量量值的测量。如用卷尺测量桌子的长度、用电流表测量线路中的电流等，相应的长度与电流称为直接测量量。直接测量按测量次数又分为单次测量和多次测量。

单次测量，只测量一次的测量称为单次测量。主要用于测量精度要求不高、测量比较困

难或测量过程带来的误差远远大于仪器误差的测量中。如在测杨氏弹性模量实验中，测钢丝长度就用的是单次测量。

多次测量，测量次数超过一次的测量称为多次测量。多次测量按测量条件，主要分为等精度测量和非等精度测量。

2. 间接测量

已知被测量与直接测量法测得的其他量值之间的函数关系，通过计算间接得到被测量量值的测量称为间接测量。例如，在测量圆柱体的密度的实验中，先测量圆柱体的高度、直径和质量，再利用公式 $\rho=\frac{4m}{\pi d^2 h}$ 计算出圆柱体的密度。

2.2　误差与不确定度

一、误差

1. 误差的定义

测量误差是指测量值与被测量的真值之差。若用 x 表示测量结果，x_0 表示真值，则误差 δ 可以表示为

$$\delta=x-x_0 \tag{2-2-1}$$

误差 δ 反映了测量值偏离真值的大小。所有的被测量在特定的条件下，理论上都有一个对应的客观、实际值存在，我们称之为“理论真值”，简称真值。真值只是一个理想的概念，由于客观实际的局限性，真值是不可知的，我们通过测量只能得到物理量的近似真值，对测量误差的量值范围也只能给予估计，国际上规定用不确定度来表征测量误差可能出现的量值范围。式(2-2-1)定义的误差 δ 又称绝对误差。

为了比较两个不同的物理量的测量优劣，引入相对误差的概念，即

$$E=\frac{\delta}{x_0}\times 100\% \tag{2-2-2}$$

相对误差也称作百分误差，是一个不带单位的纯数。它既可以评价量值大小不同的同类物理量的测量，也可以评定不同类物理量的测量，判断不同测量的优劣。例如，测得某圆柱体的长度 $L=80.20$ mm，绝对误差 $\delta_L=0.08$ mm；直径 $d=20.055$ mm，绝对误差 $\delta_d=0.005$ mm；质量 $m=215.35$ g，绝对误差 $\delta_m=0.05$ g。用相对误差可以判断，可见质量的相对误差最小，长度的相对误差最大，因此质量的测量优于直径的测量，而直径的测量又优于长度的测量。

2. 误差的分类

在测量中，存在着诸多的测量误差，这些误差均由不同的因素造成的。由于误差成因不同，误差的特征也不同。研究误差的一个重要内容就是要掌握各种误差所具有的特征，只有

这样,才能获得正确的误差处理方法。按照误差的特点与性质,可将测量误差分为系统误差、随机误差和粗大误差三类。

(1) 系统误差

在同一条件(方法、仪器、人员及环境)下,多次测量同一量值,误差的符号保持不变;或当条件改变时,误差按一定规律变化,这样的误差称为系统误差。系统误差是由固定不变的或按特定规律变化的因素造成,这些产生误差的因素是可以掌握的。系统误差的主要来源于测量装置方面的因素、环境因素、测量理论或方法的因素、测量人员方面的因素等。

产生系统误差的原因可能是各不相同的,但是他们的共同特点是确定的变化规律,这也使误差的变化具有确定规律性。各系统误差的成因不同,表现出的规律也不同。因此可以根据其产生原因,采取一定的技术措施,如校准仪器、改进实验装置和实验方法,设法消除或减小;也可以在相同条件下对已知约定真值的标准器具进行多次重复测量的方法,或者通过多次变化条件下的重复测量的方法,设法找出其系统误差的规律后,对测量结果进行修正。

(2) 随机误差

随机误差又称偶然误差,是指在对同一被测物理量进行多次测量过程中,误差的绝对值及符号均以不可预知的方式变化,这样的误差称为随机误差。实验中,即使已经消除了系统误差,但在同一条件下对某物理量进行多次测量时,仍存在差异,误差时大时小,时正时负,呈无规则的起伏,这是因为存在随机误差的缘故。随机误差是由某些偶然的或不确定的因素所引起的。如实验者受到感官的限制,读数会有起伏;实验环境(温度、湿度、风、电源电压等)无规则的变化,或是测量对象自身的涨落等。这些因素的影响一般是微小的,混杂的,并且是无法排除的。

随机误差的最主要特征是具有随机性,在重复性测量条件下,对同一被测量进行多次重复测量,单次测量的随机误差的绝对值和符号以不可预测的方式变化,没有确定的规律。但像其他随机变量一样,对无限次测量,各次测量的随机误差服从统计规律。常见的分布规律是:

① 比真值大或比真值小的测量值出现的几率相等;

② 误差较小的数据比误差较大的数据出现的几率要大得多,同时绝对值很大的误差出现概率趋于零;

③ 在多次测量中绝对值相等的正误差或负误差出现的机会是相等的,全部可能的误差总和趋于零。

这是称作正态分布(高斯分布)的一种情况。对于正态分布的随机误差,尽可能进行多次测量,增加测量次数,可以有效地减小随机误差。随机误差也存在其他分布情况,如 t 分布、均匀分布等。

(3) 粗大误差

又称为粗差,它是由于实验者使用仪器的方法不正确,粗心大意读错、记错、算错测量数据或试验条件突变等原因造成的。含有粗大误差的测量值称为坏值或异常值,正确的结果中不应包含有过失错误。在实验测量中要极力避免过失错误,在数据处理中要尽量剔除坏值。

上述各种误差在一定条件下是可以相互转化的。对某项具体误差,在此条件为系统误差，在另一条件下可能表现为随机误差，反之亦然。例如一块电表，它的刻度误差在制造时可能是随机的，但用此电表来校准一批其他电表时，该电表的刻度误差就会造成被校准的这一批电表的系统误差。又如，由于电表刻度不准，用于测量某电源的电压时必然带来系统误差，但如果采用很多块同类电表测量该电压，由于每一块电表的刻度误差有大有小，有正有负，就使得这些测量误差具有随机性。

3. 随机误差的处理

在大学物理实验中，多次独立测量得到的数据随机误差一般近似看成正态分布。在相同条件下，对某物理量 x 作 n 次独立测量，得到测量列 $x_1, x_2, \cdots, x_n$，其算术平均值为

$$\bar{x} = \frac{x_1 + x_2 + \cdots + x_n}{n} = \frac{1}{n}\sum_{i=1}^{n} x_i \tag{2-2-3}$$

可以证明，当系统误差已被消除，测量值的算术平均值最接近被测量的真值。因此常用测量列的算术平均值 $\bar{x}$ 表示测量结果，即 $\bar{x}$ 为测量最佳值。

测量值的可靠程度常用标准偏差来估计。标准偏差小，说明多次测量数据分散程序小，测量的精密度就高。满足正态分布的测量列标准偏差由贝塞尔公式定义

$$\sigma_x = \sqrt{\frac{\sum_{i=1}^{n}(x_i - \bar{x})^2}{n-1}} \tag{2-2-4}$$

由正态分布的特性可知，对于任何一次测量，其结果落在$[\bar{x}-\sigma_x, \bar{x}+\sigma_x]$区间的概率为 0.683；落在$[\bar{x}-2\sigma_x, \bar{x}+2\sigma_x]$区间的概率为 0.954；落在$[\bar{x}-3\sigma_x, \bar{x}+3\sigma_x]$区间的概率为 0.997。上述概率称为置信概率，记为 P；对应的区间称为置信区间。由于测量值落在区间$[\bar{x}-3\sigma_x, \bar{x}+3\sigma_x]$以外的可能性很小，所以将 $3\sigma_x$ 称为极限误差，作为粗差的判据，当测量值与平均值 $\bar{x}$ 的差超过 $3\sigma_x$ 时为粗差，数据处理时应予剔除。在物理实验中，置信概率一般取 0.95。

在实际工作中，人们关心的往往不是测量列的数据散布特性，而是测量结果即算术平均值 $\bar{x}$ 对真值的离散程度。误差理论可以证明，平均值 $\bar{x}$ 的标准偏差为

$$\sigma_{\bar{x}} = \frac{\sigma_x}{\sqrt{n}} = \sqrt{\frac{\sum_{i=1}^{n}(x_i - \bar{x})^2}{n(n-1)}} \tag{2-2-5}$$

其意义为待测物理量的真值落在$[\bar{x}-\sigma_{\bar{x}}, \bar{x}+\sigma_{\bar{x}}]$区间的置信概率为 0.683，真值落在$[\bar{x}-2\sigma_{\bar{x}}, \bar{x}+2\sigma_{\bar{x}}]$区间的置信概率为 0.954，真值落在$[\bar{x}-3\sigma_{\bar{x}}, \bar{x}+3\sigma_{\bar{x}}]$区间的置信概率为 0.997。

式(2-2-5)是在测量次数 n 很大，数据分布可视为正态分布的条件求得的。在物理实验教学中，测量次数往往较少(一般 $n<10$)，在这种情况下，测量结果呈 t 分布。设在正态分布的情况下，置信概率 P 对应的置信区间为$[\bar{x}-\sigma, \bar{x}+\sigma]$；对于 t 分布，相同的置信概率 P 对应的置信区间为$[\bar{x}-t_P\sigma/\sqrt{n}, \bar{x}+t_P\sigma/\sqrt{n}]$。因子 t_P 与测量次数和置信概率有关，其值可通过查表得到。表 2-2-1 给出置信概率 $P=0.95$ 时，因子 $t_{0.95}$ 和 $t_{0.95}/\sqrt{n}$ 与不同测量次数

n 的关系。

表 2-2-1　不同测量次数 n 时 $t_{0.95}$ 和 $t_{0.95}/\sqrt{n}$ 的数值

n	3	4	5	6	7	8	9	10	15	20
$t_{0.95}$	4.30	3.18	2.78	2.57	2.45	2.36	2.31	2.26	2.14	2.09
$t_{0.95}/\sqrt{n}$	2.48	1.59	1.20	1.05	0.926	0.834	0.770	0.715	0.553	0.467

4. 测量精度

精度是测量结果与真值接近的程度,与误差的大小相对应。精度可分为:

(1) 准确度,指一组测量的最佳值偏离真值的程度,反映了测量系统误差的大小。

(2) 精密度,指一组测量数据本身的离散程度,反映了测量随机误差的大小。

(3) 精确度,指一组测量数据偏离真值的离散程度,反映了测量的系统误差和随机误差的综合影响。

在科学实验中,我们希望得到精确度高的结果,即准确度和精密度都高。而对于具体的测量,精密高的,准确度不一高;准确度高的,精密度也不一定高。有人认为:重复性好的测量就是准确度高的测量。其实并不一定,因为重复性好,仅指测量结果精密度高,即测量的随机误差小。而测量结果的好坏还取决于准确度的高低,即系统误差的大小。用同一台测量仪器对同一被测物理量进行多次重复测量,仅能确定其量值的分布性和重复性,而不能断定它的准确性。

二、不确定度

1. 不确定度的基本概念

测量的目的是得到被测量的真值,但由于客观实际的局限性,真值不是准确可知的。测量结果只能得到一个真值的近似估计值和一个用于表示近似程度的误差范围,导致测量结果不能定量给出,具有不确定性。为了确切表征实验测量结果,引入了不确定度的概念,作为实验测量结果接近真实情况的量度。不确定度表征了测量结果的分散性和测量值可依赖程度,也表明了真值出现的范围。在测量方法正确的前提下,不确定度越小,测量结果越可靠;不确定度越大,测量的可靠程度越差。

需要注意的是,测量结果不确定度与测量误差是两个不同的概念,两者既有区别又有联系,不能混淆,也不能相互取代。测量误差是被测量的真值与所引用的测量结果之差,由于真值是未知的,误差是无法精确计量的。测量的不确定度则表示由于测量误差的存在,被测量值不能确定的程度,它是未知的误差数值的可能范围,是一种估计值,用来表述一种客观存在的可能性。在实际工作中遇到的几乎都是不确定度的问题。

根据国际计量局(BIPM)于 1980 年建议并推广的用不确定度来表述测量结果可信赖程度的方法[(INC－1)1980],我国计量部门也相继制定了一系列相符的规范,其范围几乎覆盖了计量系统的各个领域,这些规范中一般都采用了不确定度表示体系。

本书根据实际教学和实验的需要介绍不确定度的一种简化方案——以标准偏差表示的测量不确定度估计值,即标准不确定度。

2. 标准不确定度的两类评定

测量不确定度来源于多个因素，因而它由多个分量组成。其中一些分量可由一系列在同一条件下多次测量的结果统计分析评定，称为 A 类不确定度，记为 u_A；另一些分量由非统计分析的方法评定，称为 B 类不确定度，记为 u_B。

(1) A 类不确定度 u_A

在相同的条件下，对某物理量 x 作 n 次独立测量，得到测量列 $x_1, x_2, \cdots, x_n$，其 A 类不确定度为

$$u_A = t_P \sigma_{\bar{x}} = \frac{t_P}{\sqrt{n}} \sigma_x \tag{2-2-6}$$

式中，$t_P/\sqrt{n}$ 的取值见表 2-2-1。当置信概率 $P=0.95$、$n=6$ 或 7 时，$t_{0.95}/\sqrt{n} \approx 1$，则 A 类不确定度可近似地直接取标准偏差 σ_x 的值，即

$$u_A = \sigma_x \tag{2-2-7}$$

(2) B 类不确定度

在进行不确定度 B 类评定时，应考虑到影响测量准确度的各种可能因素，因此 u_B 通常是由多项合成的，u_B 的估计也是测量不准确度估算中的难点。在大学物理实验中，我们采用了简化方案，产生 B 类不确定度的因素主要考虑仪器误差，u_B 可简化为用仪器的最大允许误差 $\Delta_{仪}$ 表示，即

$$u_B = \frac{\Delta_{仪}}{c} \tag{2-2-8}$$

式中，c 为称为置信因子，当取置信概率 $P=0.95$ 时，$c=1.05 \approx 1$。因此

$$u_B = \Delta_{仪} \tag{2-2-9}$$

所谓最大允许误差，是指对给定的测量仪器，其有关规范、规程允许的误差极限值，以 $\Delta_{仪}$ 表示。仪器的型号不同，其最大允许误差也不同。有些仪器可以通过查询国家计量检定规程而得到，如卡尺、千分尺、天平等。有些仪器可以在其铭牌和使用说明书中查到。还有些仪器，在铭牌上给出了准确度等级，它可以换算成 $\Delta_{仪}$。表 2-2-2 给出常用仪器的最大允许误差。

表 2-2-2　常用实验仪器的最大允许误差 $\Delta_{仪}$

仪器名称	量程	最小分度值	最大允许误差
钢板尺	150 mm 500 mm 1000 mm	1 mm 1 mm 1 mm	±0.10 mm ±0.15 mm ±0.20 mm
钢卷尺	1 m 2 m	1 mm 1 mm	±0.8 mm ±1.2 mm
游标卡尺	125 mm	0.02 mm 0.05 mm	±0.02 mm ±0.05 mm

(续表)

仪器名称	量程	最小分度值	最大允许误差
螺旋测微计(千分尺)	0～25 mm	0.01 mm	±0.004 mm
七级天平(物理天平)	500 g	0.05 g	0.08 g(接近满量程) 0.06 g(1/2 量程附近) 0.04 g(1/3 量程附近)
三级天平(分析天平)	200 g	0.1 mg	1.3 mg(接近满量程) 1.0 mg(1/2 量程附近) 0.7 mg(1/3 量程附近)
普通温度计(水银或有机溶剂) 精密温度计(水银)	0～100 ℃ 0～100 ℃	1 ℃ 0.1 ℃	±1 ℃ ±0.2 ℃
电表(0.5 级) 电表(0.1 级)			0.5%×量程 0.1%×量程
数字万用电表			$\alpha\% U_x+\beta\% U_m$(其中 U_x 表示测量值即读数,U_m 表示满度值即量程,α,β 对不同的测量功能有不同的数值)

3. 合成标准不确定度

多次直接测量的合成标准不确定度为A类不确定度和B类不确定度的“方和根”

$$u_x=\sqrt{u_A^2+u_B^2} \tag{2-2-10}$$

4. 相对不确定度

以合成不确定度相对于待测量的最佳估计值所占百分比来表示,更能看出不确定度的相对大小,即把测量结果的不确定度表示为$\frac{u_x}{\bar{x}}\times 100\%$,这是不确定度的百分比表示法,又可称为相对不确定度,记为 u_{rx},即

$$u_{rx}=\frac{u_x}{\bar{x}}\times 100\% \tag{2-2-11}$$

此外,在有些实验中,待测量会有理论值或公认值的情况。此时实验结果应该表示成百分误差的形式,即将测量值与理论值或公认值相比较。如测量值为 x,其理论值为 X,则结果表示成

$$E=\frac{|x-X|}{X}\times 100\% \tag{2-2-12}$$

相对不确定度和百分误差一般保留1位或2位有效数字。

5. 不确定度的传递

在间接测量时,待测量是由直接测量量通过一定的函数关系计算而得到的。由于直接测量量存在不确定度,显然由直接测量量经过运算而得到的间接测量量也必然存在不确定

度，称作不确定的传递。

设间接测量量 N 是由直接测量量 $x, y, z, \cdots$，通过函数关系 $N=f(x, y, z, \cdots)$ 计算得到的，其中 $x, y, z, \cdots$，是彼此独立的直接测量量。则间接测量量的最佳值为

$$\overline{N}=f(\bar{x}, \bar{y}, \bar{z}, \cdots) \tag{2-2-13}$$

式中，$\bar{x}, \bar{y}, \bar{z}, \cdots$ 分别为各直接测量量的最佳值。

设 $x, y, z, \cdots$ 的不确定度分别为 $u_x, u_y, u_z, \cdots$，它们必然影响间接测量结果，使 N 也有相应的不确定度。由于不确定度相对于测量量是微小的量，相当于数学中的"增量"，因此间接测量量的不确定度的计算公式与数学中的全微分公式类似。不同之处在于：要用不确定度 u_x 代替微分 $\mathrm{d}x$，也要考虑到不确定度合成的统计性质。于是我们用以下两式简化地计算间接测量量 N 的不确定度 u_N 和相对不确定度 u_{rN}。

$$u_N = \sqrt{\left(\frac{\partial N}{\partial x}\right)^2 u_x^2 + \left(\frac{\partial N}{\partial y}\right)^2 u_y^2 + \left(\frac{\partial N}{\partial z}\right)^2 u_z^2 + \cdots} \tag{2-2-14}$$

$$u_{rN} = \frac{u_N}{\overline{N}} = \sqrt{\left(\frac{\partial \ln N}{\partial x}\right)^2 u_x^2 + \left(\frac{\partial \ln N}{\partial y}\right)^2 u_y^2 + \left(\frac{\partial \ln N}{\partial z}\right)^2 u_z^2 + \cdots} \tag{2-2-15}$$

式(2-2-14)和式(2-2-15)也称不确定度的传递公式。其中，式(2-2-14)适用于和差形式的函数，式(2-2-15)适用于积商形式的函数。

(1) 如果函数形式是若干个直接测量量相加减，则先计算间接测量量的绝对不确定度比较方便。见表 2-2-3 第一行公式。

(2) 如果函数形式是若干个直接测量量相乘除或连乘除，则先计算间接测量量的相对不确定度 u_{rN} 比较方便，然后再通过公式 $u_N=u_{rN}\overline{N}$ 求出绝对不确定度 u_N。见表 2-2-3 第二行和第五行公式。

(3) 如果间接测量量中某个直接测量量是单次测量，则直接用单次测量的结果及不确定度代入不确定度传递公式。表 2-2-3 给出一些常用函数的不确定度传递公式。

表 2-2-3 一些常用函数的不确定度传递公式

函数表达式	测量不确定度传递公式
$N=x\pm y$	$u_N=\sqrt{u_x^2+u_y^2}$
$N=xy$ 或 $N=\dfrac{y}{x}$	$u_{rN}=\dfrac{u_N}{\overline{N}}=\sqrt{u_{rx}^2+u_{ry}^2}$
$N=kx$	$u_N=ku_x, u_{rN}=u_x/\bar{x}$
$N=\sqrt[k]{x}$	$u_{rN}=\dfrac{u_N}{\overline{N}}=\dfrac{1}{k}\cdot\dfrac{u_x}{x}$
$N=\dfrac{x^k y^m}{z^n}$	$u_{rN}=\dfrac{u_N}{\overline{N}}=\sqrt{k^2u_{rx}^2+m^2u_{ry}^2+n^2u_{rz}^2}$
$N=\sin x$	$u_N=\lvert\cos\bar{x}\rvert u_x$
$N=\ln x$	$u_N=\dfrac{u_x}{x}$

6. 测量结果的表示方式和数据修约规则

完整的测量结果应给出被测量 x 的最佳值,同时按式(2-2-10)或式(2-2-14)给出合成不确定度 u_x 及相对不确定度 u_{rx},结果表示成形式,即

$$\begin{cases} x=\bar{x}\pm u_x \\ u_{rx}=\dfrac{u_x}{\bar{x}}\times 100\% \end{cases} \tag{2-2-16}$$

这表示被测量的真值以置信概率 0.95 落在$(\bar{x}-u_x,\bar{x}+u_x)$的范围内。

由于误差的存在,真值不可能获得,只能得到它的近似值,因此,无论是直接测量的仪器示值,还是通过函数关系获得的间接测量结果,不可能也没有必要记录过多的位数。数据修约的原则是能正确反映数据的可靠性,也就是按测量的不确定度来规定数据的有效位数。由于不确定度本身有一个置信概率的问题,不确定度最多保留 2 位,再多就没有意义了。我们约定:合成不确定度保留 1 位有效位数,被测量的最佳估计值的位数与合成不确定度对齐。

根据测量不确定度的大小,在对测量结果进行修约时,有效数字的末位需要做进、舍位的处理。处理办法应遵守数值修约规则:

(1) 拟舍弃数字的最左位数字小于 5 时,则舍去,即保留的各位数字不变,如下例第一行;

(2) 拟舍弃数字的最左位数字大于 5,或者等于 5 且其后数字不全为 0 时,则进 1,即拟保留的末位数字加 1,如下例中第二、三行;

(3) 拟舍弃数字的最左一位数字为 5,而后面无数字或都为 0 时,若所保留的末尾数字为奇数(1,3,5,7,9)则进一位,为偶数(2,4,6,8,0)则不变,如下例中第四、五行。

按上述舍入规则,将下列数值取成 4 位有效数字,得到

1.143428→1.143　舍去
1.143628→1.144　进位
1.143508→1.144　进位
1.143500→1.144　进位
1.142500→1.142　舍去

在要求不高时,有效数字的舍入规则亦可以按照四舍五入的方式进行。

2.3　直接测量的不确定度评估

一、单次测量的不确定度评估

有些测量比较简单,随机效应因素影响很小。例如,用天平称量物体质量,单次测量与多次测量结果几乎一致,测量误差主要是仪器的误差。在这种情况下,我们只需进行单次测

量，以仪器误差 $\Delta_{仪}$ 作为测量的不确定度即可，即

$$u_x = \Delta_{仪} \tag{2-3-1}$$

有时对某个测量结果的准确度要求不高，或在间接测量的最终结果中该分量影响较小的情况下，也可以只进行单次测量。在随机效应不能忽略的情况下，都应进行多次测量。

二、多次测量的不确定度评估

直接测量数据处理的一般程序是

1. 求出测量列 $x_1, x_2, \cdots, x_n$ 的算术平均值 $\bar{x}$ 作为测量结果的最佳估计值；
2. 计算测量的标准偏差 σ_x，根据 $u_A = \frac{t_{0.95}}{\sqrt{n}}\sigma_x$ 求出 A 类不确定度；
3. 根据测量仪器的性能得出 $u_B = \Delta_{仪}$；
4. 求出合成标准不确定度 $u_x = \sqrt{u_A^2 + u_B^2}$ 和相对不确定度 u_{rx}；
5. 最后写出结果表达式，如式(2-3-1)，并根据修约规则保留合理的位数。

下面就以两个例子来具体讨论直接测量的数据处理过程。

例 2-3-1　用 $\Delta_{仪} = 0.002$ cm 的游标卡尺测某物的高度 h，测量数据是 7.510 cm，7.506 cm，7.508 cm，7.512 cm，7.508 cm，7.512 cm，按照直接测量的数据处理程序处理数据并写出结果表达式。

解：将上面的 6 个数据输入计算器，可得该测量列的算术平均值为

$$\bar{h} = 7.50933 \text{ cm}$$

测量列的标准差为　$\sigma_h = 0.002422$ cm

由表 2-2-1 可得，当 $n=6$ 时　$\frac{t_{0.95}}{\sqrt{n}} = 1.05 \approx 1$

h 的 A 类不确定度　$u_A = \sigma_h = 0.002422$ cm

h 的 B 类不确定度　$u_B = \Delta_{仪} = 0.002$ cm

合成不确定度为　$u_h = \sqrt{u_A^2 + u_B^2} = \sqrt{0.002422^2 + 0.002^2} = 0.003141$ cm

相对不确定度为　$u_{rh} = \frac{u_h}{\bar{h}} = \frac{0.003141}{7.509} = 0.042\%$

下面对 u_h 和 $\bar{h}$ 进行修约，u_h 取一位有效数字，所以 $u = 0.003$ cm。$\bar{h}$ 末位应与 u_h 的末位对齐，u_h 的末位在千分位上，则 $\bar{h}$ 的末位也应保留到千分位，为 $\bar{h} = 7.509$ cm，则最后的结果表达式写成

$$h = (7.509 \pm 0.003) \text{cm}$$

$$u_{rh} = 0.042\%$$

例 2-3-2　在双臂电桥测电阻的实验，用螺旋测微器（$\Delta_{仪} = 0.004$ mm）测量铜棒的直径 d，7 次测量结果如下：5.528 mm，5.534 mm，5.530 mm，5.524 mm，5.526 mm，5.530 mm，

5.524 mm。按照直接测量的数据处理程序处理数据并写出结果表达式。

解:将上面的7个数据输入计算器,可得该测量列的算术平均值为

$$\bar{d}=5.528\ \text{mm}$$

测量列的标准差为 $\sigma_d=0.00365\ \text{mm}$

由表2-2-1,$n=7$时 $\dfrac{t_{0.95}}{\sqrt{n}}=0.926\approx1$

d的A类不确定度 $u_A=\sigma_d=0.00365\ \text{mm}$

d的B类不确定度 $u_B=\Delta_{仪}=0.004\ \text{mm}$

合成不确定度为 $u_d=\sqrt{u_A^2+u_B^2}=\sqrt{0.00365^2+0.004^2}=0.00542\ \text{mm}$

相对不确定度为 $u_{rd}=\dfrac{u_d}{\bar{d}}=\dfrac{0.00542}{5.528}=0.098\%$

下面对u_d和$\bar{d}$进行修约,u_d取一位有效数字,所以$u_d=0.005$ mm。$\bar{d}$的末位要与u_d的所在位对齐,u_d的末位在千分位上,则$\bar{d}$的末位也应保留到千分位,为$\bar{d}=5.528$ mm,则最后的结果表达式写成

$$d=(5.528\pm0.005)\text{mm}$$

$$u_{rd}=0.098\%$$

2.4 间接测量的不确定度评估

下面用具体例子来说明间接测量不确定度评估的一般程序。

例2-4-1 已知质量为$m=(213.04\pm0.05)$g的铜圆柱体,用分度为0.02 mm的游标卡尺测得其高度h为80.40 mm,80.38 mm,80.36 mm,80.34 mm,80.36 mm,80.40 mm;用一级千分尺测得其直径d为19.465 mm,19.476 mm,19.470 mm,19.459 mm,19.466 mm,19.468 mm。试计算该圆柱体的密度及其不确定度。

解:先写出每个变量的算术平均值(最佳值)及其合成不确定度,再按照各量之间的函数关系求出密度的最佳值及总的不确定度。

圆柱体高度h的最佳值 $\bar{h}=\dfrac{1}{n}\sum_{i=1}^{n}h_i=80.373\ \text{mm}$

h的A类不确定度 $u_{Ah}=\dfrac{t_{0.95}}{\sqrt{n}}\sigma_h=0.0242\ \text{mm}\quad\left(n=6,\ \dfrac{t_{0.95}}{\sqrt{n}}\approx1\right)$

游标卡尺的示值误差限为0.02 mm,即$\Delta_{仪}=0.02$ mm,则B类不确定度为

$$u_{Bh}=\Delta_{仪}=0.02\ \text{mm}$$

则 h 的合成不确定度为

$$u_h=\sqrt{u_{Ah}^2+u_{Bh}^2}=\sqrt{0.0242^2+0.02^2}=0.0314\ \text{mm}$$

圆柱体直径 d 的最佳值 $\bar{d}=\frac{1}{n}\sum_{i=1}^{n}d_i=19.46733\ \text{mm}$

d 的 A 类不确定度　　$u_{Ad}=\frac{t_{0.95}}{\sqrt{n}}\sigma_d=0.005645\ \text{mm}\quad\left(n=6,\ \frac{t_{0.95}}{\sqrt{n}}\approx 1\right)$

一级千分尺的仪器误差限为 0.004 mm，即 $\Delta_{仪}=0.004$ mm，则 B 类不确定度为

$$u_{Bd}=\Delta_{仪}=0.004\ \text{mm}$$

则 d 的合成不确定度为

$$u_d=\sqrt{u_{Ad}^2+u_{Bd}^2}=\sqrt{0.005645^2+0.004^2}=0.0076\ \text{mm}$$

由于质量是单次测量，直接考虑其仪器误差来计算不确定度，$\Delta_{仪}=0.05$ g，则

$$u_m=\Delta_{仪}=0.05\ \text{g}$$

密度的最佳值为

$$\bar{\rho}=\frac{4m}{\pi\bar{d}^2\bar{h}}=\frac{4\times 213.04}{\pi\times 19.467^2\times 80.373}=0.0088662\ \text{g}\cdot\text{mm}^{-3}=8.8662\ \text{g}\cdot\text{cm}^{-3}$$

密度的相对不确定度

$$\begin{aligned}u_{r\rho}&=\frac{u_\rho}{\bar{\rho}}=\sqrt{u_{rm}^2+2^2u_{rd}^2+u_{rh}^2}\\&=\sqrt{\left(\frac{0.05}{213.04}\right)^2+2^2\left(\frac{0.0314}{80.373}\right)^2+\left(\frac{0.0076}{19.467}\right)^2}\\&=0.000904=0.09\%\end{aligned}$$

密度的不确定度　　$u_\rho=\bar{\rho}\times u_{r\rho}=8.8662\times 0.09\%=0.008\ \text{g}\cdot\text{cm}^{-3}$

密度的测量结果

$$\rho=(8.866\pm 0.008)\text{g}\cdot\text{cm}^{-3}$$
$$u_{r\rho}=0.09\%$$

说明：在中间运算过程中各量的有效位数都多取几位，以防止过早的舍入而人为造成不确定度的扩大或缩小。

2.5　常用数据分析与处理方法

实验数据处理包含十分丰富的内容。例如，数据的记录、描绘，从带有误差的数据中提取参数，验证和寻找经验规律，外推试验数值等。本节结合物理实验的基本要求，介绍一些最基本的实验数据的处理方法。

一、作图法

所谓作图法，就是把实验数据用自变量和因变量的关系作成曲线，以便反映它们之间的

变化规律或函数关系。作图的优点是直观、形象,便于比较研究实验结果、求某些物理量及建立关系式等。作图时应注意以下几点:

(1) 作图一定要用坐标纸,根据函数关系选用直角坐标纸、单对数坐标纸、双对数坐标纸、极坐标纸等。

(2) 坐标纸的大小及坐标轴的比例,应当根据所测得数据的有效数字和结果的需要来确定。原则上数据中的可靠数字在图中也应是可靠的,数据中的欠准位在图中也应是估计的。要适当选取横轴和纵轴的比例和坐标分度值,使图线充分占有图纸空间,不要缩在一边或一角。坐标轴分度值比例的选取一般选择间隔为 1、2、5、10 等,以便于读数或计算。

(3) 标明坐标轴。一般是以自变量为横轴,因变量为纵轴,采用粗实线描出坐标轴,并用箭头表示方向,注明所示物理量的名称、单位。坐标轴上应标明分度值(注意有效数字)。

(4) 描点。根据测量数据,用符号"×"使其准确地落在图上相应的位置,当一张图纸上画几条实验曲线时,每条曲线应用不同的标记,如"+""⊙""△"等,以免混淆。

(5) 连线。根据不同的函数关系对应的实验数据点的分布,把点连成直线或光滑的曲线,若为校准曲线则连成折线。连线时,必须使用直尺或曲线板。当连成直线或光滑曲线时,连线并不一定通过所有的点,而是使数据点均匀地分布在连线的两侧,个别偏离很大的点应当舍去,原始数据点应保留在图中。

(6) 写图名。在图纸的空白位置或图纸下方写图的名称,一般将纵轴代表的物理量写在前面,横轴代表的物理量写在后面,中间用"-"连接。此外,还可写上实验条件、作图人、作图日期等。

下面介绍在物理实验中常用三种典型作图。

1. 直线图

图 2-5-1 为实验《线性电阻与非线性电阻的伏安特性曲线》中线性电阻伏安特性曲线。其中纵坐标表示电流,其分度值为 1 mA/cm;横坐标表示电压,分度值为 2 V/cm。也就是说在纵坐标上 1 大格(1 cm)表示 1 mA,横坐标 1 大格(1 cm)表示 2 V。如此标度后,作出的图比例适中,比较美观。可以想象如果横坐标的分度值是为 1 V/cm,那么坐标纸将会变得很宽。所以选择合适的分度值在作图中也是很重要的。

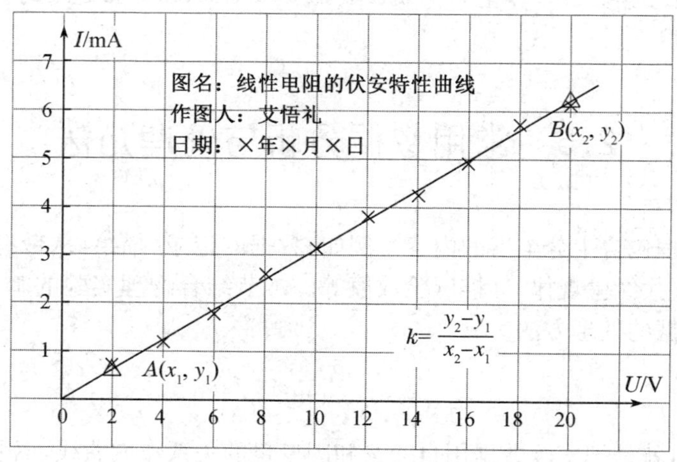

图 2-5-1　线性电阻的伏安特性曲线

作图时先在坐标纸上划上带箭头的坐标轴，再标明坐标轴上的分度值、变量符号（I、U）和单位（mA、V）。用符号"×"把所有的数据点找到对应的位置标号后，考虑各点的位置，用直尺画一直线，使直线通过的数据点尽可能多，不能通过的，使数据点均匀地分布在线的两侧。如果要求所作直线的斜率，可在所作的线上取两点 A 和 B，写下其坐标数值 $A(x_1, y_1)$，$B(x_2, y_2)$，由这些数值即可求出斜率的大小为 $k=\frac{y_2-y_1}{x_2-x_1}$。过程中应注意：$A$ 和 B 两点的位置应尽量靠近线的两端，且横坐标取整数位置以方便计算。两点取好后其标志符号应与作直线所用符号相区别，以免造成混淆。图 2-5-1 中取$\triangle$。最后在图上空白处写上图名、作图人和日期。

2. 曲线图

图 2-5-2 为晶体二极管的伏安特性曲线。需要说明的是，该实验的数据列表为两个，分别是正向和反向。把正向和反向的曲线作在同一张坐标纸上，而且在正向和反向电流和电压的标度都不相同。在纵坐标上（电流）的标度，正向为 2 mA/cm，反向为1 mA/cm；而横坐标（电压）的标度，正向为 0.1 V/cm，反向为 1 V/cm，而且不一定每个厘米格上都需要有标度。

注意：作曲线图时，对于曲线的描绘并不是要把所有的数据点都串起来，而是根据数据点的位置作出一个光滑的曲线，没有必要为个别数据点而使曲线变得曲里拐弯。作出的曲线是一个走向趋势。

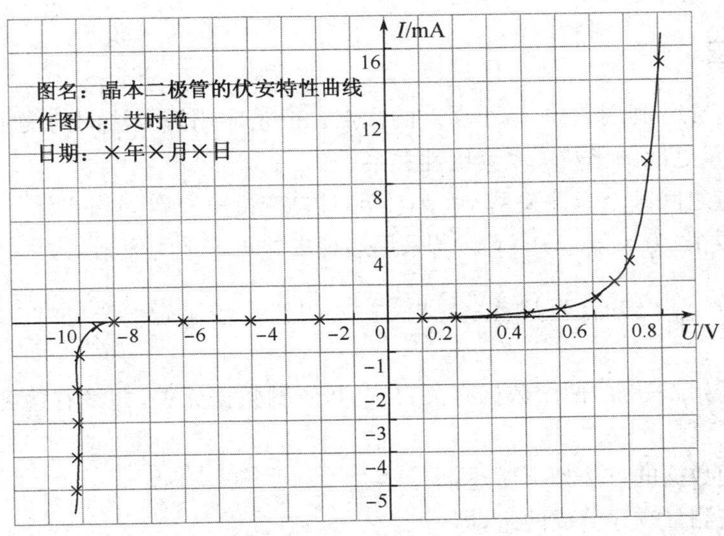

图 2-5-2　晶体二极管的伏安特性曲线

3. 折线图

在《电表的改装与校准》实验中，表头经过改装成量程为 5 V 的电压表。改装后需要对改装表进行校准，并根据校准后的数据作校准曲线。所作校准曲线如图 2-5-3 所示，纵坐标为各个刻度的绝对误差 $\Delta U'=U-U'$，横坐标为改装表的整数指示值。

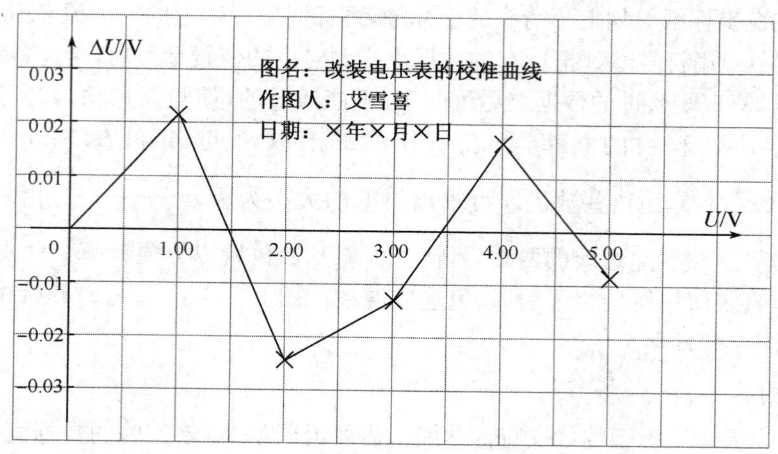

图 2-5-3　改装电压表的校准曲线图

上面的几个图中还有一个共同的特点，就是每张图中最后所做出来的曲线占全部坐标纸的比例较大，实际作图时应按照这种原则去作。坐标纸的大小是根据数据的情况及要选择的标度情况而定的。图作好后应裁下来贴在实验报告上，而不是把整张坐标纸都贴进去，既不美观又占空间，也是一种浪费。

上面介绍的是手工作图的方法，实际也可以借助计算机作图，但是同样要求标度仍取1、2、5等方法以方便阅读。

二、逐差法

逐差法具有充分利用数据、减小误差的优点。但是应用该方法的前提是自变量等间距变化，且与因变量之间的函数关系为线性关系。

设两个变量之间满足线性关系 $y=ax+b$，且自变量 x 是等间隔变化。设测量结果(测量列必须为偶数个)为 $y_1, y_2, \cdots, y_{2n}$，若采用逐项求差再求平均值结果为

$$\Delta y=\frac{(y_2-y_1)+(y_3-y_2)+\cdots+(y_{2n}-y_{2n-1})}{2n-1}=\frac{y_{2n}-y_1}{2n-1}$$

所得结果只与开始、末尾的两个数据有关，与中间所测数据无关，并没有达到多次测量减少误差的目的。

逐差法处理数据的方法如下所述：

将因变量按测量顺序分成两组，即

$$y_1, y_2, \cdots, y_n \text{ 和 } y_{n+1}, y_{n+2}, \cdots, y_{2n}$$

对应项逐项求差，即

$$\Delta y_i=y_{n+i}-y_i \quad (i=1,2,\cdots,n)$$

然后再对这 n 个 Δy_i 求平均值，考虑到每个 Δy_i 都包含 n 个测量间隔，两个相邻变量的平均间距为

$$\Delta y=\frac{(y_{n+1}-y_1)+(y_{n+2}-y_2)+\cdots+(y_{2n}-y_n)}{n^2} \tag{2-5-1}$$

逐差法的优点显而易见。但是该方法要求自变量等间距变化，且与因变量之间的函数关系为线性关系，测量列数据数目为偶数。如果自变量与因变量之间为多项式函数关系时，可采用多次逐差法，这里不做介绍。

三、最小二乘法

若两物理量 x、y 满足线性关系，并由等精度测量得到一组数据$(x_i,y_i)(i=1,2,\cdots,n)$，如何做出一条能最佳地符合所得数据的直线，以反映上述两变量间的线性关系呢？除了用作图法处理数据外，常用的还有最小二乘法。

最小二乘法认为：若最佳拟合的直线为 $y=f(x)$，则所测各 y_i 值与拟合直线上相应的各估计值 $\hat{y}_i=f(x_i)$之间的偏差的平方和为最小，即

$$s=\sum_{i=1}^{n}(y_i-\hat{y}_i)^2\rightarrow \min \tag{2-5-2}$$

因为测量总是有不确定度的存在，所以在 x_i 和 y_i 中都含有不确定度。为讨论简便起见，不妨设各 x_i 值是准确的，而所有的不确定度都只联系着 y_i。这样，如由 $\hat{y}_i=f(x_i)$所确定的值与实际测得值 y_i 之间的偏差平方和最小，也就表示最小二乘法所拟合的直线是最佳的。

一般，可将直线方程表示为：

$$y=ax+b$$

式中，k 是待定直线的斜率；b 是待定直线的截距。

如果设法确定这两个参数，该直线也就确定了，所以解决直线拟合的问题也就变成了由实验数据组(x_i,y_i)来确定 k、b 的过程。将上式代入式(2-5-2)得：

$$s(k,b)=\sum_{i=1}^{n}(y_i-kx_i-b)^2\rightarrow \min \tag{2-5-3}$$

所求的 k 和 b 应是下列方程组的解，

$$\begin{cases}\dfrac{\partial s}{\partial k}=-2\sum(y_i-kx_i-b)x_i=0\\[2ex]\dfrac{\partial s}{\partial b}=-2\sum(y_i-kx_i-b)=0\end{cases}$$

式中，$\sum$ 表示对 i 从 1 到 n 求和。

将上式展开，消去未知数 b，可得：

$$k=\frac{l_{xy}}{l_{xx}} \tag{2-5-4}$$

式中

$$\begin{cases}l_{xy}=\sum(x_i-\bar{x})(y_i-\bar{y})=\sum(x_iy_i)-\dfrac{1}{n}\sum x_i\sum y_i\\[2ex]l_{xx}=\sum(x_i-\bar{x})^2=\sum x_i^2-\dfrac{1}{n}(\sum x_i)^2\end{cases} \tag{2-5-5}$$

将求得的 k 值代入方程组,可得:

$$b=\bar{y}-k\bar{x} \tag{2-5-6}$$

至此,所需拟合的直线方程 $y=ax+b$ 就被唯一确定。

由最终结果不难得到,最佳配置的直线必然通过$(\bar{x},\bar{y})$这一点。因此在作图拟合直线时,拟合的直线必须通过该点。

为了检验拟合直线是否有意义,在数学上引入相关系数 r,它表示两变量之间的函数关系与线性函数的符合程度,具体定义为

$$r=\frac{l_{xy}}{\sqrt{l_{xx}l_{yy}}} \tag{2-5-7}$$

式中,l_{xy}的计算方法与l_{xx}类似,r 的值越接近 1,表示 x 和 y 的线性关系越好;若 r 近于 0,就可以认为 x 和 y 之间不存在线性关系。

在物理实验中,相当多的情况是所测的两个物理量 x、y 之间的关系符合某种曲线方程,而非直线方程。这时,可对曲线方程做一些变换,引入新的变量,从而将不少曲线拟合的问题转化为直线拟合问题。

如曲线方程为 $y=ax^{\alpha}$,可将等式两边取自然对数,得 $\ln y=\alpha\ln x+\ln a$。再令 $Y=\ln y$,$X=\ln x$,$b=\ln a$,即可将幂函数转化成线性函数 $Y=\alpha X+b$。

又如曲线方程 $y=a\mathrm{e}^{\alpha x}$,同样可将等式两边取自然对数 $\ln y=\alpha x+\ln a$。再令 $Y=\ln y$,$b=\ln a$,即可将指数函数转化成线性函数 $Y=\alpha x+b$。

2.6 有效数字及其运算

由于测量误差的存在,任何测量都具有确定的精确度。因此,一个物理量的测量或运算结果的位数都不应无限制地写下去。如利用分度值为 1 mm 的米尺测量一个物体的长度,以毫米为单位,测量结果的十分位是估读的,因而有一定的误差,已很不可靠,将十分位以后的数字写出来也意义不大。那么,在一般情况下测量值能准确到哪一位?从哪一位开始有误差?在数据处理的计算中应该用几位数字表示运算结果才比较合理?怎样才能做到既不损害又不夸大实际测量的精确度?这些就是有效数字所要研究的课题,下面就这些问题作简要介绍。

一、有效数字

正确有效地表示测量结果或运算结果的数字称为有效数字。有效数字由若干位准确(可靠)数字和 1 位欠准(可疑)数字组成。

有效数字=若干位可靠数字+1 位可疑位

例如,用米尺测量一块木板的长度,待测木板的一端与米尺零点对齐,另一端则处在 76～77 mm之间,如图 2-6-1 所示。以毫米为单位,读数为 76.6 mm,其中读数的准确数字

为 76 mm；超过 76 mm 的部分而需要估读，可以估计到米尺最小一格的十分之一，约 0.6 格。小数点后的 6 是估计数字，是欠准确的，但却在一定程度上反映了客观实际，表明木板的长度可能是 76.5～76.7 mm 的一个值。由于观测者在分辨能力上的差异，可疑位的估读允许有 ±0.1 格的误差。如果用最小刻度为 1 cm 的尺来测量，以厘米为单位，可靠位只能到7 cm，超过 7 cm 的部分就需要估读，读数大约为 7.6 cm 或 7.7 cm。

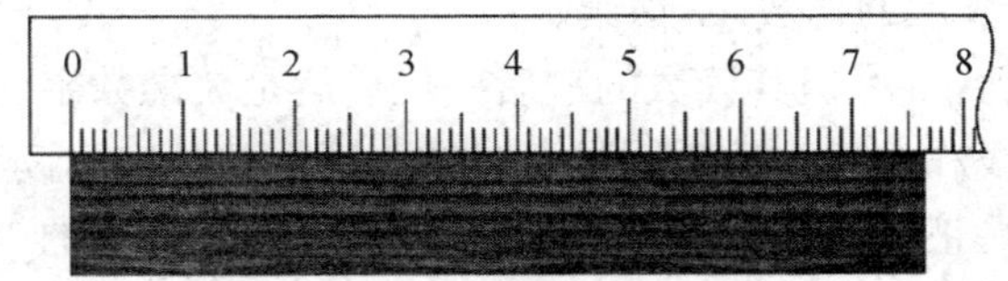

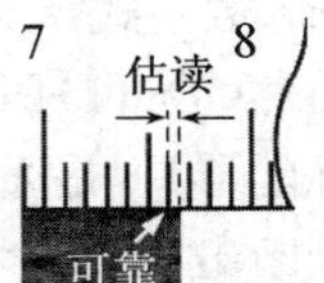

图 2-6-1　用米尺测量木板的长度

有效数字的个数称为有效数字的位数，上述测量结果 76.6 mm 为 3 位有效数字，而另一测量结果 7.6 cm 和 7.7 cm 均为 2 位有效数字。可见，有效数字的多少，表明了测量所能达到的准确程度，与测量工具的选择有关，选择不同的仪器测量相同的物理量，有效数字位数是不同的。在测量或运算中，有效数字的位数按如下方法确定：

确定被测量中有效数字的规则：

(1) 所有的非 0 数字都有效：1.234 g 有 4 个有效数字；1.2 g 有 2 个有效数字。

(2) 非零数字之间的零是有效的：1002 kg 有 4 个有效数字；3.07 mL 有 3 个有效数字。

(3) 第一个非零数字左边的零不有效，这些零仅仅表示小数点的位置：0.001 ℃ 只有 1 个有效数字；0.012 g 有 2 个有效数字。

(4) 一个数的小数点右边的零是有效的：0.023 mL 有 2 个有效数字；0.200 有 3 个有效数字。

(5) 当一个数字以非小数点右边的零结尾时，这些零不一定是有效的：190 m 可能有 2 或 3 个有效数字，5060 g 可能有 3 或 4 个有效数字。

最后一条规则的潜在歧义可以通过使用标准指数或科学符号来避免。例如，根据是否需要 3 或 4 个重要的数字，我们可以将 5060 这样写：

$$5.06\times10^{3}\ \text{g}(3\text{ 个有效数字});5.060\times10^{3}\ \text{g}(4\text{ 个有效数字})$$

可以看出，有效数字在单位转换时，位数应保持不变。

二、测读数据的取位

用测量结果表达式来报道测量结果，既给出了最佳值，又给出了不确定度，能使人们对测量值所处的范围有一个比较清楚地认识。有些场合，对被测量值不要求计算不确定度，这时，通过测量结果的有效数字位数也能大致判断不确定度的量级。一般来说，有效数字的尾数是有误差的，误差项一般只取 1 位，最多不能超过 2 位。本书约定，今后在进行有效数字运算时，所有参与运算的有效数字，其最末位具有误差，运算结果也要取到有误差的那一位为止。至于如何确定误差在哪一位，理论上应按不确定度的传播律来计算。在实际运算中为了简便与快捷，可根据有效数字运算规则来取位，或合理采用某些技巧来取位，所得到的结果与理论上的结果都比较接近。

(1) 测量仪器或仪表给出仪器的最大允许误差 $\Delta_{仪}$ 时,应读到仪器误差所在的那一位;

(2) 测量仪器的读数装置带有标尺时,应在标尺的两刻度线间估读一位数字,因为仪器的误差往往是分度值的十分之几;

(3) 有游标的器具,不估读;

(4) 数字式仪器读数一般取仪器所显示的位数,不估读。

虽然(3)、(4)两种情况没有直接进行估读,但其读数的最后一位仍然是欠准数字。比如有游标的器具,是通过判断游标上的刻线和主尺上的刻线是否对齐而读数的。实际读数时,只能判断出某根线"最"对齐,但并不是真正的对齐,因此读数最后一位仍然是欠准的。

三、运算结果的取位

1. 加减运算

运算结果的末位数的数位应与参与运算的各有效数字中末位数的数位最高者相同。

例如,100+23.643=123.643,结果应该四舍五入为124(3个有效数字)。最后一位公数是最右边的一位,是个位。

2. 乘除运算

运算结果的有效数字位数应与参与运算的各分量中有效数字位数最少者相同。

例如,3.0×12.60=37.800,结果应该四舍五入为38(2个有效数字)。因为3.0有2个有效数字,而12.60有4个有效数字,所以结果是2个有效数字。

3. 函数运算

函数运算保留有效数字的一般规则是:将自变量有效数字的最后一位变化一个单位,分别计算变化前后的函数值;从左向右比较两次函数值,结果保留到两函数值第一位不同的那一位。实际计算时,可以采用下述简化方案。

(1) 对数函数

有效数字的对数,其小数点后的位数与真数的位数相同。

例如 $y=\ln x$,式中 $x=888$,经计算器运算 $\ln 888=6.78891$。因为真数 x 是3位有效数字,所以 y 的小数点后取3位,即 $\ln 888=6.789$。

(2) 指数函数

指数函数运算后的有效数字的位数可与指数的小数点后的位数相同(包括紧接小数点后的零)。

例如 $y=10^x$,$x=2.08$,经计算器运算 $10^{2.08}=120.2264$,因为2.08小数字点后只有2位,所以 y 也只有2位有效数字,即 $10^{2.08}=1.2\times10^2$。

(3) 三角函数

三角函数的取位由角度的有效数字而定。一般用分光计测量角度时,应读到1分。此时,应取4位有效数字。

4. 乘方开方运算

有效数字在乘方或开方时,若乘方或开方的次数不太高,其结果的有效数字位数与原底数的有效数字位数相同。

例如 $A=4.25$，求 $y=A^2=?$ 计算器运算得 $4.25^2=18.0625$，原底数 4.25 为 3 位有效数字，所以最后取 $A^2=18.1$，结果为 3 位有效数字。

5. 正确数

在运算过程中，我们还可有碰到一种特定的数，称为正确数。例如，将半径化为直径 $d=2r$ 时的倍数 2，它不是由测量得出的；实验测量次数 n，它总是正整数，没有可疑部分。有效数字的运算规则不适用于这些正确数，在计算时只需由测量值的有效数字的多少来决定运算结果的有效数字，正确数对运算结果的有效数字倍数没有影响。

在运算过程中，我们还可能碰到一些常数，如 π、g 和 e 之类，一般我们取这些常数与测量的有效数字的位数相同。例如：圆周长 $l=2\pi r$，当 $r=2.356$ mm 时，此时 π 至少应取为 3.142。实际运算时多用计算器或电子表格软件，将计算结果按有效数字运算规则及修约规则保留到合适有效数字位数，而不是算出结果的位数越多越好！

有效数字的位数多少取决于测量仪器，而不是运算过程。因此，选择测量仪器时，应使其所给出的位数不少于应有的有效位数，否则将使测量结果精度降低。

Chapter Ⅱ　Measurement Error, Uncertainty and Data Processing

2.1　The Definition of Measurement and the Basic Knowledge

2.1.1　The Measurement

1. The Definition of Measurement

Measurement is a comparison process of obtaining the multiple between the physical quantity of interest and a similar known physical quantity that is selected as the measurement standard. The value of the multiple is indeed the value of the physical quantity to be measured, and the known quantity chosen as the measurement standard is called a unit. Therefore, the measured value of a physical quantity should be composed of two parts, namely numerical value and its unit, both being indispensable.

2. The Results of Measurement

The result of measurement, i.e. the value of what is obtained by the measurement, gives measured physical quantity. The word "give" here means that the result of the measurement is an estimated result of the true value rather than a "true value" (the true value of the measured physical quantity). For direct measurements, if only a single measurement is made, this value of measurement can be used as a result of a measurement; if multiple repetitive measurements are made on the same physical quantity, and multiple observations are obtained, the arithmetic mean of multiple measurements is taken as the result of the measurement.

2.1.2　The Unit of Physical Quantity

According to the Metrology Law of the People's Republic of China, the physical units are based on the International System of Units (SI). Specifically, units for the length, mass, time, current, thermodynamic temperature, the amount of substance and luminous intensity are the basic units. Alternatively, the derived units refer to the units derived from these basic units.

2.1.3 The Classification of Measurement

There are numerous types of classification criteria for measurement, and this textbook only introduces the classification based on the way of obtaining the measured values, namely direct measurement and indirect measurement.

1. Direct Measurement

Direct measurement is a measurement that directly reaches the value of the measured physical quantity without no auxiliary calculation of the function relation of the measured values and other values. Good cases in point are the length-measuring of a table with a tape measure and the current-measuring with an ammeter, where the corresponding length and current are called as direct measurements. The direct measurements are sorted into single and multiple measurements according to the number of measurements.

Single measurement: single measurement refers to the only one performance of measuring. Single measurement is mainly applied in the occasions with lower accuracy of the measurement, the higher difficulty of measuring, or the possibly greater errors in the measuring process than those caused by the instruments. The wire length-measuring in the test of Yang's Elastic Modulus is a case of single measurement.

Multiple measurements: The repeated measurements, usually more than one measuring, are called multiple measurements. Multiple measurements, according to the measuring conditions, are mainly divided into equal precision measurement and non-equal precision measurement.

2. Indirect Measurement

Suppose we have known the functional relationship between the measured quantity and the values of other quantities that measured directly. Once the values of the directly measured quantities are known, we can calculate the measured quantity by applying this relationship. Different from direct measurements, the indirect measurement refers to measuring the values of measured quantities in an indirect way. For instance, in the experiment of measuring the density of cylinders, the height, diameter and quality of the cylinder are first measured, and then the formula is applied $\rho=\frac{4m}{\pi d^2 h}$ to calculate the density of the cylinder. Such measuring procedures can be called as indirect measurements.

2.2 Error and Uncertainty

2.2.1 Error

1. Definition of Error

The error of measurement refers to the difference between the measured value and the true value of the measured quantity. Suppose the result of the measurement is x and the true value is x_0, the error can be expressed as

$$\delta = x - x_0 \tag{2-2-1}$$

δ reflects the deviation of the measured value from the true value. Theoretically speaking, all the values of a given measured quantity have a corresponding objective and practical value under given conditions. We call this value the "theoretical truth value", also referred to as the true value. Confined to the limitation of objective reality, the true value is only an ideal concept and the true value is unattainable. Through the measurement, we can only obtain the approximate true value of the physical quantity within the estimated range of the measurement error. It is internationally acknowledged that the uncertainty should be utilized to characterize the possible value range of the measurement error. The error δ defined by Equation (2-2-1) is also known as an absolute error.

In order to compare the effects of the measurements of two different physical quantities, it is necessary to introduce the concept of relative error as follows:

$$E = \frac{\delta}{x_0} \times 100\% \tag{2-2-2}$$

The relative error is also known as the percentage error, which is a pure number without units. The relative error can be used to evaluate the measurement of the same physical quantity in different sizes and of different physical quantities. Here is an example of comparing the merits and demerits of different measurements for a cylinder: the length of a cylinder is $L=80.20$ mm, and the absolute error is $\delta_L=0.08$ mm; the diameter is $d=20.055$ mm and the absolute error is $\delta_d=0.005$ mm; the mass is $m=215.35$ g and the absolute error is $\delta_m=0.05$ g. Through calculation, it is discovered that the relative error of the mass is the smallest, and the relative error of the length is the largest. Therefore, it's safe to conclude that the measurement of the mass is better than that of the diameter, and the measurement of the diameter is better than that of the length.

2. The Classification of Errors

There are numerous errors caused by different factors in the practical measurement.

As the cause differs, the characteristics of the error are also different. An important part of the error study is to grasp the characteristics of various errors, and only in this way can we obtain the correct method of processing the error. Based on the characteristics and features of the errors, three types of the measurement error can be classified as systematic error, random error and coarse error.

(1) Systematic error

The repeated measurements of the same quantity, under the same conditions (methods, instruments, personnel and the environment), can produce the unchanged signs of the error, whereas the error changes with certain patterns are generated under different conditions. And errors with such features are generally known as the systematic errors. The systematic errors can be attributed to some consistent factors or rule-based changing factors, which can be mastered. The main contributing factors of systematic error include the measurement device, the environment, the theory or method of measurement, the measuring personnel and etc.

The causes for the systematic error may differ, but there are fixed changing regularities, resulting in certain changing rules for the change of the error. Different causes for the systematic error are responsible for the different rules. Thus, with the consideration of the possible causes for the systematic errors, some technical measures, include calibrating the equipment, improving the experimental device and experimental methods, can be adopted to eliminate or reduce the systematic error. Further, the repeated measurements of the standard equipment with the true agreed value under the same conditions can be adopted. Alternatively, the repeated measurements under different conditions can be utilized to identify the rules of the systematic error and then to modify the measuring results.

(2) Random error

Random error, also known as accidental error, refers to the error whose absolute value and sign changes in an unpredictable way during the process of repeatedly measuring the same physical quantity. The random errors can account for the fact that various fluctuations of errors, big or small, positive or negative, can occur in repeatedly measuring a physical quantity under the same conditions, despite the elimination of the systematic errors.

Random errors are caused by some accidental or uncertain factors, including the experimenter's sensory limitations for the reading fluctuation, the irregular changes of the experimental environments (temperature, humidity, wind, power supply voltage, etc.) or the rising and falling of the measured object etc. The possible impacts of these factors may be small and mixed, yet they cannot be ruled out.

The most important characteristic of the random error is its randomness. The absolute value and the sign of the random error of a single measurement vary in an unpredictable way and there is no definite law. In an infinite number of measurements, the random errors of each measurement obey the statistical law. The common distribution is:

① The probability of the measured value being larger than the true value and the probability of the measured value being smaller than the true value are equal;

② The occurrence probability of data with smaller error is much higher than that of data with larger error, while the occurrence probability of errors with larger absolute value tends to be zero;

③ The chance of positive or negative errors with equal absolute values is equal during the multiple measurements, and the sum of all possible errors tends to zero.

This is what is called a normal distribution (Gaussian distribution). For a random error of the normal distribution, we should measure as many times as possible since the increasing number of measurements can effectively reduce the random error. Random errors also have other types of distribution, such as the t distribution, the uniform distribution and etc..

(3) Rough error

Rough error is also known as gross error. And it is caused by the following reasons: the incorrect experiment method, the experimenter's carelessness reading, recording or miscalculating of measurement data, or the sudden change of test conditions etc. The correct results should not contain the negligent errors, and the value of measurement which contains a rough error is known as a bad value or an exception value. In the experimental measurement, we should attempt to avoid mistakes and errors. Thus, in the progress of data processing, the bad value should be removed.

The above errors can be inter-transformed under certain conditions. A specific error is likely to belong to a systematic error under one condition, but it may be expressed as a random error under other condition, and vice versa. For example, a meter may have a random scale error from manufacturing. However, if we calibrate a number of other meters with this meter, the scale error of the meter will cause the systematic error of a batch of calibrated meters. In another example, the inaccurate meter scale is sure to result in a systematic error when measuring the voltage of a power supply. However, if a lot of meters with the same type are used to measure the voltage of a power supply, the errors of each meter scale will range from large to small or positive to negative, which means these measurement errors are random.

3. The Processing of a Random Error

In a university physics experiment, the random errors of the multiple independent measurements can be approximated as a normal distribution. Under the same conditions where a physical quantity x is independently measured many times, we can obtain the measurement as x_1, x_2, $\cdots$, x_n, and the arithmetic mean is

$$\bar{x} = \frac{x_1 + x_2 + \cdots + x_n}{n} = \frac{1}{n}\sum_{i=1}^{n} x_i \tag{2-2-3}$$

It can be illustrated that when the systematic error has been eliminated, the arithmetic mean of the measured value is closest to the true value of the measured physical quantity. Consequently, the arithmetic mean of the commonly used measurement sequence x represents the measurement result, that is, $\bar{x}$ is the optimum value.

The reliability of the measured values is often estimated by the standard deviation. A small standard deviation indicates that the degree of dispersion of the data measured many times is small, and the precision of the measurement is high. The standard deviation of the measurement sequence satisfying the normal distribution is defined by the Bessel formula as

$$\sigma_x = \sqrt{\frac{\sum_{i=1}^{n}(x_i - \bar{x})^2}{n-1}} \tag{2-2-4}$$

As can be seen from the characteristics of the normal distribution, for any measurement, the probability that the result falls in the interval $[\bar{x} - \sigma_x, \bar{x} + \sigma_x]$ is 0.683, and the probability that the result falls in the interval $[\bar{x} - 2\sigma_x, \bar{x} + 2\sigma_x]$ is 0.954, and the probability that the result falls in the interval $[\bar{x} - 3\sigma_x, \bar{x} + 3\sigma_x]$ is 0.997. The probability is known as the confidence probability, denoted by P. And the corresponding interval is known as the confidence interval. Because the probability that the value of measurements falls outside the interval $[\bar{x}-3\sigma_x, \bar{x}+3\sigma_x]$ is small, the $3\sigma_x$ is called the limit error, which can be applied as the criterion of a gross error. An error which is larger than $3\sigma_x$ between the value of measurements and the average value is known as a gross error, which should be removed when processing data. In physical experiments, the probability of confidence is generally taken as 0.95.

In practical work, one focuses more on the measurement results rather than the data distribution characteristics of the measuring sequence, specifically, the degree of dispersion of the arithmetic mean $\bar{x}$ from the true value. The error theory proves that the standard deviation of the mean $\bar{x}$ is

$$\sigma_{\bar{x}} = \sqrt{\frac{\sum_{i=1}^{n}(x_i - \bar{x})^2}{n(n-1)}} \tag{2-2-5}$$

Notably, its meaning is the confidence probability that the true value of the measured physical quantity falls in the interval $[\bar{x} - \sigma_{\bar{x}}, \bar{x} + \sigma_{\bar{x}}]$ is 0.683, the confidence probability that it falls in the interval $[\bar{x}-3\sigma_{\bar{x}}, \bar{x}+2\sigma_{\bar{x}}]$ is 0.954, and the confidence probability that falls in the interval $[\bar{x}-3\sigma_{\bar{x}}, \bar{x}+3\sigma_{\bar{x}}]$ is 0.997.

Equation (2-2-5) can be obtained under the condition where the data distribution with a large number of measurements can be considered as the normal distribution. In physical experiments, the number of measurements is often small (general $n<10$). And in this case, the measurement results behave as t distribution. Assuming that in the case of a normal distribution, the confidence interval for which the confidence probability P corresponds is $[\bar{x}-\sigma_{\bar{x}}, \bar{x}+\sigma_{\bar{x}}]$, then for the t distribution, the confidence interval that the same confidence interval P corresponds to is $\left[\bar{x}-\frac{t_P\sigma}{\sqrt{n}}, \bar{x}+\frac{t_P\sigma}{\sqrt{n}}\right]$. The value of t_P is related to the number of measurements and the confidence probability, and it can be obtained through referencing Table 2-2-1, which illustrates the relationship between the factors $t_{0.95}$, $t_{0.95}/\sqrt{n}$ and the different measuring times n with the confidence probability $P=0.95$.

Table 2-2-1 The value of $t_{0.95}$ and $t_{0.95}/\sqrt{n}$ with different measurements n

n	3	4	5	6	7	8	9	10	15	20
$t_{0.95}$	4.30	3.18	2.78	2.57	2.45	2.36	2.31	2.26	2.14	2.09
$t_{0.95}/\sqrt{n}$	2.48	1.59	1.20	1.05	0.926	0.834	0.770	0.715	0.553	0.467

4. The Measurement Precision

Precision is the closeness of the measurement results relative to the true value and it corresponds to the size of the error. The precision can be divided into:

(1) Accuracy: accuracy refers to the extent to which the optimal value of a set of measurements deviates from the true value, reflecting the size of the measured systematic error.

(2) Punctuality: punctuality refers to the degree of dispersion of a group of the measured data, reflecting the size of measurement random error.

(3) Definition: definition refers to the degree of dispersion of a group of measurement data deviated from the true value, reflecting the combined effects of the measured system error and random error.

In the scientific experiment, we aim to obtain a result with high precision, accuracy and punctuality. For a specific measurement, the truth is that the high punctuality may not necessarily bring the high accuracy and that the high accuracy may not necessarily produce the high punctuality either.

It is believed that the measurement with good repeatability must have high accuracy. Yet it is not necessarily true in that good repeatability only refers to the measurement results with high punctuality, namely the measured random error is small.

The quality of measurement results depends on the measurement accuracy, that is, the size of the systematic error. Conducting repeated measurements on the same measured physical quantity with the same measuring instrument can only determine the distribution and repeatability of the value, but it cannot determine its accuracy.

2.2.2 Uncertainty

1. The Basic Concept of Uncertainty

The purpose of the measurement is to obtain the true value of the measured value, but the true value is not accurate due to the limitation of objective reality. The measurement result can only give an approximate estimate of the true value and an error range used to represent the approximate degree. Therefore, the experimental results cannot be obtained quantitatively. In other words, they possess uncertainty. In order to accurately characterize the experimental measurement, we introduced the concept of uncertainty to measure the closeness of the experimental results to a real situation. The uncertainty not only characterizes the dispersion and the dependability of the measurement results, but also illustrates the range of the true value. Under the premise of the correct measurement method, the smaller the uncertainty is, the more reliable the measurement results will be, while the greater the uncertainty is, the less reliable the measurement will be.

It should be noted that the uncertainty of the measurement results and the measurement error are two different concepts. There are difference and correlation between two concepts. They can neither be confused nor substituted with each other. The measurement error is caused by the error between the measured true value and the quoted measurement results. The error cannot be accurately measured because of the unknown true value. The uncertainty of the measurement indicates that the measured value cannot be determined due to the existence of the measurement error. Additionally, it is the possible range of the unknown error value and an estimated possibility of an objective existence. In practical work, all problems encountered are about the uncertainty.

Based on the methods of utilizing uncertainty to describe the degree of reliability of the measurement [(INC - 1) 1980] (which was recommended and promoted by the Bureau International des Poids et Mesures (BIPM) in 1980), Chinese Measurement Departments have also developed a series of consistent norms. With an uncertainty representation system, these norms nearly cover all areas in the metering system.

This book introduces a simplified solution to uncertainty based on the needs of practical teaching and experiments—the estimated value of the measurement uncertainty represented by the standard deviation, namely the standard uncertainty.

2. Two Types of Assessment of Standard Uncertainty

The measurement uncertainty can be derived from a number of factors and thus it consists of multiple components. Some of the components can be evaluated through a series of statistical analyses of the results of multiple measurements under the same conditions. They are known as the type A uncertainties, denoted as u_A; other components are evaluated by non-statistical methods, known as the type B uncertainties, denoted by u_B.

(1) Type A uncertainties u_A

Under the same conditions, n times of an independent measurement on a physical quantity x results in the measurement sequence $x_1, x_2, \cdots, x_n$. In this case, the type A uncertainties u_A is

$$u_A = t_P \sigma_{\bar{x}} = \frac{t_P}{n} \sigma_x \tag{2-2-6}$$

Where the value of $t_P/\sqrt{n}$ is illustrated in Table 2-2-1. When the confidence probability is $P = 0.95$, $n = 6$ or 7, $t_{0.95}/\sqrt{n} \approx 1$, the type A uncertainties can be directly approximated to take the value of the standard deviation σ_x, which is

$$u_A = \sigma_x \tag{2-2-7}$$

(2) Type B uncertainties u_B

In the case of assessing the type B uncertainties, various possible factors affecting the accuracy of the measurement should be taken into account. Therefore, u_B is usually synthesized by multiple items. And the estimate of u_B is also a challenge when estimating the inaccuracy of the measurement.

The error that produce type B uncertainties is mainly considered as the instrument error. Additionally, u_B can be simplified as the maximum allowable error of instrument, that is

$$u_B = \frac{\Delta_{\text{instrument}}}{C} \tag{2-2-8}$$

Table 2-2-2 Maximum allowable error of the commonly used lab instruments

The name of equipment	Range	Minimum scale value	Maximum allowable error
Steel plate ruler	150 mm 500 mm 1000 mm	1 mm 1 mm 1 mm	±0.10 mm ±0.15 mm ±0.20 mm
Steel tape	1 m 2 m	1 mm 1 mm	±0.8 mm ±1.2 mm
Vernier caliper	125 mm	0.02 mm 0.05 mm	±0.02 mm ±0.05 mm

(Continue)

The name of equipment	Range	Minimum scale value	Maximum allowable error
Spiral micro caliper (micrometer)	0～25 mm	0.01 mm	±0.004 mm
Seven-level balance (physical balance)	500 g	0.05 g	0.08 g(close to full scale) 0.06 g(near the range of 1/2) 0.04 g(near the range of 1/3)
Three-level balance (analytical balance)	200 g	0.1 mg	1.3 g(close to full scale) 1.0 g(near the range of 1/2) 0.7 g(near the range of 1/3)
Common thermometer(mercury or organic solvent) Precisionthermometer (Mercury)	0 ℃～100 ℃ 0 ℃～100 ℃	1 ℃ 0.1 ℃	±1 ℃ ±0.2 ℃
Meter(0.5 level) Meter(0.1 level)			0.5%×Range 0.1%×Range

In the teaching of physics experiments in university, a simplified scheme is applied, whereby the factor c is a confidence factor, and when the confidence probability P is 0.95, $c=1.05\approx 1$, thus

$$u_B=\Delta_{\text{instrument}} \tag{2-2-9}$$

The so-called maximum allowable error, for a given measuring instrument, is the error limit allowed by its relevant norms and procedures of the instrument, and it is represented by $\Delta_{\text{instrument}}$. Different models of instrument have different maximum allowable error. The error of some instruments can be obtained by querying the national measurement procedures, such as calipers, micrometers, balances, etc. The error of some instruments can be identified on their nameplate and instruction manuals. Additionally, some instruments have an accuracy rating on the nameplate, which can be converted into $\Delta_{\text{instrument}}$. Table 2-2-2 provides the maximum allowable error of the frequently used instruments.

3. The Synthetic Standard Uncertainty

The synthetic standard uncertainty of multiple direct measurement is the "square root of the square sum" of type A and type B uncertainties, that is

$$u_x=\sqrt{u_A^2+u_B^2} \tag{2-2-10}$$

4. The Relative Uncertainty

The relative uncertainty is expressed as the percentage of the synthetic uncertainty relative to the optimal estimated measurement, thus the relative size of uncertainty can be easily idenfied. The percentage representation of uncertainty can be expressed as $\frac{u_x}{\bar{x}}\times 100\%$, also known as the relative uncertainty, represented as u_{rx}, that is

$$u_{rx}=\frac{u_x}{\bar{x}}\times 100\% \tag{2-2-11}$$

Additionally, in some experiments, a situation will occur where the value to be measured has theoretical or recognized value. Under such condition, the experimental results should be expressed in the form of a percentage error, which means comparing the measured value with the theoretical value or recognized value. For example, if the measured value is x, and the theoretical value of it is X, the results can be expressed as

$$E=\frac{|x-X|}{X}\times 100\% \tag{2-2-12}$$

Relative uncertainty and percentage error generally retain 1 or 2 significant digits.

4. The Delivery of Uncertainty

In the indirect measurement, the quantity to be measured is obtained through the direct measurement quantity calculation on the calculation of certain function relationships. Due to the uncertainty of direct measurement, the uncertainty must arise in the indirect measurement quantity (which is obtained by the direct measurement quantity with calculation), which is known as the delivery of uncertainty.

Suppose that the indirect measurement quantity N is calculated by the direct quantities $x, y, z, \cdots$ with the function relation $N=f(x, y, z, \cdots)$, where $x, y, z, \cdots$ are direct measurements independent of each other. The optimum value of the indirect measurement is

$$\overline{N}=f(\bar{x}, \bar{y}, \bar{z}, \cdots) \tag{2-2-13}$$

Where $\bar{x}, \bar{y}, \bar{z}, \cdots$ is the optimum values of direct measurement respectively.

The uncertainties of $x, y, z, \cdots$, namely, $u_x, u_y, u_z, \cdots$, inevitably affect the results of the indirect measurement, thus N also has the corresponding uncertainty. Compared with the value of measurement, the uncertainty is a small amount, which is equivalent to the "incremental" in mathematics. Therefore, the calculation formula of the uncertainty of the indirect measurement is similar to the all-differential formula in mathematics.

The differences lie in the fact that the uncertainties of u_x, u_y, u_z, $\cdots$ are used instead of the differential $\mathrm{d}x, \mathrm{d}y, \mathrm{d}z, \cdots$, and the statistical properties of the uncertainty synthesis is also taken into account as well. Then, the following two formulas are applied in order to simplify the calculation of the uncertainty u_N and the relative uncertainty u_{rN} derived from the indirect measurement, that is,

$$u_N=\sqrt{\left(\frac{\partial N}{\partial x}\right)^2 u_x^2+\left(\frac{\partial N}{\partial y}\right)^2 u_y^2+\left(\frac{\partial N}{\partial z}\right)^2 u_z^2+\cdots} \tag{2-2-14}$$

$$u_{rN}=\frac{u_N}{\overline{N}}=\sqrt{\left(\frac{\partial \ln N}{\partial x}\right)^2 u_x^2+\left(\frac{\partial \ln N}{\partial y}\right)^2 u_y^2+\left(\frac{\partial \ln N}{\partial z}\right)^2 u_z^2+\cdots} \tag{2-2-15}$$

Equation (2-2-14) and Equation (2-2-15) are also called transfer formula of uncertainty, in which Equation (2-2-14) is applied to the function in the form of the addition and subtraction, and Equation (2-2-15) is applied to the function in the form of the multiplication and division.

(1) If the form of the function is the mutual addition and subtraction of a number of quantities obtained from direct measurements, it is more convenient to initially calculate the absolute uncertainty of the indirect measurement, which is displayed in the first line of the formula of Table 2-2-3.

(2) If the form of function is the mutual or continuous multiplication and division of a number of the quantities obtained from direct measurements, it is more convenient to calculate the relative uncertainty u_{rN} of the quantity of indirect measurement. Then, the absolute uncertainty u_N can be identified through Equation $u_N=u_{rN}\overline{N}$, which can be seen in the second line and fifth line of the formulas in Table 2-2-3.

(3) If a direct measurement of an indirect measurement is a single measurement, the transfer formula of uncertainty can be directly substituted with the results and the uncertainty of the single measurement. Table 2-2-3 provides the uncertainty transfer formulas of some common functions.

Table 2-2-3　The uncertainty transfer formulas of some common functions

The expression of function	The uncertainty transfer formulas of the common functions
$N=x\pm y$	$u_N=\sqrt{u_x^2+u_y^2}$
$N=xy$ or $N=\frac{y}{x}$	$u_{rN}=\frac{u_N}{\overline{N}}=\sqrt{u_{rx}^2+u_{ry}^2}$
$N=kx$	$u_N=ku_x$, $u_{rN}=\frac{u_x}{\overline{x}}$
$N=\sqrt[k]{x}$	$u_{rN}=\frac{u_N}{\overline{N}}=\frac{1}{k}\cdot\frac{u_x}{\overline{x}}$
$N=x^k y^m z^{-n}$	$u_{rN}=\frac{u_N}{\overline{N}}=\sqrt{k^2 u_{rx}^2+m^2 u_{ry}^2+n^2 u_{rz}^2}$
$N=\sin x$	$u_N=\lvert\cos\overline{x}\rvert u_x$
$N=\ln x$	$u_N=\frac{u_x}{\overline{x}}$

5. The Representation of the Measurement Results and the Data Rounding-off Rules

The complete measurement results should illustrate the optimum value of the measured x, and the synthetic uncertainty u_x and the relative uncertainty u_{rx} are demonstrated in Equation (2-2-10) or (2-2-14). The result is expressed in the following form:

$$\begin{cases} x=\bar{x}\pm u_x \\ u_{rx}=\dfrac{u_x}{\bar{x}}\times 100\% \end{cases} \tag{2-2-16}$$

This means that the true value of the measured quantity falls within the range of $\bar{x}-u_x$, $\bar{x}+u_x$ with a confidence probability of 0.95.

Because of the existence of the error, it is impossible for the true value to be obtained, so its approximation is the only factor that can be obtained. Therefore, whether it is directly measured via instrument indication, or the indirect measurement results obtained through the relationship of function, it is impossible and unnecessary to record too many bits. The principle of rounding off data is to correctly reflect the reliability of the data, i. e., to determine the effective number of bits of the data according to the uncertainty of the measurement. Given the confidence probability of the uncertainty, two bits of uncertainty is enough with more bits being meaningless. We agree on the following stipulation:

The synthetic uncertainty preserves the 1-bit significant digit, and the optimal experimental measurement or result should be rounded to the same decimal place as the synthetic uncertainty.

When calculating the results of the measurement, we should comply with the rules for rounding off numbers:

(1) If the leftmost digit to be dropped is less than 5, the last remaining digit is left as it is. See the following example in the first line.

(2) If the leftmost digit to be dropped is greater than 5, or equal to 5 with the following digits not being 0, the last retained digit is increased by one. See the following example in the second and third line.

(3) If the leftmost digit to be dropped is 5 with no following digit or the following digit being 0, the last remaining digit is increased by one. See the following example in the fourth line.

(4) If the leftmost digit to be dropped is 5 and is followed only by zeroes, the last remaining digit is increased by one if it is odd, but left as it is if even. See the following example in the fifth line.

The rules for rounding off numbers can be shown in the following example, each rounding a value to four significant digits:

1.143428→1.143 Round down
1.143628→1.144 Round up
1.143508→1.144 Round up
1.143500→1.144 Round up
1.142628→1.142 Round down

2.3 Uncertainty Assessment of Direct Measurement

2.3.1 Uncertainty Assessment of Single Measurement

Some measurements are quite simple and the effect of random factors is small. Take weighting the mass of objects with scales for example. The results of single measurement and multiple measurements are almost identical whereas the measurement errors are mainly caused by the errors of the instrument. In this case, we only need a single measurement with instrument error $\Delta_{\text{instrument}}$ as the uncertainty of measurement. That is

$$u_x = \Delta_{\text{instrument}} \tag{2-3-1}$$

When the required accuracy for a measurement result is not that high, or when the measurement result has little effect on the final result of indirect measurement, a single measurement is acceptable. If random effects cannot be ignored, repeated measurements are necessary.

2.3.2 Uncertainty Assessment of Repeated Measurements

The general procedures of processing the data of direct measurement are listed as follows:

(1) To calculate the arithmetic mean $\bar{x}$ of the measurement sequence $x_1, x_2, \cdots, x_n$ as the best estimate of the measurement results.

(2) To calculate the standard deviation σ_x of the measurement, and then identify the uncertainty of the type A in accordance with $u_A = \frac{t_{0.95}}{\sqrt{n}}\sigma_x$.

(3) To obtain $u_B = \Delta_{\text{instrument}}$ based on the performance of the measuring instrument.

(4) To find the synthetic standard uncertainty $u_x = \sqrt{u_A^2 + u_B^2}$ and the relative uncertainty u_{rx}.

(5) Finally, to write the result expression, such as Equation (2-2-15), complying with the rules for rounding numbers.

Here are two examples to discuss the procedure of processing the data of direct measurement.

Case 1: Using the Vernier caliper with $\Delta_{instrument}=0.002$ cm to measure the height h of a cylinder, the measurement results were 7.510, 7.506, 7.508, 7.512, 7.508, and 7.512 cm. The data should be disposed of in accordance with the data processing program of the direct measurement and the result expression should be written.

Solution: Entering the above 6 into the calculator, we can get the following results:

The arithmetic mean of the measurement column is $\bar{h}=7.50933$ cm.

The standard deviation of the measurement column is $\sigma_h=0.002422$ cm.

From Table 2-2-1, it can be seen that, when $n=6$, $\frac{t_{0.95}}{\sqrt{n}}=1.05\approx 1$.

The uncertainty of type A of h is $u_A=\sigma_h=0.002422$ cm.

The uncertainty of type B of h is $u_B=\Delta_{instrument}=0.002$ cm.

The synthetic uncertainty is $u_h=\sqrt{u_A^2+u_B^2}=\sqrt{0.002422^2+0.002^2}=0.003141$ cm.

The relative uncertainty is $u_{rh}=\frac{u_h}{\bar{h}}=\frac{0.003141}{7.509}=0.042\%$

Therefore, $u_h=0.003$ cm, and $\bar{h}$ should be rounded to 3 decimal places, thus the final resultant expression is

$$\bar{h}=(7.509\pm 0.003)\text{cm},\ u_{rh}=0.042\%$$

Case 2: In the experiment of measuring the resistance with Kelvin bridge, the micrometer screw ($\Delta_{instrument}=0.004$ mm) is required to measure the diameter d of a copper bar, and the seven readings are 5.528, 5.534, 5.530, 5.524, 5.526, 5.530 and 5.524 mm. The data in accordance with the data processing program of the direct measurement should be disposed of and the result expression should be written.

Solution: Entering the above 7 data into the calculator, we get the following results:

The arithmetic mean of the measurement column is $\bar{d}=5.528$ mm

The standard deviation of the measurement column is $\sigma_d=0.00365$ mm

From Table 2-2-1, it can be seen that, when $n=7$, there is $\frac{t_{0.95}}{\sqrt{n}}=0.926\approx 1$.

The uncertainty of type A of d is $u_A=\sigma_d=0.00365$ mm

The uncertainty of type B of d is $u_B=\Delta_{instrument}=0.004$ mm

The synthetic uncertainty is $u_d=\sqrt{u_A^2+u_B^2}=\sqrt{0.00365^2+0.004^2}=0.00542$ mm

The relative uncertainty is $u_{rd}=\frac{u_d}{\bar{d}}=\frac{0.00542}{5.528}=0.098\%$

Therefore, $u_d=0.005$ mm, and $\bar{d}$ should be rounded to 3 decimal places, thus the final results can be written as

$$\bar{d}=(5.528\pm 0.005)\text{mm}$$

$$u_{rd}=0.098\%$$

2.4 Uncertainty Assessment of Indirect Measurement

A specific example is given below to illustrate the general procedure for the uncertainty assessment of indirect measurement.

Case 3: A copper cylinder has a mass measured to be $\bar{m}=(213.04+0.05)\text{g}$. Its height h measured with a vernier caliper of resolution 0.02 mm reads 80.40, 80.38, 80.36, 80.34, 80.36, 80.40 mm; and its diameter d measured with a micrometer reads 19.465, 19.476, 19.470, 19.459, 19.466, 19.468 mm. Please calculate the density of the cylinder and its uncertainty.

Solution: Firstly, the arithmetic mean (optimal value) of each variable and its synthetic uncertainty is written, and then the optimum value of the density and its total uncertainty according to the function relation between each quantity is calculated.

For the height h

$$\bar{h}=\frac{1}{6}\sum_{i=1}^{6}h_i=80.373\ \text{mm}$$

$$u_{\text{A}h}=\frac{t_{0.95}}{\sqrt{6}}\sigma_h=0.0242\ \text{mm}\quad\left(n=6,\ \frac{t_{0.95}}{\sqrt{n}}\approx 1\right)$$

$$u_{\text{B}h}=\Delta_{\text{vernier}}=0.02\ \text{mm}$$

$$u_h=\sqrt{u_{\text{A}h}^2+u_{\text{B}h}^2}=\sqrt{0.0242^2+0.02^2}=0.0314\ \text{mm}$$

For the diameter d

$$\bar{d}=\frac{1}{6}\sum_{i=1}^{6}d_i=19.46733\ \text{mm}$$

$$u_{\text{A}d}=\frac{t_{0.95}}{\sqrt{6}}\sigma_d=0.005645\ \text{mm}\quad\left(n=6,\ \frac{t_{0.95}}{\sqrt{n}}\approx 1\right)$$

$$u_{\text{B}d}=\Delta_{\text{micrometer}}=0.004\ \text{mm}$$

$$u_d=\sqrt{u_{\text{A}d}^2+u_{\text{B}d}^2}=\sqrt{0.005645^2+0.004^2}=0.0076\ \text{mm}$$

$$u_m=\Delta_{\text{balance}}=0.05\ \text{g}$$

$$\bar{\rho}=\frac{4\bar{m}}{\pi d^2 h}=4\times\frac{213.04}{\pi\times 19.467^2\times 80.373}=0.0088662\ \text{g/mm}^3=8.8662\ \text{g/cm}^3$$

$$u_{r\rho}=\frac{u_\rho}{\rho}=\sqrt{u_{rm}^2+2^2u_{rd}^2+u_{rh}^2}=\sqrt{\left(\frac{0.05}{213.04}\right)^2+4\times\left(\frac{0.0314}{80.373}\right)^2+\left(\frac{0.0076}{19.467}\right)^2}=0.09\%$$

$$u_\rho=\bar{\rho}\,u_{r\rho}=8.8662\times 0.09\%=0.008\ \text{g/cm}^3$$

Therefore, the result of the measurement of the density of the cylinder is reported as

$$\rho=(8.866\pm 0.008)\text{g/cm}^3$$

$$u_{r\rho}=0.09\%$$

Note: In a long calculation involving mixed operations, we should carry as many digits as possible through the entire set of calculations and then round off the final result appropriately.

2.5 Common Methods of Data-Analyzing and Data-Processing

Experimental data processing contains incredibly rich contents, such as recording data, depicting graphs, extracting parameters from the data with error, validating and identifying the rule of experience, extrapolating test values and etc. In this section, we will introduce some basic methods of experimental data processing with the consideration of the basic requirements of physics experiments.

2.5.1 Graphic Method

The purpose of the graphic method is to make the curve with the experimental data on the basis of the correlation between the dependent and independent variables to reflect the changing rules or the function relationship between them. The advantage of graph lies in its intuitive image, and it is convenient to compare and display the experimental results, and to percieve some physical quantities so as to establish a relation, and etc. The following points should be paid attention to when constructing the graphic:

(1) The diagram must be drawn in the coordinate paper. The rectangular, single logarithmic, double logarithmic, or polar coordinate paper can be selected given the function relations.

(2) The size of the coordinate paper and the proportion of the coordinate axis should be set with regards to the requirements of the results and the significant numbers of measured data. In principle, the reliable numbers of the data should also be dependable in the diagram, and the unreliable digits of data should also be estimated in the graphic.

The proportion of the horizontal axis and the vertical axis division value should be properly selected so that the figure line could fully occupy the drawing space and not be shrunk into one side or in the corner. The scale of the axis division value generally selects the intervals of 1,2,5,10, and etc. in order to facilitate reading and calculation.

(3) To indicate the coordinate axis. Generally, the independent variables should be chosen as the horizontal axis and the dependent variable as the longitudinal axis. Outline axis with thick arrows indicating the direction, and specify the name and units of the shown physical quantities. Meanwhile, the dividing value must be marked on the axis

(note the valid number).

(4) Tracing points. Based on the measured data, the symbol "x" should be marked to make it accurately fall in the corresponding location on the diagram. The experimental curves on the graphics should be labeled with different marks, such as "+", "⊙", "◇", in order to avoid confusion.

(5) Connection. Based on the distributions of different experiment data points corresponding to their functional relationship, the points are to be connected into a straight line, a smooth curve or a broken line in terms of the calibration curve. It is necessary to use a straight edge or a curve plate for connecting. The connection of a straight line or a smooth curve does not necessarily run through all the points. However, to ensure that all data points are evenly distributed on both sides of the connection, individual points with large deviations should be deleted while the original data points should be retained in the graphics.

(6) To entitle the figure. Write the title of the above figure on the blank space of the graphics or below. The physical quantity represented by the vertical axis should be written in the front while the horizontal axis representing the physical quantity should be written at the back. The two physical quantities are connected with "—". Additionally, the experimental condition, the drafter and the date of the drawing can be marked as well.

The following are three typical diagrams used in physics experiments.

1. Straight Line Graph

Fig. 2-5-1 is the current-voltage characteristic curve for a linear resistor in experiment. The vertical coordinate is current, the unit value being 1 mA/cm; the Horizontal axis is voltage, unit value being 2 V/cm. That is to say, 1 big grid on the ordinate (1 cm) represents 1 mA while 1 big grid on the abscissa (1 cm) represents 2 V. With such a scale, the figure illustrated has moderate and beautiful proportions. Suppose the division value of abscissa being 1 V/cm, then the coordinate paper will widen a lot. Thus an appropriate division value is also very important in the drawing.

When drawing, the axis with arrows should be initially drawn on the graph paper. Then, the unit value, variable symbol (I, U) and unit (mA, V) on the axis are to be indicated. Symbol "×" is used on all the data points identified corresponding to the location. Considering the position of each point, a straight line can be drawn with a ruler, running through the data points as many as possible. The data points not passed by the straight line must be evenly distributed on both sides of the straight line. If the slope of the straight line is requested, we can take two points (A and B) from the line, writing down its coordinate numeric values $A(x_1, y_1)$, $B(x_2, y_2)$. With these values, the size of the slope can be calculated as $k=\frac{y_2-y_1}{x_2-x_1}$. In this process, it needs to be noticed that the locations of the two points should be near both ends of the line and the abscissa should take a rounded position for easy calculation. The marks of the two points should be distinguished from the symbols of the linear phase. In order to avoid confusion, we use the symbol ⚠ in Fig. 2-5-1. Finally, the caption of the drawing is written down in the blank space along with the name of the person who constructed the drawing and the date.

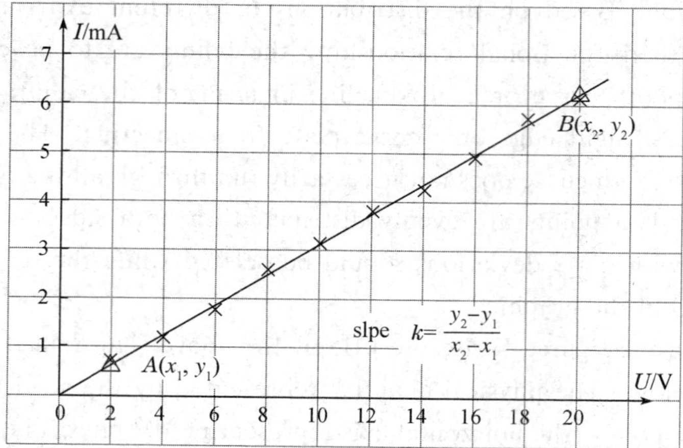

Fig. 2-5-1 Current-voltage characteristic curve for a linear resistor

2. The Curve Graph

Fig. 2-5-2 illustrates the current-voltage curve for a diode. The left part represents for negative voltage across the diode while the right part represents for positive voltage across the diode. The scales for positive voltage and negative voltage and the scales for positive current and negative current are different.

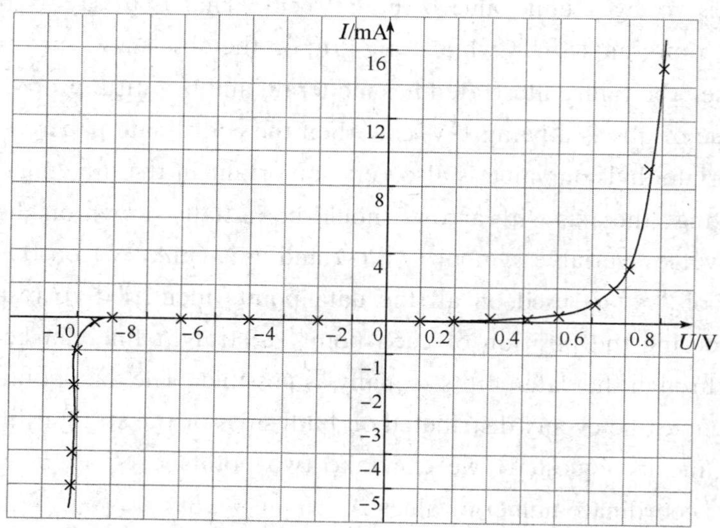

Fig. 2-5-2 Current-voltage characteristic curve for a diode

Note: When composing the graph, it is unnecessary to connect all the data points to depict the curve. We will make a smooth curve based on the position of data point rather than making the curve wind and turn because the curve direction of individual data points only stands for a trend.

3. The Line Chart

In the experiment "Converting a Galvanometer to a Voltmeter", a galvanometer is converted into a voltmeter of range 5V. The voltmeter needs to be calibrated, therefore a calibration curve needs to be plotted based on the calibrated data. The calibration curve is illustrated in Fig. 2-5-3, and the coordinate is the absolute error of each calibration $\Delta U = U - U'$, where the abscissa stands for the readings of the voltmeter.

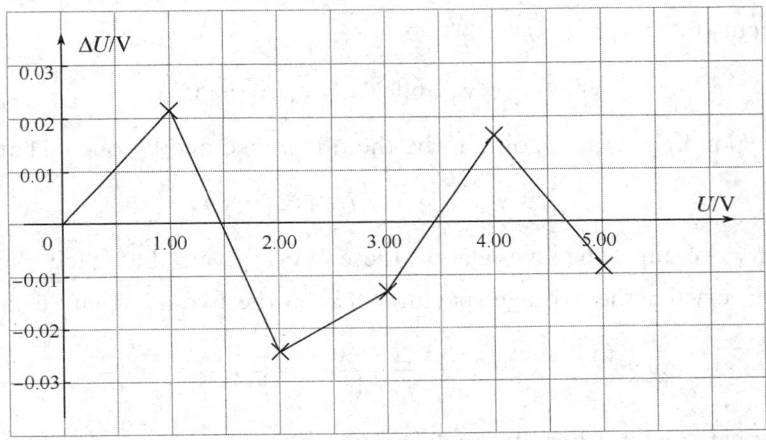

Fig. 2-5-3　Illustration of the calibration curve of the modified voltmeter

The above figures demonstrate a common characteristic, namely, the end curve in each figure occupies a bigger part of the entire paper. This principle should be allowed in the practical drawing. The size of the paper should be determined on the basis of the data and the scale to be selected. The figure should be cut after being completed and then be attached to on the experiment report. Making the entire piece of graph paper sticker is a kind of waste, since it is neither beautiful nor space-saving.

We have introduced the methods of manual drawing above. The computer drawing is also acceptable while it should follow the same scale taken of the 1, 2, 5, … lattices so as to provide the reader-friendly instructions.

2.5.2 The Successive Difference Method

This method has the advantage of making full use of the data and reducing the number of errors. However, the premise of applying this method is that the independent variables are equally spaced, and that the linear functional relation between independent variables and the dependent variables should be identified.

Suppose that the two variables satisfy the linear relationship $y=ax+b$, and the independent variable x changes through an equal interval. Assume that the measurement results (there must be an even number of variables in the measured sequence) are y_1, $y_2, \cdots, y_{2n}$, the average result with the differential method can be obtained as follows:

$$\Delta y=\frac{(y_2-y_1)+(y_3-y_2)+\cdots+(y_{2n}-y_{2n-1})}{2n-1}=\frac{y_{2n}-y_1}{2n-1}$$

The final result is only related to the two sets of data at the beginning and in the end while being independent of the data measured in the middle. As a result, the goal of reducing the error through multiple measurements can not be achieved.

The procedures of the successive difference method can be illustrated as follows: first split the independent and dependent variable sequences into two groups according to the measurement order. That is

$$y_1, y_2, \cdots, y_n \text{ and } y_{n+1}, y_{n+2}, \cdots, y_{2n}$$

The corresponding term is going to be the difference one by one. That is

$$\Delta y_i=y_{n+i}-y_i \quad (i=1,2,\cdots,n)$$

After that, we take the average of these Δy_i, given that each Δy_i contains n measurement intervals, the average spacing between the two adjacent variables is

$$\overline{\Delta y}=\frac{(y_{n+1}-y_1)+(y_{n+2}-y_2)+\cdots+(y_{2n}-y_n)}{n^2} \tag{2-5-1}$$

The difference method has clear advantages, this method requires the independent variables spacing interval to be equal. Additionally, there should be the linear function relationship between independent variable and the dependent variable, and the measurements sequence data should have even number. If the function relationship between the independent variable and the dependent variable is polynomial functional, the many-time difference method can be utilized, which will not be introduced in this section.

2.5.3 Least Square Method

If the two physical quantities x and y satisfy the linear relationship, and we receive a set of data (x_i, y_i), $i=1, 2, \cdots n$, according to an equal accuracy measurement, how to draw a straight line that is best suited the data received so as to reflect the linear relationship between the two variables? In addition to processing the data through drawing, the least square method is commonly used.

According to the least square method means: if the best-fit line is $y=f(x)$, the sum of the squares of deviation between the measured y_i value and the corresponding estimated value $\hat{y}_i=f(x_i)$ on the fitting line is the smallest. This is

$$s = \sum_{i=1}^{n}(y_i - \hat{y}_i)^2 \rightarrow \min \tag{2-5-2}$$

Because of the uncertainty of measurement, the uncertainty is also contained in x_i and y_i. To simplify the discussion of this method, we assume that each x_i value is accurate, and all the uncertainty are connected only with y_i. In this way, if the sum of the squares of deviation between the value determined by $\hat{y}_i = f(x_i)$ and the measured value y_i is the least, the straight line fitted by the least square method is the optimal line.

In general, the equation of the line can be expressed as: $y = kx + b$. Where k is the slope of the undetermined line; b is the intercept of the undetermined line. If the two parameters are determined, the line can be set. So, the process of setting the fitting line means determining k and b by using the experimental data group (x_i, y_i). Putting the upper equation into Equation (2-5-2), we can obtain the following formulas:

$$s(k,b) = \sum_{i=1}^{n}(y_i - kx_i - b)^2 \rightarrow \min \tag{2-5-3}$$

k and b are the solutions of the following equations to be obtained:

$$\begin{cases} \dfrac{\partial s}{\partial k} = -2\sum\limits_{i=1}^{n}(y_i - kx_i - b)x_i = 0 \\ \dfrac{\partial s}{\partial b} = -2\sum\limits_{i=1}^{n}(y_i - kx_i - b) = 0 \end{cases}$$

where $\sum$ represents the sum of i from 1 to n. After the expansion of upper equation and the elimination of the unknown number b, we obtain the following formulas:

$$k = \frac{l_{xy}}{l_{xx}} \tag{2-5-4}$$

where

$$\begin{cases} l_{xy} = \sum\limits_{i=1}^{n}(x_i - \bar{x})(y_i - \bar{y}) = \sum\limits_{i=1}^{n}(x_i y_i) - \dfrac{1}{n}\sum\limits_{i=1}^{n}x_i\sum\limits_{i=1}^{n}y_i \\ l_{xx} = \sum\limits_{i=1}^{n}(x_i - \bar{x})^2 = \sum\limits_{i=1}^{n}x_i^2 - \dfrac{1}{n}\left(\sum\limits_{i=1}^{n}x_i\right)^2 \end{cases} \tag{2-5-5}$$

When k is placed into the equation, we obtain:

$$b = \bar{y} - k\bar{x} \tag{2-5-6}$$

Until now, the line equation to be fitted $y = kx + b$ is uniquely identified.

According to the above equation, it's not difficult to obtain the final result, and the optimal configured line must pass through $(\bar{x}, \bar{y})$. Therefore, when fitting the line graph, the fitted line must pass through this point.

In order to check out the meaningfulness of the fitted line, the correlation coefficient r is introduced, which demonstrates the degree of conformity of the function relation between two variables with the linear function, which is defined as:

$$r=\frac{l_{xy}}{\sqrt{l_{xx}l_{yy}}}$$

In the formula, the calculation method of l_{yy} is similar to that of l_{xx}, and the much closer r is to 1, the better linear relationship between x and y is. If r goes to 0, it can be inferred that there is no linear relationship between x and y.

In physics experiments, there are some situations where the relationship between two measured physical quantities x, y conforms to a certain curve equation rather than to a straight line. Based on the curve equation, various transformations can be made and new variables can be introduced. As a result, some curve fitting problems can be converted to a linear fitting problem.

For example, if the curve equation is $y=ax^{\alpha}$, the natural logarithm on both sides of the equation can be taken, obtaining $\ln y=\alpha\ln x+\ln a$. Then make that $Y=\ln y$, $X=\ln x$, $b=\ln a$, and the power function can be also converted into a linear function $Y=\alpha X+b$.

Similarly, for a curve described by equation $y=a\mathrm{e}^{\alpha x}$, we can take natural logarithm on both sides and obtain $\ln y=\alpha x+\ln a$. Then make that $Y=\ln y$, $b=\ln a$, the exponential function can be converted into a linear function $Y=\alpha X+b$.

2.6 Significant Figures and Rules for Calculation

Due to the presence of measurement errors, any measurement has its certain accuracy. Therefore, the measurement of a physical quantity or the digits of computing results should not be stated indefinitely. If a metric scale of resolution 1 mm is used to measure the length of an object, the one tenth position of its measurement result is estimated. Since the possible errors make it unreliable, it would be meaningless to write the digits after the one tenth position. Then in common cases, to what level can the measurements be accurately determined? Where may the initial errors occur? How many digits should be used in the calculation of data processing to ensure that the reasonability of the calculation results? How can we achieve the accuracy of the actual measurement without damaging or exaggerating it? These are the questions to be discussed in this section of significant numbers.

2.6.1 Significant Figures

The digits what correctly and effectively represents measurement results or calculation results are called significant figures. The significant figure is made up of an accurate (reliable) digit and an inaccurate (suspicious) digit.

The significant figure = a number of reliable digits + a suspicious digit.

For example, when measuring the length of a board with a metric scale, the left side of the board to be measured aligns with the metric scale zero and the other end is between 76 mm and ~77 mm, as shown in Fig. 2-6-1. Taking millimeter as the unit, the reading is 76.6 mm, the exact number of the readings is 76 mm; and the over part of the 76 mm needs to be estimated up to one tenth of the minimum spacing of the meter ruler, namely around 0.6 gap. The number 6 after the decimal point is the estimated number. Despite of the inaccuracy, the estimated digit, within a certain extent, can reflect the objective reality, suggesting that the length of the board may be around 76.5~76.7 mm. Due to the difference of the observers in the resolution, the estimation of a suspicious digit can accept the error of about ±0.1 spacing. When using a ruler with smallest scale 1 cm to make the measurement and set centimeter the unit, the reliable measurement can only reach 7 cm, and the parts over 7 cm need to be estimated. The reading is about 7.6 cm or 7.7 cm.

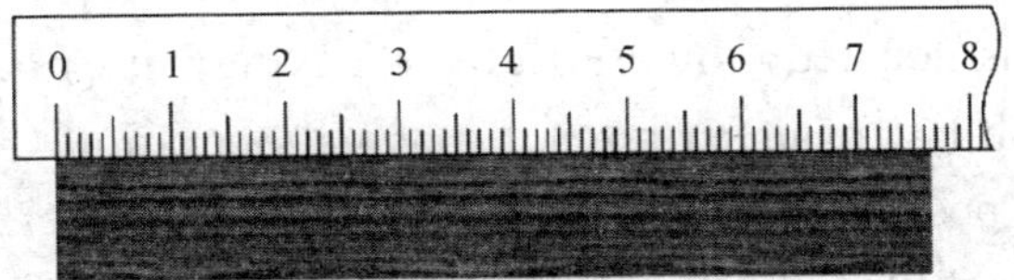

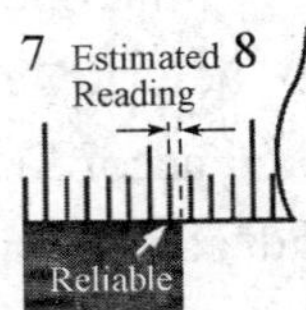

Fig. 2-6-1　The use of metric scale to measure the length of the board

The measurement result of 76.6 mm has three significant digits, and 7.6 cm and 7.7 cm have two significant digits. The number of significant digits shows that the accuracy of the measurement is associated with the seleeted measurement tools. Consequently, choosing a different instrument to measure the same quantity would result in different numbers of significant digits in measurements or operations.

Rules for deciding the number of significant digits in a measured quantity:

(1) All nonzero digits are significant: 1.234 g has 4 significant digits. 1.2 g has 2 significant digits.

(2) Zeroes between nonzero digits are significant: 1002 kg has 4 significant digits. 3.07 mL has 3 significant digits.

(3) Zeroes to the left of the first nonzero digits are not significant; such zeroes merely indicate the position of the decimal point: 0.001 ℃ has only 1 significant digit. 0.012 g has 2 significant digits.

(4) Zeroes to the right of a decimal point in a number are significant: 0.023 mL has 2 significant digits. 0.200 g has 3 significant digits.

(5) When a number ends in zeroes that are not to the right of a decimal point, the

zeroes are not necessarily significant: 190 m may have 2 or 3 significant digits, 5060 g may have 3 or 4 significant digits. The potential ambiguity in the last rule can be avoided by the use of standard exponential, or scientific notation. For example, depending on whether 3 or 4 significant digits are required, we could write 5060 g as:

5.06×10^3 g (3 significant digits); 5.060×10^3 g (4 significant digits)

In unit conversion, the number of significant digits should remain the same.

2.6.2 Choosing Digits of the Measurement Data

To report measurement results, both the optimal values and the uncertainty are given. This can provide people with a clear understanding on the range of measured values. On some occasions, where the uncertainty of measured values is not required to be calculated, measuring the digits of effective numbers can approximately determine the level of uncertainty. In general, the valid numbers recorded have an error, and the only one digit for the error term is usually taken, no more than two. This book sets the convention that when all the effective digits participate in the effective digital operation, the operation result should be retained to the bottom digit with error. As for how to determine where an error is located, theoretically speaking, it should be calculated according to the propagation law of uncertainty. In actual operation, for the easy and convenient calculation, the position can be taken according to the effective numerical algorithm or some rational skills and the results obtained are close to the theoretical results.

(1) When the measuring instrument or meter gives the maximum permissible error $\Delta_{\text{instrument}}$ of the instrument, the reading of the digits should be set where the instrument error is located;

(2) When a ruler is used as the measuring instrument, the reading of the digits should be estimated between the two calibration lines of the ruler since the error of the instrument is usually a few tenths of the dividing value;

(3) The digit reading of the equipment with a cursor is not estimated;

(4) Reading digital instruments generally takes the number of digits displayed by the instrument rather than estimating the reading.

Although case (3) and case (4) are not directly required to be estimated, the last digit of its reading is still the less accurate value. For the instrument with vernier scale, the reading is determined by whether the scale lines of the main scale are aligned with the vernier scale. The aligned scale lines are the merely superficial aligned instead of being really aligned.

2.6.3 Rules for Mathematical Operations

1. Addition and Subtraction

In addition and subtraction, the result is rounded off to the last common digit occurring furthest to the right in all components.

For example, $100+23.643=123.643$, which should be rounded to 124 (3 significant digits). The last common digit is furthest to the right of 100 and 23.643 is one digit.

2. Multiplication and Division

In multiplication and division, the result should be rounded off so as to have the same number of significant digits as in the component with the least number of significant digits.

For example, $3.0\times12.60=37.800$, which should be rounded off to 38 (2 significant digits). Because 3.0 has 2 significant digits while 12.60 has 4, the result of 3.0×12.60 has 2 significant digits.

3. Function Operation

The general rule for function operation can be described as the following steps: to change the last number of significant digits of the independent variable by 1 unit, and calculate the pre-change and post-change function values respectively; to compare the two function values from left to right and the results are saved to the first difference between the two function values. In actual calculation, the following simplified scheme can be adopted.

(1) Logarithmic function

All logarithmic function have two parts. The decimal part is called the mantissa. The digits in the mantissa are the only significant digits in a logarithm. The mantissa should have the same number of digits as the number from which it was derived.

For example, $x=888$ (3 significant digits). Calculating with a calculator, we obtain $\ln 888=6.78891$. Therefore, $\ln 888=6.789$ (The mantissa 0.789 has 3 significant digits, the same as 888).

(2) Exponential function

The number of digits in the mantissa gives the number of significant digits in the resulting number.

For example, $x=2.28$, the mantissa 0.28 has 2 significant digits. $10^x=10^{2.28}=190.54607$ has 2 significant digits too. Therefore, $10^{2.28}=1.9\times10^2$.

(3) Trigonometric functions

The position of the trigonometric function depends on the significant number of angles. When measuring angle with spectrometer generally, should read 1 minute. In this case, we should take four significant digits.

4. Power and Square Root Operation

The resulting number of significant digits is the same significant digits as the measurement if the degree of power or square root is not too high.

For example, A=4.25 (3 significant digits), $A^2=4.25^2=18.0625$ should have 3 significant digits. Therefore, $A^2=4.25^2=18.1$.

5. The Exact Number

Some numbers are exact because they are known with complete certainty. For example, the formula for radius to diameter is $d=2r$, 2 is the exact number. The experimental measure is n, which is always a positive integer. The calculation rules of significant digits do not apply to these exact numbers. In calculation, the significant number of the calculation result is determined only by the significant number of the measured value, and the correct number has no effect on the significant number multiple of the calculation result.

In the process of calculation, we may encounter some constants, such as g and e, which we usually take to be the same as the number of significant digits measured. For example, the circumference of a circle should be at least 3.142. In actual calculation, calculator or spreadsheet software is used to keep the calculation results to the appropriate number of significant digits according to the calculation rules and modification rules, rather than the number of digits calculated as much as possible.

The number of significant digits depends on the measuring instrument, not the calculation process. Therefore, when choosing the measuring instrument, the number of digits given by it should not be less than the significant digits, otherwise the accuracy of the measurement results will be reduced.

第 3 章　基本测量方法及常用仪器介绍

3.1　物理实验中基本测量方法及基本仪器介绍

一、物理实验中基本测量方法

物理学是一门实验科学，凡物理学的概念、规律及公式等都是以客观实验为基础的。实验是发现规律、形成理论的基础，而理论又对实验有着指导和预见的作用，两者的关系是密不可分的。正是两者之间相互促进的关系造就了物理学的非凡成绩。因此物理实验的重要性是不言而喻的。

物理实验是以测量和观测为基础的，任何物理实验都离不开物理量的测量。在实验物理学中，对各种物理量的研究和测量已经形成了自身的理论和卓有成效的测量方法。这些基本测量方法不仅对物理学科是必需的，而且由于这些理论和方法还有其基本性和通用性，对其他有实验的学科也是必不可少的。物理实验种类繁多，相应的待测物理量也非常广泛。它包括力学量、热学量、电磁学量和光学量等等。因此，以实验装置和实验技术为手段的测量方法也非常多。下面对物理实验中最常用的几种基本测量方法作简单的介绍。

1. 比较法

实验中将相同类型的被测量与标准单位进行比较，测出其大小的方法称为比较法。比较法是物理实验中最基本、最常用、最重要的测量方法。主要过程为：首先确定与被测量为同类量的一个单位量，将此单位量作为标准，然后将被测量与此单位量进行比较，求出它们的倍率，得到测量结果。一般可分为直接比较法和间接比较法。

直接比较中，标准单位一般可选标准量具，这样测量的准确度主要决定于标准量具的准确度。比如用钢直尺测量长度、用等臂天平测量物体质量时，长度测量结果的准确度取决于钢直尺的准确度等级、质量测量结果的准确度决定于砝码的准确度等级。

在一些物理量难以用直接比较法测量时，可以利用物理量之间的函数关系将待测物理量与同类标准量进行比较。例如，用惠斯登电桥测电阻时，就是利用电桥电阻间的函数关系将待测电阻与标准电阻进行比较而得到结果。

2. 放大法

在物理实验中常遇到一些微小量或微小变化量，这些量难以直接测量或直接测量时精度不高。这时,为了方便测量或为提高测量精度,常采用合适的放大方法,选用相应的测量装置将被测量放大后再进行测量。根据放大方式的不同又分为累计放大、机械放大、电子学放大和光学放大。

当被测量能够进行简单叠加,或能多次连续重复时,可以使用累计放大的方法。例如在用秒表测量一个单摆摆动的周期时,为了提高准确度,一般不是测一个周期,而是测量 50 个周期或 100 个周期。在光的干涉实验中测量干涉条纹的间距或宽度时,出于同样的原因,一般测量十条或几十条条纹之间的间距。这些都是累计放大的方法。50 或 100 个周期的时间。这就是累计放大。若所用的秒表的仪器误差为 0.1 s,设某单摆的周期约为 1 s,则测量单个周期时间间隔的相对误差为

$$E=\frac{0.1}{1.0}=10\%$$

若累计测量 100 个周期的时间间隔,则相对误差为

$$E=\frac{0.1}{100.0}=0.1\%$$

可见,测量精度大为提高。

机械放大是利用传动原理将测量仪器的读数机构细分,从而提高测量精度的方法。其中螺旋测微放大法是一种典型的机械放大法。利用螺旋结构把螺杆上的 1 mm 分成 100 份可在螺旋筒上,相当于把毫米间隔放大了 100 倍。使用这种放大方法的仪器有,螺旋测微器(千分尺)、声速测量仪、读数显微镜、迈克尔孙干涉仪等。

电子学放大是利用电子学的方法把较弱的电信号先放大再测量的方法。使测量灵敏度得到提高。该方法亦可用于把其他非电学量经传感器转换为电信号放大后的测量。

光学放大是将需要测量的微小量或观察到的微小效应,应用光学器件按一定的规律将它们“放大”显示,以便于更准确地观察和测量。主要是利用光学的方法来放大视角以提高测量精度。使用这种方法的仪器主要有放大镜、显微镜、望远镜等。

3. 补偿法

补偿法是通过采用一个可以变化的附加装置,用以补偿实验中某部分损失或变化,使系统处于补偿状态或平衡状态。简而言之,补偿法就是将因种种原因使测量状态受到的影响尽量加以弥补。例如,用电压补偿法弥补因用电压表直接测量电压时而引起被测支路工作电流的变化;用温度补偿法弥补因某些物理量(如电阻)随温度变化而对测试状态带来的影响;用光程补偿法弥补光路中光程的不对称等等。

4. 模拟法

模拟法是利用某些现象或过程存在的相似性,人为地制造一个与研究对象的物理现象或过程相似的模型,使所研究的物理现象重现、延缓或加速,通过测量这个模型来获得测量结果的方法。例如,在设计飞机外形时,通常按比例设计一个飞机模型,放入风洞中进行试验。通过测量模型的动力学参量而相应的得到飞机原尺寸的动力学参量。在静电场的描绘实验中,由于稳恒电流场的与静电场的等位线都满足相同的数学方程式,则可采用模拟法测

量静电场的分布。另外,由于计算机仿真技术的发展,有时采用计算机模拟法,即采用适当的数学模型可以把一个物理系统用计算机程序来代替,从而改为在计算机上进行实验。

5. 转换法

转换法是将一种物理量转化为另一种物理量进行观测的方法。在物理量的测量中,常常会遇到一些不可直接测量或不易测准。为了使这些物理量变得可测、易测或为了提高测量的精度、减少测量的误差,常常将被测量转换为其他的物理量来进行测量,然后再根据转换关系求得被测量。

在转换法中应用最多的是将非电学量转化为电学量来测量。其中的关键是传感器,传感器是一种能以一定的精确度把被测量转化为与之有确定对应关系的、便于应用的某种物理量的器件。我们常用到的有以下几种:

(1) 热电式传感器,包括热电阻传感器和热电偶传感器。热电阻传感器是根据电阻随温度变化的特性制成,热电偶是根据热电效应将温度转换为电势差的测量。

(2) 压电式传感器,是利用某些固体电介质材料的压电效应制成。当这种材料受到外力而发生机械形变时,其内部晶体点阵中会产生极化,在表面产生束缚电荷,形成与极化方向一致的电场,于是可以将压力的测量转换为对电场的测量。反之,对压电晶体加上电场,材料会发生机械形变。这两个过程分别成为压电效应和逆压电效应。

(3) 磁电式传感器,主要是霍尔传感器。将通有电流的半导体材料霍尔片放在磁场中,并在两相对的薄边加上电压使内部通以电流,如果电流方向与磁场方向垂直,则在与磁场和电流方向都垂直的方向上会产生横向电势差,该电势差称为霍尔电压,这个现象称为霍尔效应。利用霍尔传感器可以实现非接触测量,并且不从磁场中获取能量。

(4) 光电式传感器,是将光信号转化成电信号进行测量的一种传感器件。这种测量方法具有结构简单、非接触性、高可靠性、高精度和反应快等优点,被广泛用于自动控制检测技术中。该类传感器的基础是光电效应。如利用外光电效应制出的光电管、光电倍增管,利用内光电效应制出的光敏电阻、光导管,利用光生伏特效应制出的光电池、光敏晶体管等。

3.2 物理实验的常用仪器

物理实验中,需要进行物理量的定量测量和定性观测。而测量和观测必须要借助适当的实验仪器来进行。物理实验仪器种类繁多,有力学、热学、声学、电磁学、光学等种种仪器。各种实验仪器的使用将在相关实验中具体介绍。本节对物理实验常用仪器作一些介绍,以期事先了解仪器的性能、掌握仪器的原理、布局及线路连接的常识。

一、力学、热学实验仪器

1. 游标卡尺

游标卡尺是由主尺和可沿主尺滑动的游标尺(副尺)构成。游标尺是为了帮助实验者比较准确地对主尺最小刻度后面的读数进行估读。游标可分为直游标和角游标,直游标就是常见的游标卡尺,角游标应用于分光计。尽管各种游标的形状、长度、游标的分度格数各不相同,但是基本原理和读数方法都是一样的,游标卡尺读数为主尺 0 刻线到游标 0 刻线间的距离。

下面以图 3-2-1 所示的分度值为 0.02 mm 的直游标卡尺为例来介绍其读数方法。

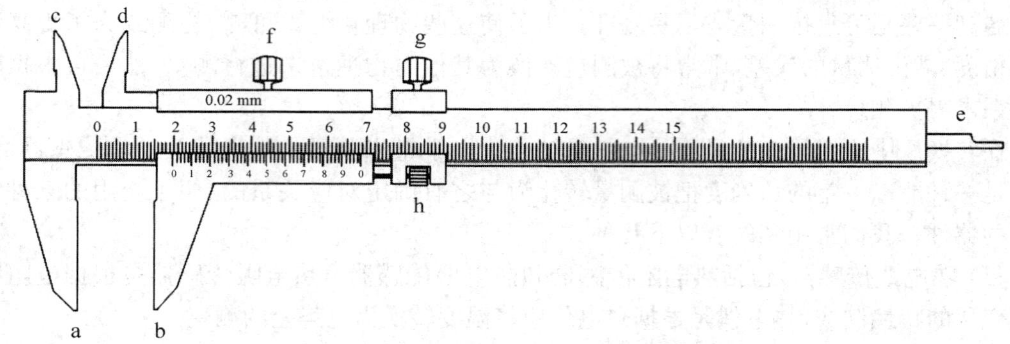

图 3-2-1　游标卡尺构造

a, b—外径卡口; c, d—内径卡口; e—测深端; f, g—固定螺钉; h—微调螺母。

该型号游标卡尺主尺上的最小刻度为 $a=1$ mm。游标共有格数 $N=50$,游标总长度为游标的 49 格的长度,为 49 mm(图 3-2-2)。则游标 1 格的长度 $b=49/50=0.98$ mm,则主尺 1 格与游标 1 格的差值 $\delta=a-b=0.02$ mm,δ 称为游标的分度值,即主尺最小分度与游标格数的比值

$$\delta=\frac{\text{主尺最小分度 } a}{\text{游标格数 } N}=\frac{1}{50}=0.02\ \text{mm}$$

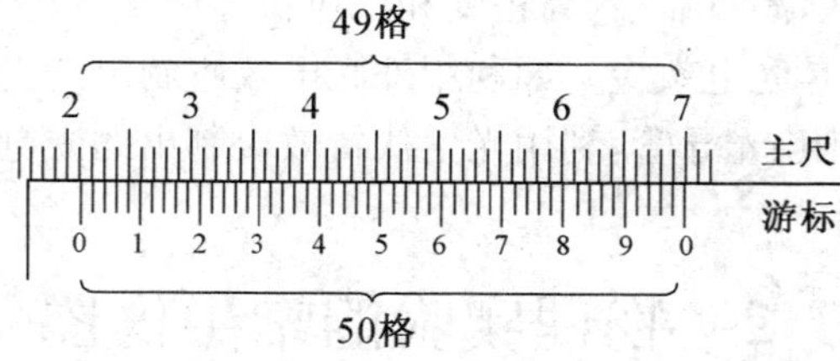

图 3-2-2　游标卡尺的游标

当游标的 0 刻度线与主尺上的刻度线对齐时,最后一条线亦与主尺一条线对齐,见图 3-2-2。此时游标的第一条线与主尺上右边最近的线相距 δ,第二条线与主尺上右边最近的线相距 2δ,以此类推,第 n 条即为 $n\delta$。由此,当游标上的第 n 条线和主尺上的线对齐时,则游标的 0 刻度线与主尺上左边最近的线相距 $n\delta$,这样我们就知道游标的读数。

实际读数时,游标 0 刻线左边整数毫米部分在主尺上读出,不足 1 mm 部分在游标上读出,具体方法如下:

将游标总长看成主尺最小 1 格长度,50 分度的游标卡尺为 1 mm,这个 1 mm 被等分成了 50 小格,游标中的数字所在线为 1 mm 的 10 等分点。在游标中找到与主尺中某条刻线最对齐的一根线,根据游标上标注的数字,很容易读出这条线左边部分在 1 mm 中占多少。

如图 3-2-3 所示，游标 0 刻线左边整数毫米部分读数为 102 mm，游标上不足 1 mm 部分的读数为 0.66 mm，所以被测物长度为 102.66 mm。

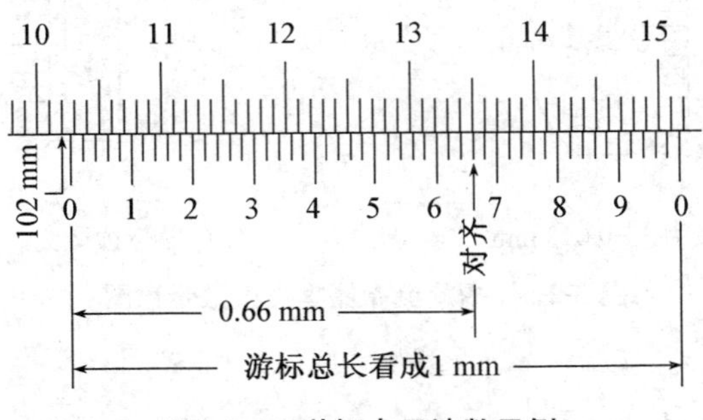

图 3-2-3 游标卡尺读数示例

实验中常用的 50 分度游标卡尺的仪器误差为 $\Delta_{仪}=0.02$ mm。

2. 螺旋测微器

螺旋测微器又叫千分尺，用于精确测量较短的长度，例如金属丝的直径、薄板的厚度等。实验室常用的螺旋测微器外形如图 3-2-4 所示。其结构的主要部分是精密测微螺杆和套在螺杆上的螺母套管以及与测微螺杆固定上的微分套筒。螺母套管上的主尺有两排刻线，毫米刻线和半毫米刻线。微分套筒圆周上刻有 50 个分度，螺距为 0.5 mm，则微分套筒每旋转一周，螺杆移动(前进或后退)一个螺距 0.5 mm，则套筒每转动一个分度，螺杆移动的距离为$\frac{0.5\ \text{mm}}{50}=0.01$ mm，即螺旋测微器的分度值为 0.01 mm。

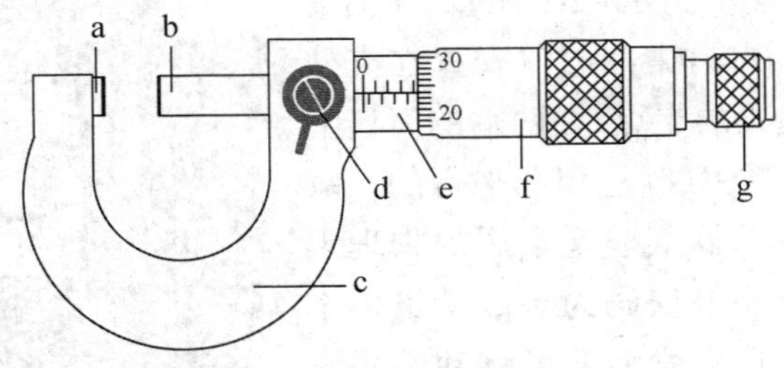

图 3-2-4 螺旋测微器

a—测砧；b—螺杆；c—支架；d—锁紧装置；e—螺母套管；f—微分套筒；g—棘轮。

使用螺旋测微器测量前，应先记录其零点读数(零点误差)，即测微螺杆与测砧合上之后，螺旋测微器的读数。测量物体长度时要将直接读数减去零点读数，才是待测长度。零点读数的大小是微分套筒上的 0 刻度线与螺母套管横线间的读数。当微分套筒上的 0 刻度线在螺母套管横线上方时，零点读数为负值，如图 3-2-5(a)所示；反之，零点读数为正值，如图 3-2-5(b) 所示。

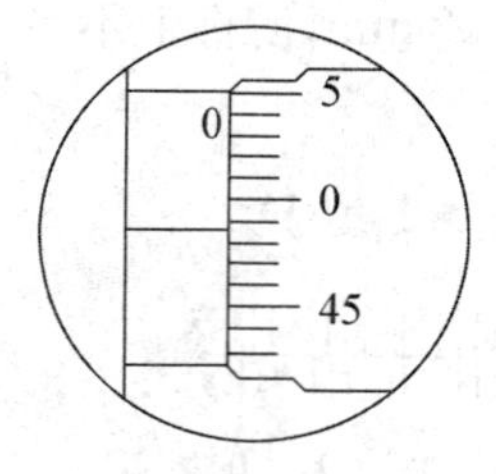

(a) 零点读数为：−0.013 mm

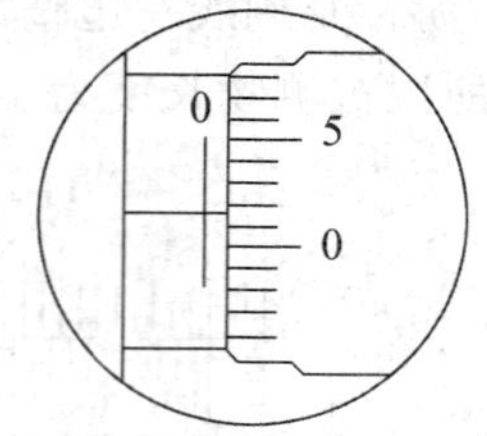

(b) 零点读数为：+0.018 mm

图 3-2-5　螺旋测微器零点读数示意图

使用注意事项：

(1) 手握螺旋测微器的支架部分，被测工件也尽量少用手接触，以免因热膨胀影响测量精度。

(2) 测量时须用棘轮。测量者转动螺杆时对被测物所加压力大小，会直接影响测量的准确。为此，在结构上加一棘轮作为保护装置。当测微杆端面将要接触到被测物之前，应改用旋转棘轮，直至接触上被测物时，它就自行打滑，并发出“咔咔”声，此时即应停止旋转棘轮，进行读数。

(3) 使用完毕放入盒内时，应将螺杆退回几圈，留出空隙，以免热膨胀使螺杆变形。

实验中常用的螺旋测微器的仪器误差为 $\Delta_{仪}=0.004$ mm。

3. 物理天平

物理天平是根据等臂杠杆的原理制成的测量质量的仪器。物理天平的构造如图 3-2-6 所示，在横梁的中点及两端共有三个刀口，中间刀口安置在支柱顶端的玛瑙垫上，作为横梁的支点，在两端的刀口上悬挂两个等重的秤盘，三个刀口是相互平行的。每架物理天平都配有一套砝码，实验室常用的物理天平最大称量为 1000 g，最小的砝码为 1 g，质量中小于 1 g 的部分，则可通过在横梁上的可移动游码测出。横梁上每格分度值为 0.05 g。游码在横梁上向右移动一个分度，就相当于在右盘中加上 0.05 g 的砝码，横梁下部装有读数指针。立柱上装有标尺。根据指针刻度标尺上的数来判断天平是否平衡。物理天平的操作步骤：

图 3-2-6　物理天平

(1) 调节水平螺钉使支柱铅直，这可利用底座上的水准泡是否在中间来检查。

(2) 调整零点。把游码拨到刻度 0 处，将秤盘吊钩挂在两端刀口上，转动止动旋钮，支起天平横梁，观察指针的摆动情况，判断天平是否平衡；当指针在标尺的中线左右作等刻度摆动时，天平即达到平衡，否则，可以调整平衡螺母置至平衡。

(3) 称量，将待测物体放在左盘内，砝码放在右盘内进行称量。每次加减法码，应止动天平以便保护刀口，全部称完后将秤盘摘离刀口。

物理天平的操作规则：

（1）天平的负载量不得超过其最大称量，以免损坏刀口或横梁。

（2）为了保护刀口，必须切记：在取放物体、加减砝码、调节平衡螺母以及不使用天平时，都必须将天平止动，只有在判断天平是否平衡时，才能启动天平。无论是启动还是止动时，动作要轻，止动最好在天平指针接近标尺中间刻度时进行。

（3）砝码不得直接用手拿取，只准用镊子夹取，从秤盘上取下砝码应放入砝码盒中。

（4）天平的各部分以及砝码盘易锈蚀，应注意避免直接称量高温物体、液体以及带腐蚀性的化学药品。

实验室中物理天平的仪器误差可按表 3-2-1 考虑。

表 3-2-1　常用天平的仪器误差

天平型号	载荷 0～250 g 的 $\Delta_{仪}$	载荷 250～500 g 的 $\Delta_{仪}$	载荷 500～1000 g 的 $\Delta_{仪}$
TW05	0.05 g	0.1 g	
TW1	0.1g		0.2g

4. 电子天平

实验室所用电子天平，其构造如图 3-2-7 所示，它的最大量程为 200 g，操作简易程序为：

（1）调平：调整地脚螺丝使水平仪内空气泡位于圆环中央；

（2）开机：接通电源，按开关键 ON/OFF 进入待测状态；

（3）预热：初次通电，至少要预热 30 min 以取得理想的工作状态。

图 3-2-7　电子天平

（4）称量：使用去皮键，去皮清零。放置样品进行称量，自动显示重量。

5. 温度计

温度表明物质的冷热程度，是表征物质微观热运动程度的一个状态参量。测量温度的方法很多，其测量原理是利用对温度敏感元件的某一物理特性随温度的显著变化而设计的。如根据热膨胀规律制成的液体温度计，是利用感温液体受热膨胀、遇冷收缩的原理进行温度测量的。一根粗细均匀的毛细玻璃管连接一贮液泡，贮液泡内装有液体（水银、酒精等），在一定温度范围内，液体在毛细玻璃管中升降的位置和温度成线性，依此可读出温度值。此外，根据电阻随温度变化可以制成金属电阻温度计、半导体电阻温度计，利用不同金属温差电动势的变化可以制成热电偶温度计。测量高温时可利用辐射规律可以制成光测高温计等等。

二、电磁学实验仪器

1. 直流稳压电源

直流稳压电源是将 220 V 交流电，经降压（或升压）、整流、滤波、稳压之后，得到的稳定的直流电压源，也就是说虽然输出电流可以在一定范围内随负载变化，输出电压却保持不变。通常直流稳压电源的输出电压值都是可调的,转动调节旋钮,可以选取所需要的直流电压值。有的直流稳压电源面板上装有电压表和电流表,直接显示出输出的电压值和电流值,使用时若超过额定电流,输出电压将会不稳定,而且会烧断保险丝,或因内部自动保护而切断电路。我们所用的直流稳压电源内阻一般在零点几欧姆以下,输出相当稳定。使用时注意切不可将电源正负极用导线短路,这时电流非常大,会烧断保险丝。短路或接错极性,都是电学实验的大忌,应务必注意。

WYK－303B_2 型双路直流电源简介。WYK 系列稳压、稳流电源是由变压器、整流滤波电路、大功率三极管等组成,输入交流电由变压器降压,经整流滤波、大功率三极管调整及基准取样放大后输出。双路电源可单独输出,也可串联或并联使用。串联时从路输出电压跟踪主路输出电压;并联时最大输出电流可达到两路单独输出电流之和。本系列稳压电源,采用全桥斩波提高整机功率,输出电压可从零伏起调,输出电流可以从零安设置。WYK－303B_2 型双路直流电源如图 3-2-8 所示。

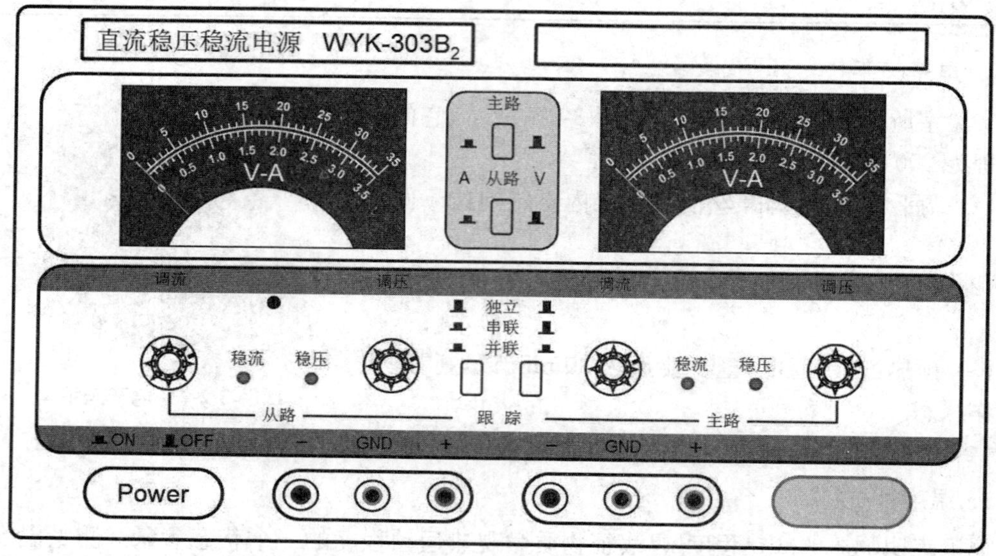

图 3-2-8 WYK－303B_2 型直流稳压、稳流电源面板

(1) 电源开关

按下电源开关时,两路输出端均有输出;主路和从路的电压电流值可分别在面板上的"V-A"表上显示。两表中间有两个按钮开关,有按下和弹出两种状态,面板上有提示:上边的按钮选择按下右边的表显示主路电流,弹出显示主路电压;下边的按钮按下左边的表显示从路电流,弹出显示电压。

(2) 输入输出

输入为 220±10%V、50 Hz 交流电。输出有 2 路:主路和从路通过"＋""－"分别由红、黑接线柱供给,而"GND"(⊥)接线柱的通常空置。

(3) 工作方式选择开关

在主路和从路之间有两个按钮开关,有按下和弹出两种状态,面板上有提示:当两个按钮均在弹出状态，此时两路（主、从）电源为各自独立工作；当从路的按钮按下而主路的按钮弹出时，为两路电源串联输出；当两个按钮均按下，此时两路（主、从）电源为并联工作。

（4）稳压使用

如使用主路电压源输出，要将主路电源的“稳流”调节旋钮置最右端（即电流输出最大），此时主路的“稳压”调节旋钮边的“稳压”指示灯亮时，就可以通过调节“稳压”旋钮，可得到稳压输出。同理，从路的稳压输出调节类似，要避免电压源外接的负载短路。

（5）稳流使用

如使用主路电流源输出，要将主路电源的“稳压”调节旋钮置最右端（即电压输出最大），此时主路的“稳流”调节旋钮边的“稳流”指示灯亮时，就可以通过调节“稳流”旋钮，可得到稳流输出。同理，从路的稳流输出调节类似，要避免电流源外接的负载开路。

（6）注意事项

在使用电源的过程中，如果因为外接电路的短路或者开路等错误操作时，电源自身的保护报警装置会以长时间蜂鸣声来提示，操作者应该尽快关闭 WYK－303B_2 型双路直流电源的。

2. 电表

电表是量度电学量的仪器，分为交流和直流两种，而在普通物理实验中所接触的电表都是直流电表。

（1）直流磁电式模拟指针电表

图 3-2-9 为磁电式毫安表面板示意图，电流表和电压表是由一只表头根据需要并、串联上合适的电阻而得到。该类型电表读数时首先要考虑所选取的量程，电表指针满偏即为所选量程的值，再根据指针所在的位置而读取实际读数。电表使用时应注意：

线路连接：电流表必须与要测量电流的电路串联，而伏特表必须与需要测量电压的电路并联。

正负极性：作为直流电表使用时必须注意电表的极性问题，如果极性反接，则电表指针反偏，容易损坏仪表。正极表示电流从这个极流入，而负极表示从这个极流出。通常电表的正极标志为“＋”，负极标志为“－”。

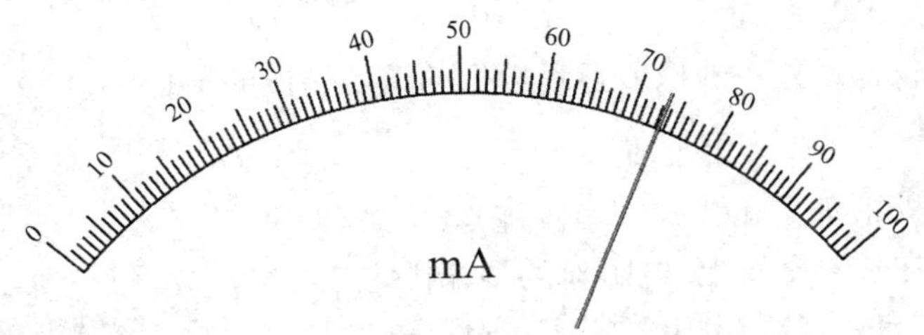

图 3-2-9　磁电式毫安表面板示意图

量程：量程指仪器允许量度的最大数值。当测量的电流或电压值超过电表量程，就得更换量程大的电表来进行测量，反之，测量的电流或电压值远小于电表的量程，就得用小量程的电表。如果测量的电流或电压值事先不能估计，一般情况下，应先用较大量程的电表测试，然后改用适当量程的电表，避免打坏电表指针甚至烧坏电表。

电表的误差与电表的等级:在测量电学量时,由于电表本身结构及测量环境的影响,如温度、外界电场和磁场等环境影响均能引起测量结果有误差,而电表本身结构缺陷(如摩擦、游丝残余形变、装配不良及标尺刻度不准确等)产生的误差则为仪表基本误差,它不因使用者不同而变化,因而基本误差也就决定电表所能保证的准确程度。常用仪表的准确度用等级来定义。

电表的等级是反映电表的准确度的一个参数,由电表的测量误差决定的。为最大绝对误差与仪表量程比值的百分数。我国规定电表分为7个等级,它们是0.1,0.2,0.5,1.0,1.5,2.5,5.0级。数值越小,电表的精确度越高(电表的等级标在表盘上)。知道了电表的准确度等级即可求得仪器误差

$$\Delta_{仪}=量程\times准确度等级$$

电表的实际操作:在实际选用电表时,在待测量不超过所选量程的前提下,应尽量使电表指针有较大的偏转,一般至少偏转2/3以上。

(2) 数字电表

图3-2-10为数字电表面板示意图。数字仪表是直接用数码管显示数据的仪表,在使用中有它的优越性,测量准确度高,读数方便,即使极性输入接反、输入超量程的电量,一般也不会对仪表造成损坏。尽管精度较高,在用数字仪表测量的过程中,如果使用不当,也可能产生很大的误差。在使用时为了减少测量误差,应使显示的数位尽可能得多,这样就可以减小误差,这需要引起使用者注意。

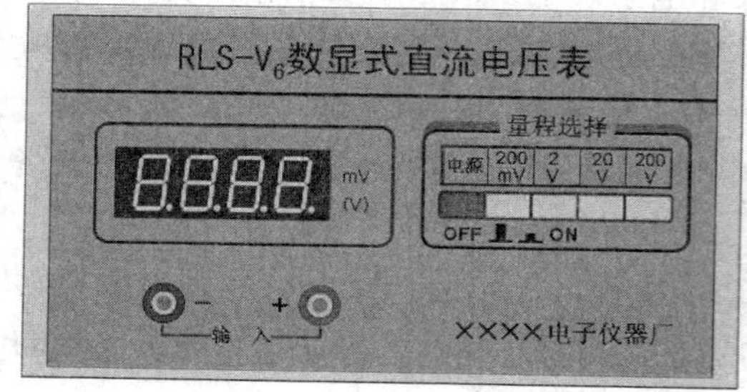

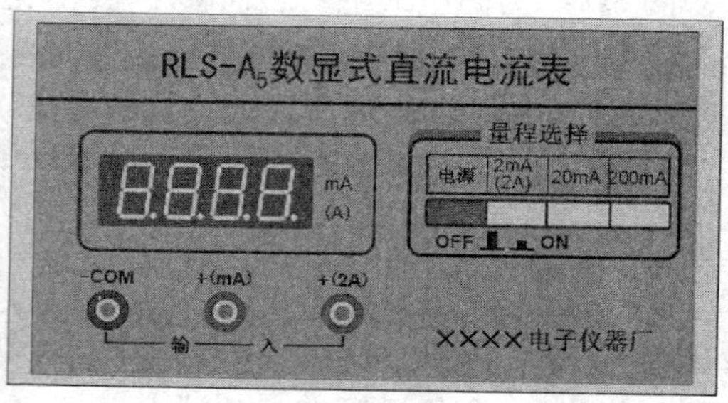

图 3-2-10　数字电表面板图

数字式电表的仪器误差一般取所显示数据的最后一位的±1。

3. 滑线变阻器

滑线变阻器的外形如图3-2-11所示。它是将涂有绝缘层的电阻丝绕在长直瓷管上,电阻丝的两端固定在接线柱 A、B 上。滑动头 C 与电阻丝紧密接触,滑动时能改变引出电阻值的大小。

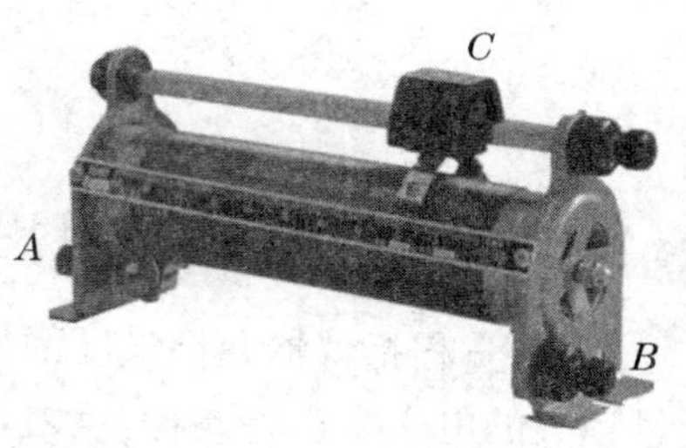

图 3-2-11　滑线变阻器

滑线变阻器在电路中有两种用法:

(1) 限流接法。如图3-2-12(a)所示,将变阻器中任一固定端,比如 A 与滑动端 C 串联在电路里。当滑动头 C 向 A 移动时,AC 间电阻变小;当滑动头向 B 移

动时，AC 间电阻变大。在这个过程中改变了电路中的总电阻，从而使电路中的电流发生变化，由此达到了在电路中的限流目的。

(2) 分压接法。如图 3-2-12(b)所示，将变阻器的两个固定端分别与电源的两极相连，滑动端 C 和任一固定端与负载相接将电压输出。移动 C 时，输出电压在 0 到 U_{AB} 范围内连续变化。应注意，在接通电源前，应将滑动头 C 移至 A 处，这样接通电源后 $U_{AC}=0$。当 C 点向 B 点移动时，AC 间电压 U_{AC} 即负载上的电压增大。当 C 点向 A 点移动时，U_{AC} 减小。可见改变活动头 C 的位置就改变了 AC 间的电压，由此达到了分压的目的。开始实验前，在限流接法中，变阻器的活动头 C 应放电阻最大处为宜，在分压接法中，变阻器的滑动头应放在输出电压较小的位置。

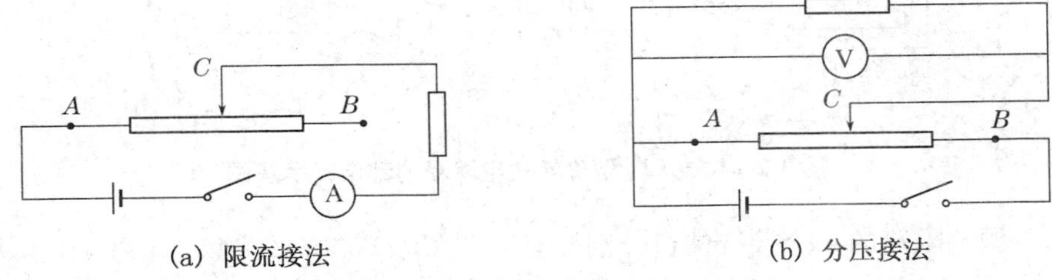

(a) 限流接法　　(b) 分压接法

图 3-2-12　滑线变阻器的两种接法

4. ZX21 型电阻箱

电阻箱是由若干个准确的固定电阻元件，按照一定的组合接在特殊的变换开关装置上构成的，组成电阻箱的元件组装在一个匣子里，阻值标在匣子的面板上，用旋钮可以选择不同的阻值，实验室中常把电阻箱作为标准电阻使用。如图 3-2-13 所示为常用的 ZX21 型旋转式电阻箱面板，图 3-2-14 为其内部接线图。面板上有 6 个旋钮，每个旋钮控制一组等值标准电阻，每组等值标准电阻的个数一般为 9 个；每个旋钮下标明的倍率(×0.1、×1、×10、×100、×1000、×10 000)就是该组各等值电阻的大小(例如，×100 表示该组每只电阻的阻值是 100 Ω)。旋转各旋钮即可选取所需阻值，旋钮下方箭头所指读数乘以相应的倍率即得该旋钮所对应的电阻值，各旋钮所选取的电阻值的总和即为所取总电阻。如图3-2-13 所示阻值为 6935.5 Ω(接“99 999.9 Ω”和“0”两个接线柱时)

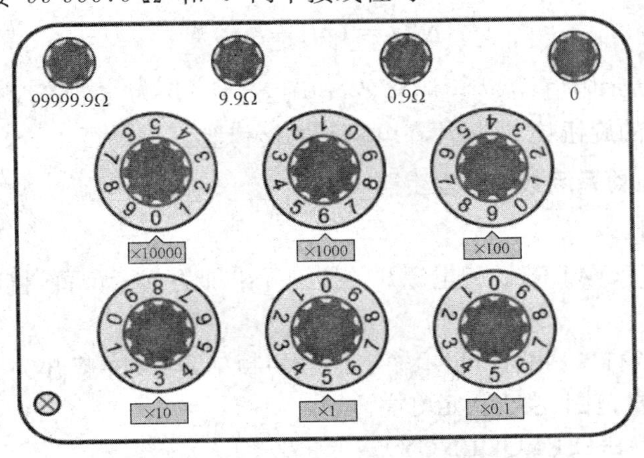

图 3-2-13　ZX21 型旋转式电阻箱面板

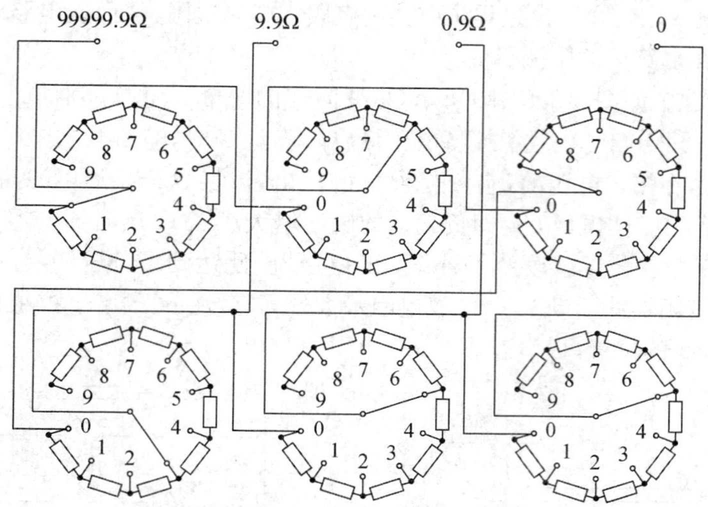

图 3-2-14 ZX21 型旋转式电阻箱内部电路示意图

ZX21 型电阻箱的 4 个接线柱边分别标有"0""0.9 Ω""9.9 Ω""99 999.9 Ω"。从其内部电路图可以看出,连接"0.9 Ω"与"0"两接线柱,阻值调整范围为 0.1～0.9 Ω;连接"9.9 Ω"与"0"两接线柱,阻值调整范围为 0.1～9.9 Ω;连接"99 999.9 Ω"与"0"两接线柱,阻值调整范围为 0.1～99 999.9 Ω。实际使用时,如只需要 0.1～0.9 Ω 或 9.9 Ω 的阻值变化,则将导线接到"0"和"0.9 Ω"或"9.9 Ω"两接线柱,可以避免电阻箱其余部分的接触电阻和导线电阻对低阻值带来不可忽略的误差。电阻箱各挡电阻允许通过的最大电流是不同的,ZX21 型电阻箱各档电阻允许通过的最大电流如表 3-2-2。

表 3-2-2 ZX21 型电阻箱各挡电阻允许通过的最大电流

旋钮倍率	×0.1	×1	×10	×100	×1000	×10 000
容许负载电流/A	1.5	0.5	0.15	0.05	0.015	0.005

电阻箱按其准确度可分为 0.02、0.05、0.1、0.2 和 0.5 级,电阻箱的仪器误差通常可用下式计算

$$\Delta R_{仪}=(aR+bm)\%$$

式中,R 为电阻箱所示阻值;α 为电阻箱的准确度等级;b 为与准确度等级有关的系数;m 是所使用的电阻箱的旋钮数,均标注在电阻箱的铭牌上。

5. YB1601P 型功率函数信号发生器

(一) 面板说明

(1) 电源开关(POWER)。将电源开关按键弹出即为"关"位置,将电源线接入,按电源开关,以接通电源。

(2) LED 显示窗口。此窗口指示输出信号的频率,当"外测"开关按下,显示外测信号的频率。如超出测量范围,溢出指示灯亮。

(3) 频率调节旋钮(FREQUENCY)。

（4）占空比(DUTY)。占空比开关，占空比调节旋钮，将占空比开关按下，占空比指示灯亮，调节占空比旋钮，可改变波形的占空比。

（5）波形选择开关(WAVE FORM)。按对应波形的某一键，可选择需要的波形。

（6）衰减开关(ATTE)。电压输出衰减开关，二挡开关组合为20dB、40dB、60dB。

（7）频率范围选择开关(并兼频率计闸门开关)。根据所需要的频率，按其中一键。

（8）计数、复位开关。按计数键，LED显示开始计数，按复位键，LED显示全为0。

（9）计数/频率端口。计数、外测频率输入端口。

（10）外测频开关。此开关按下LED显示窗显示外测信号频率或计数值。

（11）电平调节。按下电平调节开关，电平指示灯亮，此时调节电平调节旋钮，可改变直流偏置电平。

（12）幅度调节旋钮(AMPLITUDE)。顺时针调节此旋钮，增大电压输出幅度。逆时针调节此旋钮可减小电压输出幅度。

（13）电压输出端口(VOLTAGE OUT)。电压输出由此端口输出。

（14）TTL/CMOS输出端口。由此端口输出TTL/CMOS信号。

（15）功率输出端口。功率输出由此端口输出。

（16）扫频。按下扫频开关，电压输出端口输出信号为扫频信号，调节速率旋钮，可改变扫频速率，改变线性/对数开关可产生线性扫频和对数扫频。

（17）电压输出指示。3位LED显示输出电压值，输出接50Ω负载时应将读数÷2。

（18）功率按键。按下按键上方，左边绿色指示灯亮，功率输出端口输出信号，当输出过载时，右边红色指示灯亮。

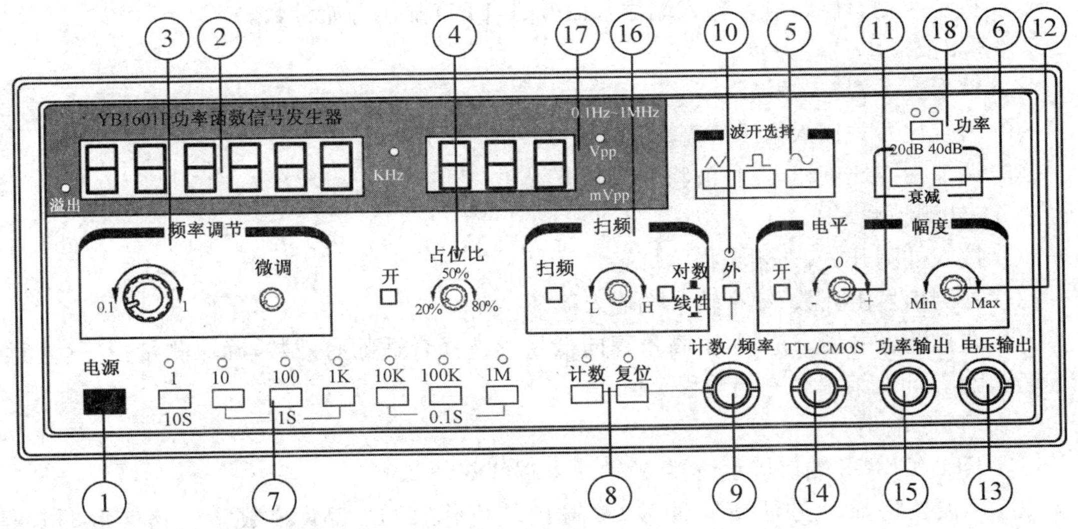

图3-2-15　YB1601P型功率函数信号发生器

（二）使用方法

打开电源开关之前，首先如表3-2-3所列示设定各个控制键。

表 3-2-3　YB1601P 型功率函数信号发生器初始按键设置

电源(POWER)	弹出	电平	弹出
衰减开关(ATTE)	弹出	扫频	弹出
外测频(COUNTER)	弹出	占空比	弹出

所有的控制键如上设定后,打开电源。函数信号发生器默认 10K 挡正弦波,LED 显示窗口显示本机输出信号频率。

(1) 将电压输出信号由幅度(VOLTAGE OUT)端口通过连接线送入示波器 Y 输入端口。

(2) 三角波、方波、正弦波产生。

① 将波形选择开关(WAVE FORM)分别按正弦波、方波、三角波。此时示波器屏幕上将分别显示正弦波、方波、三角波。

② 改变频率选择开关,示波器显示的波形以及 LED 窗口显示的频率将发生明显变化。

③ 幅度旋钮(AMPLITUDE)顺时针旋转至最大,示波器显示的波形幅度将$\geqslant 20V_{p\text{-}p}$。

④ 将电平开关按下,顺时针旋转电平旋钮至最大,示波器波形向上移动,逆时针旋转,示波器波形向下移动,最大变化量±10 V 以上。注意:信号超过±10 V 或±5 V(50 Ω)时被限幅。

⑤ 按下衰减开关,输出波形将被衰减。

(3) 计数、复位。

① 按复位键、LED 显示全为 0。

② 按计数键、计数/频率输入端输入信号时,LED 显示开始计数。

(4) 斜波产生。

① 波形开关置“三角波”。

② 占空比开关按入指示灯亮。

③ 调节占空比旋钮,三角波将变成斜波。

(5) 外测频率。

① 按入外测开关,外测频指示灯亮。

② 外测信号由计数/频率输入端输入。

③ 选择适当的频率范围,由高量程向低量程选择合适的有效数,确保测量精度(注意:当有溢出指示时,请提高一挡量程)。

(6) TTL 输出。

① TTL/CMOS 端口接示波器 Y 轴输入端(DC 输入)。

② 示波器将显示方波或脉冲波,该输出端可作 TTL/CMOS 数字电路实验时钟信号源。

(7) 扫频(SCAN)。

① 按入扫频开关,此时幅度输出端口输出的信号为扫频信号。

② 线性/对数开关,在扫频状态下弹出时为线性扫频,按入时为对数扫频。

③ 调节扫频旋钮,可改变扫频速率,顺时针调节,增大扫频速率,逆时针调节,减慢扫频

速率。

(8) VCF(压控调频)。

由 VCF 输入端口输入 0～5 V 的调制信号。此时,幅度输出端口输出为压控信号。

(9) 调频(FM):由 FM 输入端口输入电压为 10 Hz～20 kHz 的调制信号,此时,幅度端口输出为调频信号。

(10) 50 Hz 正弦波:由交流 OUTPUT 输出端口输出 50 Hz 约 2 $V_{p\text{-}p}$的正弦波。

(11) 功率输出:按入功率按键,上方左边指示灯亮,功率输出端口有信号输出,改变幅度电位器输出幅度随之改变,当输出过载时,右侧指示灯亮。

三、光学实验仪器

1. 望远镜

望远镜是用来观察和测量远距离物体的。它由长焦距的物镜和短焦距的目镜组成,在目镜和物镜之间安装有叉丝或分划板,作为测量的参考点。实验室常用的望远镜是开普勒望远镜,它的物镜和目镜部是会聚透镜,远处的物体经过物镜在其焦平面附近成一倒立的缩小实像,再经目镜得到放大、倒立的虚像。使用时,首先调节目镜,使眼睛贴近目镜时能看到清晰的叉丝,然后再转动调焦手轮,使物体成像清晰,并且和叉丝没有视差。所谓视差,是指观察者改变位置,像和叉丝的相对位置有明显的移动。由于视差的存在会影响测量的精确度,所以在实验过程中一定要尽可能地消除视差,然后再测量。图 3-2-16 所示是杨氏模量实验仪中使用的望远镜。

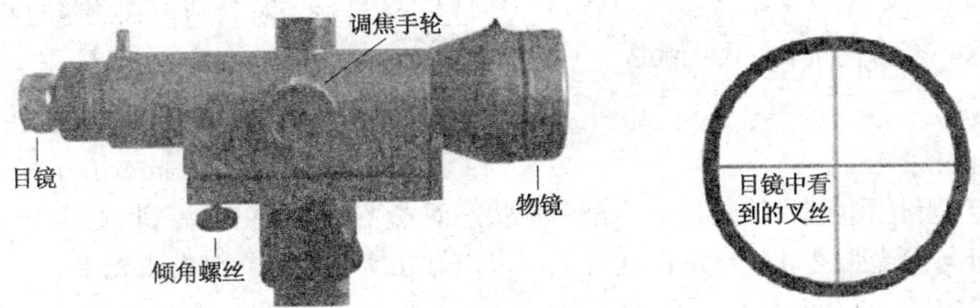

图 3-2-16　杨氏模量实验仪中的望远镜

2. 读数显微镜

读数显微镜是由显微镜和读数移动装置组成,用于测量长度的精密仪器,如图 3-2-17 所示。显微镜镜筒由目镜、物镜和分划板组成,调节调焦手轮,镜筒可上下移动,使观察者看到清晰的像。同时镜筒固定在测微螺杆上,旋转测微鼓轮,就可推动测微螺杆前进或后退,从而带动显微镜左右移动,可以从读数装置读取移动的距离。所以读取数显微镜是利用光学放大原理测量长度的精密仪器。

读数显微镜使用方法和步骤为:

(1) 将被测工件放置在工作台面上,用弹簧片压紧;

(2) 旋转目镜,看清楚叉丝;

(3) 调节物镜调焦手轮,使物像成像在叉丝平面上并和叉丝无视差;

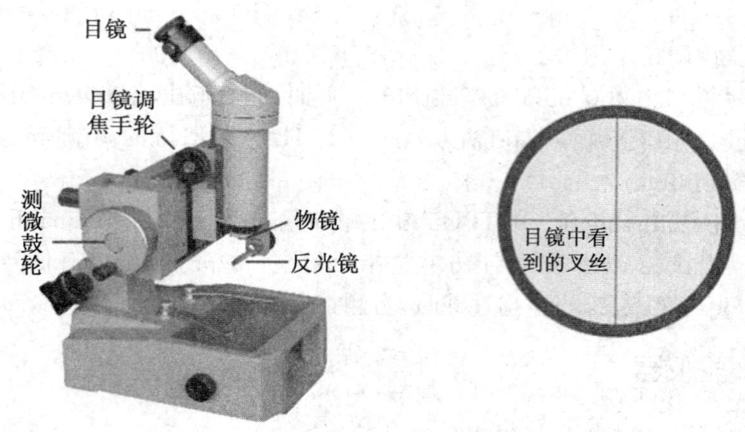

图 3-2-17 读数显微镜

(4) 转动测微鼓轮,使叉丝的交点与被测物的像重合,即可读数。沿同一个方向继续转动测微鼓轮,使叉丝的交点与被测物的像的另一端重合,又可得到另一个读数。两者之差为被测物的尺寸;

(5) 读数方法与螺旋测微器相同,主尺的分度值为 1 mm,测微鼓轮共有 100 个刻度,则分度值为 0.01 mm,可估读到 0.001 mm;

(6) 读数显微镜筒能在左右方向移动是由测微鼓轮通过螺纹推动测微螺杆引起的,而螺纹之间由于存在间隙,所以测微鼓轮当一个方向向反方向转动较小时,螺杆并没有发生移动,产生空转,这种空转会造成读数误差,称为回程差或螺旋空程。所以在测量过程中,测微鼓轮要按同一个方向旋转,不可倒退。

3. 光源

(1) 白炽灯

白炽灯是利用电能将钨灯丝加热至白炽状态而发光的热辐射光源,其光谱为连续光谱,光谱的成分与发光强度和灯丝的温度有关。白炽灯在实验室可作为白光光源。

(2) 钠光灯

钠光灯是在灯泡内充有钠蒸汽的气体放电光源,可发出两条波长相近的光谱线,其波长分别为 589 nm 和 589.6 nm。因颜色为橙黄色,所以常称钠灯为钠黄灯。由于两条光谱线非常接近,钠光灯可作为较好的单色光源,通常取 589.3 nm 作为平均波长。

(3) 汞灯

汞灯是汞蒸汽放电光源。在石英玻璃管内充有汞蒸汽,按汞蒸汽压强的大小分为低压、高压和超高压汞灯。实验室常用的是低压汞灯,这种灯常在一个大气压或小于一个大气压下工作。汞灯点亮后一般需 5～15 min 才稳定,在可见光区域内有十几条不同波长的谱线。汞灯辐射紫外线较强,眼睛不要直视汞灯。使用钠光灯和汞灯时应注意尽量避免反复开关电源。由于灯管在启动发光过程的损耗远大于连续点燃的损耗,因此灯管一旦点燃,应做完实验再关灯。灯管熄灭后,则应等灯管冷却后再点燃,以增加灯管的使用寿命。

（4）氦氖激光器

激光器是受激辐射光源。与普通光源相比，激光器具有单色性好、方向性强、相干性好、能量集中等优点。大学物理实验中常用的激光器为He-Ne激光器，波长为632.8 nm，功率从几到几十毫瓦。使用He-Ne激光器时应注意以下内容：

① He-Ne激光器两端接有高压，使用时谨防触电；

② He-Ne激光器的辉光电流不得超过额定电流，但不能低于阈值，否则激光闪烁；

③ He-Ne激光束能量高度集中，切勿裸眼直视激光束，以免损伤眼睛。

Chapter Ⅲ Introduction of Basic Measurement Methods and Common Instruments

3.1 Basic Measurement Methods in Physics Experiments

As an experimental science, the concepts, laws and formulas of physics are all based on objective experiments. Experiments are the basis of law-discovering and theory-forming, and theories play a guiding and foreseeing role in experiments. There is an inseparable relationship between experiments and theories and it is just such a mutually reinforcing relationship between them that contributes to the remarkable achievements of modern physics. Therefore, the importance of physical experiments is self-evident.

Physics experiments are based on measurement and observation. Any physics experiment cannot do without the measurement of physical quantity. In experimental physics, the research and measurement of various physical quantities have contributed to the formation of distinctive theories and effective measurement methods. As the essential elements of physics, these basic theories and experiments also play necessary roles for other scientific subjects. A great variety of physical experiments correspond to the extensive range of physical quantities to be measured, including mechanical quantity, thermal quantity, electromagnetism quantity, and optical quantity and so on. As a result, many measurement methods by means of experimental devices and techniques can be found in physics. This section provides a brief introduction to the most commonly used measurement methods in physics experiments.

3.1.1 Comparison Method

Comparison method in the experiment means comparing the same type of measured unit with the standard unit and then measuring its size. The comparison method is the most basic, common and important measurement method in physics experiments. The main procedures are as follows: first, to determine a unit quantity of the same kind as the measured quantity; next, to set the unit quantity as the standard and compare the measured unit quantity with this unit quantity; finally, to calculate their multiple ratios and obtain the measurement result. Generally, comparison method can be divided into direct comparison and indirect comparison.

In the direct comparison, the standard measuring tool is generally selected as the standard unit, thus the accuracy of measurement mainly depends on the accuracy of the standard measuring tool. For example, when a steel ruler is used to measure length, the accuracy of length measurement results will depend on the accuracy level of the steel ruler; if the equal arm balance is selected to measure object mass, the accuracy of quality measurement results will depend on the accuracy level of the weights.

When it is difficult to measure some physical quantities with direct comparison method, the physical quantities to be measured can be compared with the same standard quantities via the functional relationship between physical quantities. For example, when the Wheatstone bridge is used to measure resistance, the functional relationship between the bridge resistance can be adopted to compare the resistance to be measured with the standard resistance so as to get the results.

1. Amplification Method

In physics experiments, it is difficult to directly measure some small quantities or small changes because the direct measurement only has a lower accuracy. On such occasions, amplification method can be used to select the corresponding measurement device to amplify the objects to be measured so as to facilitate measurement or to improve measurement accuracy. According to different amplification modes, amplification method can be further classified into cumulative amplification, mechanical amplification, electronic amplification and optical amplification.

When the measurement can be simply superimposed or repeated, the cumulative amplification method can be used. For example, when measuring the period of a simple pendulum with a stopwatch, it is generally to measure 50 cycles or 100 cycles instead of one cycle in order to improve the accuracy. Similarly, the distance between ten or dozens of fringes is generally measured for the distance or width of interference fringes in the light interference experiment. A period of 50 or 100 cycles is the cumulative amplification. If the instrument error of the stopwatch is 0.1 s and the period of a pendulum is about 1s, the relative error of measuring the time interval of a single cycle is 0.1 s, namely

$$E=\frac{0.1}{1.0}=10\%$$

If the time interval of 100 cycles is measured, the relative error is

$$E=\frac{0.1}{100.0}=0.1\%$$

It can be seen that the measurement accuracy has greatly improved.

Mechanical amplification is a transmission-principle-based method to subdivide the reading mechanism of the measuring instrument so as to improve the measurement accuracy. Spiral micrometer magnifying method is a typical mechanical amplification method. The 1 mm on the screw is divided into 100 parts with the spiral structure, which is equivalent to magnifying the millimeter interval by 100 times. The instruments suitable for this method include the micrometer, sound velocity meter, reading microscope, Michelson interferometer, and etc.

Electronic amplification is a kind of electronic method to amplify weak electrical signals so as to enhance the measurement sensitivity. This method can also be used to convert other non-electrical quantities into amplified electrical signals for measurement.

Optical amplification is to magnify the tiny quantities or observed effects to be measured through optical devices according to certain rules, so as to achieve more accurate observation and measurements. The optical method is mainly used to enlarge the angle of view to improve the measurement accuracy. The instruments using this method include the magnifying glass, microscope, telescope, and etc.

2. Compensation Method

The compensation method is the approach to compensate up for the partial loss or change of some parts in the experiment through using a variable attaching device, so as to keep the system in the compensation state or equilibrium state. In short, the compensation method means making up for the unsatisfied results of various possible impacts on the measurement state. For example, the voltage compensation method is used to compensate for the change of the working current of the measured branch resulting from the direct measurement of voltage with a voltmeter; the temperature compensation method is used to compensate for the influence of some physical quantities (such as resistance) on the test state due to the change of temperature; the optical path compensation method is used to compensate for the asymmetry of optical path in the optical path, etc.

3. Simulation Method

On the foundation of the similarities between some physical phenomena or processes, the simulation method aims to artificially create a model similar to the physical phenomenon or process of the research object, so as to reproduce, delay or accelerate given physical phenomenon and then obtain the final results by measuring the artificial model. For example, when designing the shape of an aircraft, a model aircraft is usually made in proportion and tested in a wind tunnel. By measuring the dynamic parameters of the model, the dynamic parameters of the original size of the aircraft can be obtained. In the experiment of describing electrostatic fields, the simulation method can be used to measure the distribution of electrostatic fields, given that the equipotential line of steady current fields and electrostatic fields satisfy the same mathematical equation. Additionally, thanks to the development of computer simulation technology, computer simulation method can be applied in physics experiments. With appropriate mathematical model, physical systems can be replaced by computer programs and thus the physics experiment can be conducted on computer.

4. Conversion Method

The convention method refers to the process of transforming one physical quantity into another for observation. In the measurement of physical quantities, some physical quantity cannot be directly measured or accurately measured. To make these physical quantities measurable and easy to be measured, or to improve the measurement accuracy and reduce the measurement error, it is a common practice to convert the physical quantities to be measured into other physical quantities, and then to obtain the measured values according to the conversion relationship.

The conversion method is most frequently used to convert the non-electrical quantities into electrical quantities for measurement. The sensor is the critical device for the convention. The sensor is the device to convert the measured quantity to a corresponding physical quantity with certain accuracy and easy application. The most frequently used sensors include:

Thermoelectric sensors include a thermal resistance sensor and a thermocouple sensor. The thermal resistance sensor is made according to the characteristics of resistance changing with temperature, and the thermocouple sensor is based on the thermoelectric effect to convert temperature into potential difference for measurement.

The piezoelectric sensor is made by using the piezoelectric effect of some solid dielectric materials. When the material is mechanically deformed by external force, polarization will be generated in the crystal lattice inside the material, and bound charges will be generated on the surface to form an electric field consistent with the polarization direction. Therefore, the measurement of pressure can be converted into the measurement of electric field. On the contrary, the piezoelectric crystal will deform when the electric field is applied. These two processes are called the piezoelectric effect and inverse piezoelectric effect respectively.

Magneto electric sensor mainly refers to Hall sensor. When placing the semiconductor material-hall plate in the magnetic field and applying the voltage to the two opposite thin edges of the Hall plate, the internal electric current will be generated. If the current direction is perpendicular to the direction of the magnetic field, a transverse potential difference will be generated in the direction perpendicular to both the magnetic field and the current direction. Such a magneto electric effect is called the Hall effect, also named as Hall voltage. Hall sensor can realize the non-contact measurement without obtaining energy from magnetic fields.

The working principle of photoelectric sensor is to convert optical signal into electrical signal for measurement. With the advantages of simple structure, non-contact, high reliability, high accuracy and quick response, this measurement method has been widely used in automatic control detection technology. Photoelectric sensors are produced on the basis of the photoelectric effect. For example, the external photoelectric effect can be applied to produce the photocell and photomultiplier tube, the internal photoelectric effect can be used to make the photosensitive resistor and light guide, while the photocell and photosensitive transistor are made with the photo voltage effect.

3.2 Basic Instruments for Physics Experiments

In physics experiments, it is necessary to carry out the quantitative measurement and qualitative observation of physical quantity. Yet physical measurements and observations must be conducted via appropriate experiment instruments. A great variety of physics experiment instruments can be identified, including mechanical instruments, thermal instruments, acoustic instruments, electromagnetic instruments, optical instruments and etc. The usage of various instruments will be described in each experiment in detail. This section will provide the basic knowledge about some commonly-used instruments in the laboratory, such as the instrument performance, instrument principle, and instrument layout structure and circuit connection.

3.2.1 Mechanical and Thermal Experimental Instruments

1. Vernier Caliper

The vernier caliper is composed of the main ruler and vernier rulers (sub-rulers) that can slide along the main ruler. The purpose of the cursor is to help the experimenter accurately evaluate the reading after the minimum scale of the main ruler. Cursors can be divided into straight cursors and angle cursors. Straight cursors are common calipers while angle cursors can be used in spectrometers. Although the shape, length and dividing grid number of cursors are different, the basic principle and reading methods are the same. The caliper's reading mainly refers to the distance between the zero line of main ruler and the zero line of the cursor.

As shown in Fig. 3-2-1, a straight vernier caliper with a dividing value of 0.02 mm is taken as an example to introduce the reading method.

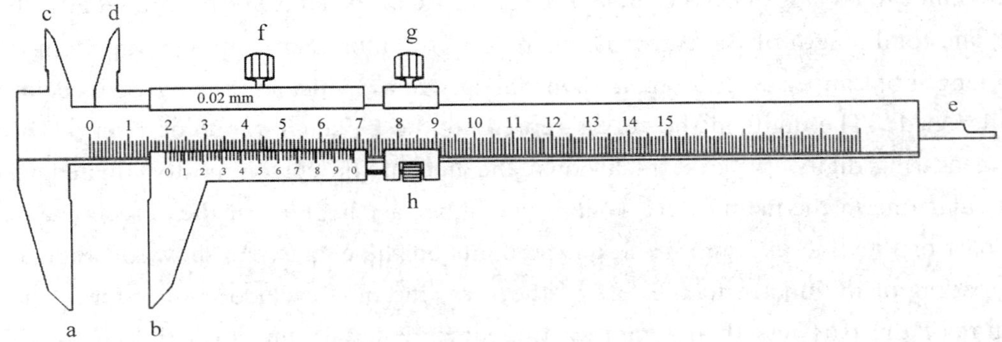

Fig. 3-2-1　Component diagram of a vernier caliper

a, b—bayonet of outer diameter; c, d—bayonet of inner diameter; e—depth measuring end; f, g—fixing screw; h—trimming nut.

The minimum scale on the main caliper of this model is $a=1$ mm. The number of cells of a cursor N is 50. The total length of the cursor is 49 lattice length of the cursor, or 49 mm (see Fig. 3-2-2.) The length of the cursor is denoted by b, where $b=49/50=0.98$ mm, the difference in value between the main ruler and the cursor is $\delta=a-b=0.02$ mm, where δ is the index value of the cursor, that is, the ratio of the minimum index of the master ruler to the number of cursor cells.

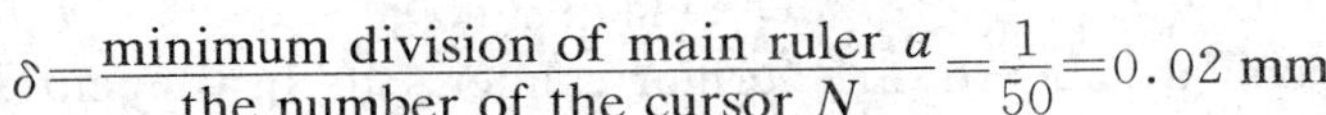

$$\delta=\frac{\text{minimum division of main ruler } a}{\text{the number of the cursor } N}=\frac{1}{50}=0.02 \text{ mm}$$

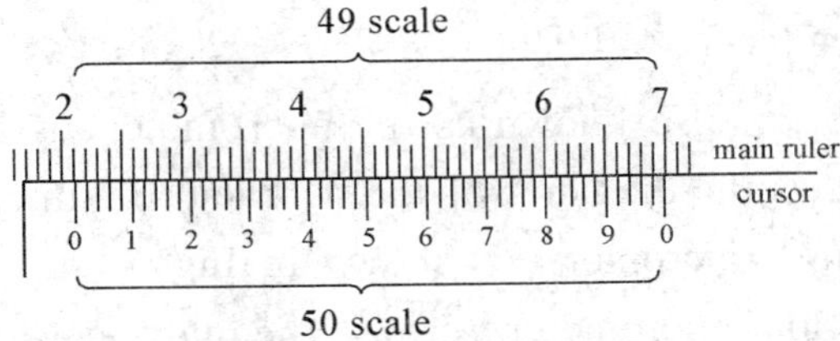

Fig. 3-2-2　The cursor of a vernier caliper

As is shown in Fig. 3-2-2, the 0 scale line of the cursor is aligned with the scale line of the main ruler and the last line is aligned with the same line on the main ruler. At the same time, the first line of the cursor is δ away from the nearest line on the right side of the main ruler, and the second line of the cursor is 2^{δ} away from the nearest line on the right side of the main ruler. Similarly, it can be judged that the nth line is n^{δ}. As a result, when the nth line on the cursor is aligned with the line on the main ruler, the zero scale of the cursor is n^{δ} away from the nearest line on the left of the main ruler, then the cursor reading can be obtained.

In an actual reading, the reading for integer of millimetre to the left of the zero scale of the cursor can be obtained on the main ruler while the readings for the part less than 1mm can be judged on the cursor. The specific procedures are illustrated as follows:

The total length of the cursor is regarded as the minimum length of the main ruler. The length of caliper with 50 scale divisions is 1 mm, which is equally divided into 50 smaller grids. The digits in the cursor represents the 10 equal points of 1 mm. To read the measuring digits, we can track down the line in the cursor mostly aligned with a particular line in the main ruler. Then it would be easy to find out the percentage of the left part of this line in 1 mm based on the digits on the cursor. As shown in Fig. 3-2-3, the reading of millimetre to the left of the zero line of the cursor is 102 mm, and the reading of the part less than 1 mm on the cursor is 0.66 mm, thus the length of the measured object is 102.66 mm.

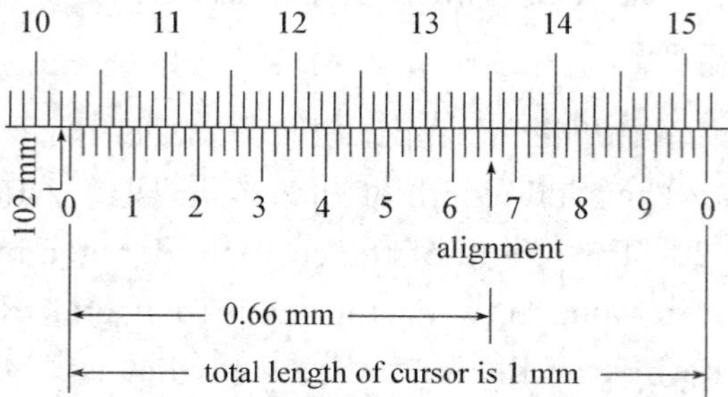

Fig. 3-2-3 Example of vernier caliper reading

For the commonly used 50 - degree vernier caliper, the instrument error is $\Delta_{\text{instrument}} = 0.02$ mm.

2. Screw Micrometer

The screw micrometer, also known as the micrometer, is used to measure shorter lengths accurately, such as the diameter of the wire and the thickness of the sheet, etc. The frequently-used spiral micrometer is shown in Fig. 3-2-4. The main parts of the structure are the precision micrometer screw, the nut sleeve on the screw and the differential sleeve fixed on the micrometer screw. There are two rows of lines on the main ruler of the nut sleeve, namely millimeter line and half millimeter line. The circumference of the differential sleeve is engraved with 50 scale division, and the pitch of screw is 0.5 mm. For each rotation of the differential sleeve, the screw moves (forward or backward) with a pitch of 0.5 mm. Thus, for each rotation of the sleeve, the distance the screw moves is $\frac{0.5\text{ mm}}{50}=0.01$ mm, that is to say, the dividing value of the screw micrometer is 0.01 mm.

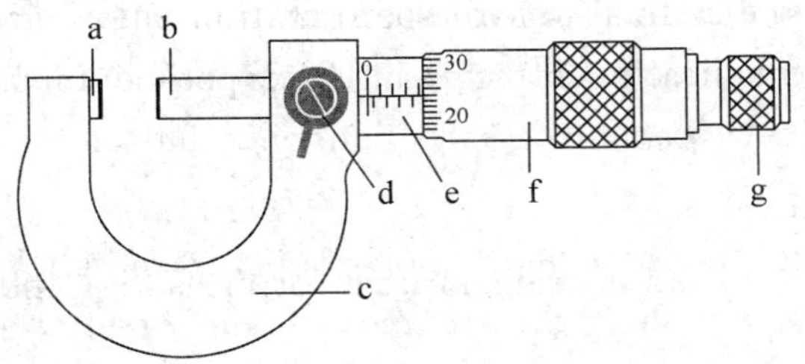

Fig. 3-2-4 Screw micrometer

a—anvil; b—screw; c—stent; d—locking device;
e—nut casing; f—differential sleeve; g—ratchet.

Before using the screw micrometer to measure, we should first record its zero reading (zero error), namely the reading of screw micrometer after the combination between micrometer screw and anvil. The length of the object to be measured is the final result of direct reading minus zero reading. The value of the zero reading lies in the digits between the zero calibration on the differential sleeve and the horizontal line of the nut sleeve. When the zero calibration on the differential sleeve is above the horizontal line of the nut sleeve, the zero reading is negative, as is shown in Fig. 3-2-5 (a). Conversely, the zero reading is positive, as is shown in Fig. 3-2-5(b).

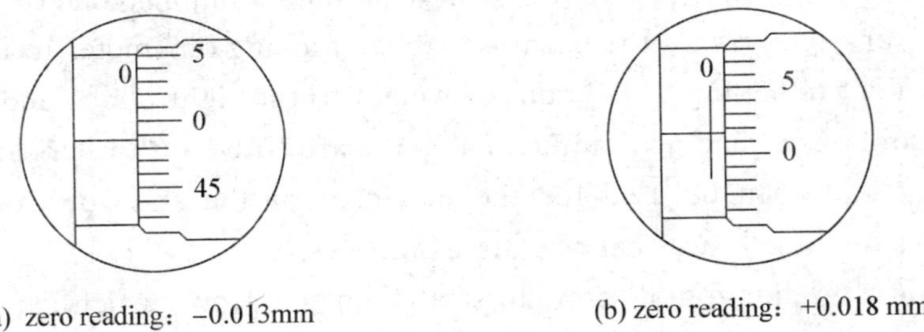

(a) zero reading: −0.013mm

(b) zero reading: +0.018 mm

Fig. 3-2-5 Diagram of zero reading of screw micrometer

Notes for use:

1) The supporting part of the screw micrometer is the only part to be held by hand, and the workpiece under the test should be touched as little as possible, so as not to affect the measurement accuracy due to thermal expansion.

2) A ratchet should be used for measurement. The measurement accuracy will directly depend on the pressure applied to the measured object when the tester rotates the screw. Thus, a ratchet wheel can be attached as a protective device. Before the contact of the ending face of the measuring rod with the measured object, the rotary ratchet should be adopted. On contacting the object to be measured, the rotary ratchet will go slipping by itself and make a click sound. At this time, the rotary ratchet should be stopped and the measurement should be read.

3) When the screw is put into the box after its usage, the screw should be returned

several times to leave space, so as to avoid deformation caused by thermal expansion. As for the commonly used helical micrometer in the experiment, the instrumental error is $\Delta_{instrument}=0.004$ mm.

3. Physical Balance

The physical balance is an instrument used for measuring the mass according to the principle of equal arm lever.

The structure of the physical balance is shown in Fig. 3-2-6. The physical balance has three parallel knife edges, with one being at the middle point and the other two being located in the ends of the beam. As the fulcrum of the beam, the middle knife edge is placed on the agate cushion at the top of the pillar. Two equal weight scales are hung on the knife edge at both ends. Each physical balance is equipped with a set of weights. The maximum weight of the physical balance commonly used in the laboratory is 1000 g, and the minimum weight is 1 g. The part of the mass less than 1 g can be measured by moving the swimming code on the beam. The dividing value of each grid on the beam is 0.05 g. Moving one meter to the right on the beam is equivalent to adding 0.05 grams of weight to the right plate, and the lower part of the beam contains a reading pointer. The posts are fitted with rulers and the readings on the pointer scale can be used for the judgment of the state of the balance. The operating steps of the physical balance are as follows:

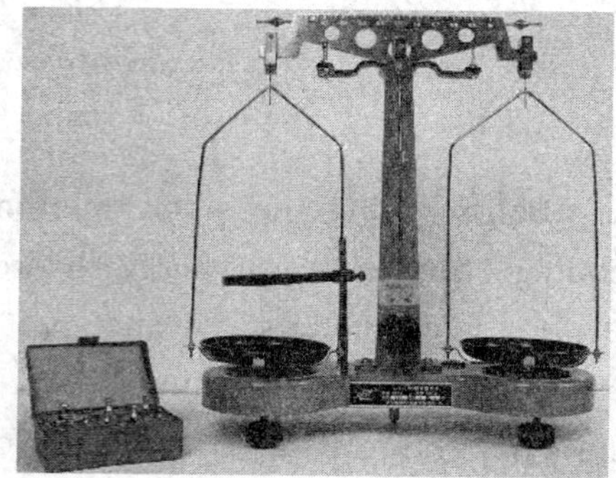

Fig. 3-2-6 The physical balance

1) To adjust the horizontal screw to straighten the prop, which can be checked by whether the horizontal bubble on the base is in the middle.

2) To adjust the zero point. First dial the swimming code to 0, then hang the lifting hook of the scale plate on both ends of the knife edge, next turn the stop knob and support the beam of the balance. After observing the swing of the pointer, whether the balance is balanced or not can be judged. When the pointer swings to the left and right of the center line of the scale, the balance is balanced. If not, the balance nut can be adjusted to achieve balance.

3) To weigh the objects, put the object to be measured on the left plate and the weights on the right plate. When adding and subtracting weights, try to stop the balance in order to protect the knife edge. The scale should be removed from the knife edge after weighing is completed.

Operating rules of physical balance are as follows:

1) The load of the balance must not exceed its maximum weight so as not to damage the knife edge or beam.

2) In order to protect the blade, it must be remembered that the scales must stay balanced whenever taking or putting objects, adding or subtracting silicon yards, adjusting the balance nut or not using it. The balance can be started only to judge whether the balance is balanced or not. Light operations are necessary whenever starting or stopping the balance. It is highly suggested to stop the balance only when the balance pointer is close to the scale center.

3) It is forbidden to directly take the weights by hand. Weights can only be clipped with tweezers. Weights must be put back into the weight box after being taken from the scale plate.

4) All parts of the balance and the weight plate can easily rust. Attention should be paid to avoid directly weighing high-temperature objects, liquids and corrosive chemicals.

The common instrumental errors of the physical balance in the laboratory are illustrated in Table 3-2-1.

Table 3-2-1　Instrumental error of commonly used balances

Balance model	Instrument with load of 0～250 g $\Delta_{\text{instrument}}$	Instrument with load of 250～500 g $\Delta_{\text{instrument}}$	Instrument with load of 500～1000 g $\Delta_{\text{instrument}}$
TW05	0.05 g	0.1 g	
TW1	0.1 g		0.2 g

4. Electronic Balance

The structure of the electronic balance used in the laboratory is shown in Fig. 3-2-7. Its maximum range is 200 g. The simple operation procedures are as follows:

Fig. 3-2-7　Electronic balance

1) Leveling: Adjust the anchor screw so that the air bubble in the level is located in the center of the ring;

2) Power ON: press ON/OFF to enter the state to be tested;

3) Preheat: Preheat for at least 30 minutes after the first power on to get the ideal working state.

4) Weighing: Use the peeling key to clear the skin. Place the sample for weighing and automatically display the weight.

5. Thermometer

Temperature is a state parameter that describes the degree of microscopic thermal motion of a substance. There are many ways to measure temperature, and the thermometers are designed on the basis of the principle that a given physical characteristic of the temperature-sensitive element change significantly with temperature. For example, a liquid thermometer measures temperature according to the law of thermal expansion. A capillary glass tube of uniform thickness is connected to a storage vacuole, which are filled with liquid (mercury, alcohol, etc.). Within a certain range of temperature, the rising and falling of liquid in the capillary glass tube is linear with the temperature. Thus, the temperature value can be read out according to the liquid level. Alternatively, metal resistance thermometers and semiconductor resistance thermometers are made on the basis of the resistance change of electrical resistivity with temperature. Thermocouple thermometers are produced according to the fact the electromotive force changes result from the metal temperature difference. Besides, an optical pyrometer for high temperature is designed by taking advantage of the radiation law.

3.2.2 Electromagnetic Experiment Instrument

1. DC Regulated Power Supply

DC regulated power supply is a stable DC voltage source obtained by reducing (or boosting), rectifying, filtering and stabilizing the voltage of 220 V AC. That is to say, the output voltage remains unchanged despite of the output current change within the load range. Usually, the output voltage value of the DC regulated power supply is adjustable, and the required DC voltage value can be selected by turning the adjusting knob. Some DC regulated power supply panels are equipped with voltmeters and ammeters, which can directly display the output voltage and current values. If the rated current is exceeded, the output voltage will be unstable, and the fuse will be blown, or the circuit will be cut off due to internal automatic protection.

However, the internal resistance of the DC regulated power supply is generally below a few tenths of an ohm, and the output is quite stable. Attention should be paid not to short circuit the positive and negative poles of the power supply with wires. On such occasion, the high current will cause the blow out of fuse. Short circuiting or wrong polarity are big taboos in electrical experiments, and thus much caution should be taken for the electrical experiments.

Brief introduction of the WYK－$303B_2$ dual-channel DC power supply

The WYK series of stabilized voltage and current power supplies are mainly composed of a transformer, a rectifying and filtering circuit and a high-power triode, etc.. The input AC is depressurized by the transformer and then output after rectification filtering, high-power triode adjustment and reference sampling amplification. Notably, the dual power supply can be output separately or used in series or parallel. When it is connected in series, the output voltage of the slave circuit follows that of the main circuit. Meanwhile, when connected in parallel, the maximum output current can reach the sum of two separate output currents. This series of regulated power supplies adopt the full-bridge chopper to improve the power of the entire machine. The output voltage can be adjusted from zero volt, and the output current can be set from zero ampere. The WYK－303B_2 dual-channel DC power supply is illustrated in Fig. 3-2-8.

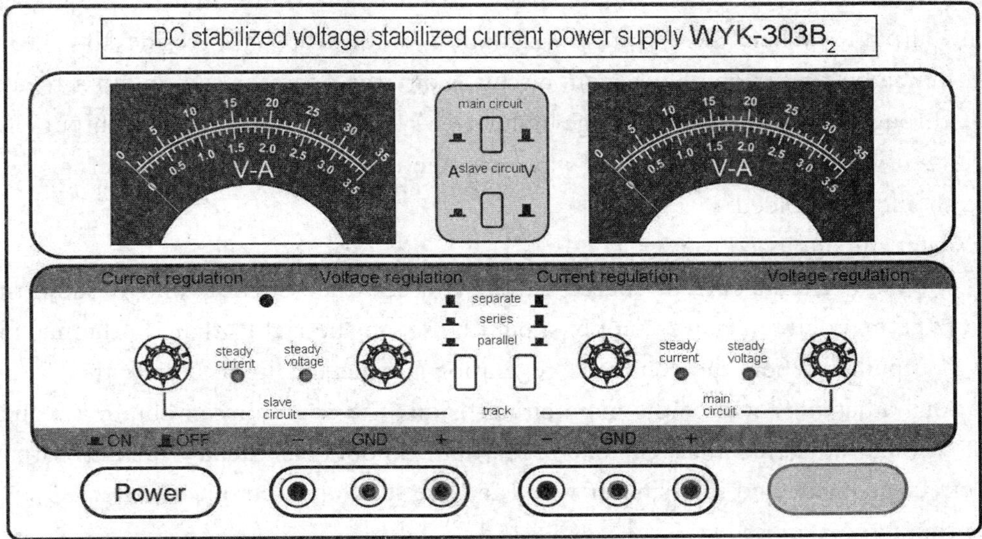

Fig. 3-2-8　WYK－303B_2 DC stabilized voltage current power supply panel power switch

1) Power switch

When the power switch is pressed, both output terminals have outputs; the voltage and current values of the main circuit and the slave circuit are displayed on the "V-A" table on the panel. There are two button switches in the middle of the two tables, which can be pressed and popped up. There are prompts on the panel: press the button on the upper side to select and press the meter on the right side to display the main circuit current, and pop up to display the voltage of the main circuit; press the button at the bottom and press the meter on the left to display the current of the secondary circuit, and pop up the display voltage.

2) Input and output

The input is 220±10% V, 50 Hz AC. There are two output channels: the main circuit and the slave circuit, which are supplied by red and black terminals through "＋" and "－", while the "GND" (⊥) terminal is usually vacant.

3) Working mode selection switch

There are two button switches between the main circuit and the slave circuit and there are two states of pressing and popping. There are prompts on the panel: when both buttons are in the popping state, the two (master-slave) power supplies work independently; when the button of the slave circuit is pressed and the button of the main circuit pops up, the two power supplies are output in series; when both buttons are pressed, the two (master-slave) power supplies work concurrently.

4) Voltage stabilized use

If the main circuit voltage source output is used, the "steady current" adjustment knob of the main circuit power supply should be set to the right (that is, the maximum current output). When the "voltage regulation" indicator light beside the "voltage regulation" adjusting knob of the main circuit is on, the operator can obtain a regulated output through adjusting the "voltage regulation" knob. The regulated output of the slave circuit is conducted in a similar way, and the load short-circuit of external voltage source should be avoided.

5) Current stabilized use

If the main circuit current source output is used, the "steady voltage" adjustment knob of the main circuit power supply should be set to the right (that is, the maximum voltage output). When the "current regulation" indicator light beside the "current regulation" adjusting knob of the main circuit is on, the operator can obtain a regulated output through adjusting the "current regulation" knob. The steady current output of slave circuit is conducted in a similar way, and the load open-circuit of external current source should be avoided.

6) Notes

While using the power supply, if a short circuit or an open circuit of the external circuit happens, the protection alarm device of the power supply will be prompted by a long beep. On such occasion, the operator must turn off the WYK - 303B_2 dual-channel DC power supply as soon as possible.

2. Electricity Meter

Electric meter is an instrument for measuring electrical quantity. Electric meter can be classified into AC and DC, and the DC meter is most frequently used in common physical experiments.

7) DC magneto electric analog pointer meter

Fig. 3-2-9 illustrates the schematic diagram of magneto electric millimeter panel. Ammeter and voltmeter are obtained through connecting a meter head with appropriate resistance in series or in parallel as required. To read this type of meter, the selected range should be considered first. The full deviation of the meter pointer is the value of the selected range, and then the actual reading can be obtained according to the position of the pointer. There are some points to be noticed for the use of electric meter as follows:

Line connection: the ammeter must be connected in series with the circuit to be measured, and the voltmeter must be connected in parallel with the circuit to be measured.

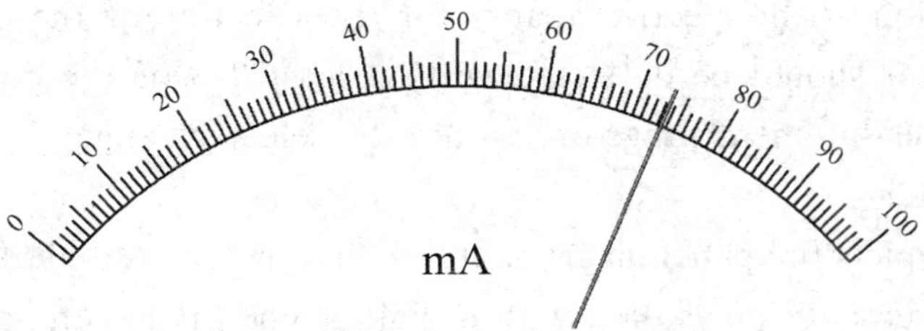

Fig. 3-2-9　Schematic diagram of mA meter panel

Positive and negative polarity: the polarity of the meter is the top priority for the use of a DC meter. If the polarity is reversed, the pointer of the meter will be reversed, resulting in the damage to the meter. The positive pole means that the current flows in from this pole, while the negative pole indicates that the current flows from this pole. Generally, the positive sign of the meter is "+", and the negative sign is "−".

Range: range refers to the maximum value allowed by the instrument. When the measured current or voltage value exceeds the range of the meter, the operator must choose the meter with larger range for measurement. On the contrary, if the measured current or voltage value is far less than the range of the meter, a small range meter is necessary. If the measured current or voltage value cannot be estimated in advance, the common practice is to try the meter with larger range and then the meter with the appropriate range can be used to avoid damaging the pointer of the meter or even burning the meter.

Error and grade of the electric meter: in the measurement of electrical quantity, the errors of measurement results may often occur resulting from the meter structure and the measuring environments, including temperature, external electric field and magnetic field.

The errors caused by the structural defects of the meter (such as friction, residual deformation of the hair spring, poor assembly and inaccurate scale, etc.) belong to the basic errors of the instrument, which will not change with different operators. As a result, the basic error can determine the accuracy of the meter and the accuracy of common instruments can be defined by grade.

The grade of the meter is a parameter reflecting the accuracy of the meter, which is determined by the measurement error of the meter. It is the percentage of the ratio of the maximum absolute error to the instrument range. In China, there are seven grades of electrical meters, which are 0.1, 0.2, 0.5, 1.0, 1.5, 2.5 and 5.0. The smaller the value is, the higher the accuracy of the meter (the grade of the meter is marked on the dial) will be. The instrument error can be specifically obtained through learning the accuracy level of the meter.

$$\Delta_{\text{Instrument}} = \text{range} \times \text{accuracy level}\%$$

Actual operation of the electric meter: for the selection of the electric meter, the pointer of the meter should be deflected with at least 2/3 of the deflection under the premise that the measurement range are within the selected range.

8) Digital meter

Fig. 3-2-10 depicts the panel diagram of a digital meter. A digital meter is a kind of instrument that directly displays data with a digital tube. It has the advantages of high measurement accuracy and convenient reading. Despite the high accuracy, the improper operation may also cause great errors in the measuring process. One point needs to be noticed that the operator should try to display the number of digits as many as possible so as to reduce the measurement error. Generally, the plus-or-minus one among the last numbers of the displayed digit is taken as the instrumental error of the digital meter.

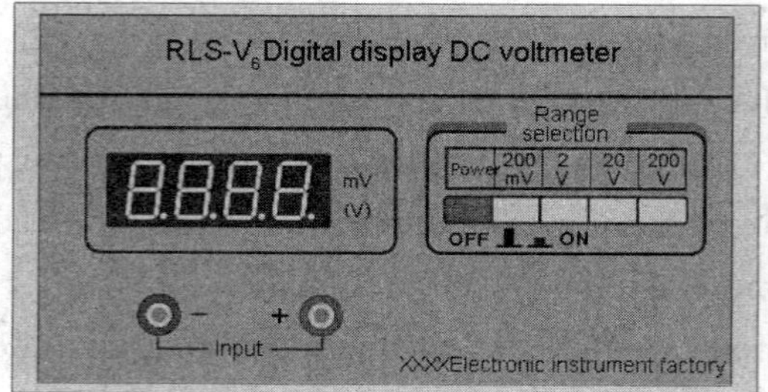

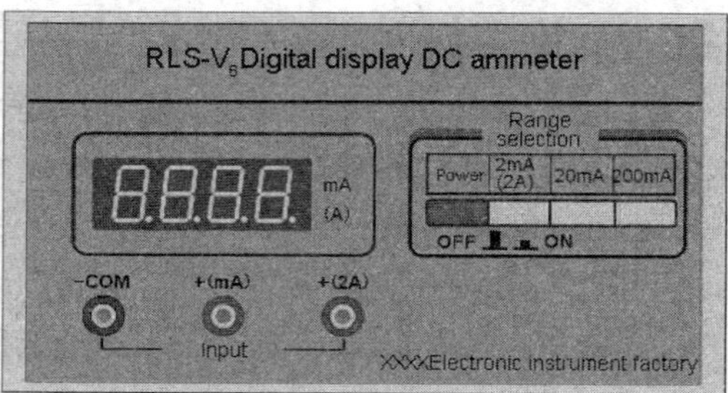

Fig. 3-2-10 Digital meter panel diagram

3. Slide Wire Rheostat

The shape of slide wire rheostat is illustrated in Fig. 3-2-11. The resistance wire coated with an insulating layer is wound on the long straight porcelain tube, and the two ends of the resistance wire are fixed on terminal A and terminal B. The sliding head C is in close contact with the resistance wire and the resistance value can be changed when sliding.

Fig. 3-2-11 Slide wire rheostat

There are two uses of a slip wire rheostat in a circuit:

1) Current limiting connection: As shown in Fig. 3-2-12(a), the operator can connect any fixed ends of the rheostat, for example, A can be connected with sliding end C in series in the circuit. When the slider C moves to A, the resistance between AC will become smaller. Meanwhile, when the slider C moves to B, the resistance between AC will become larger. In this process, the current in the circuit changes along with the total resistance, resulting in the current limitation.

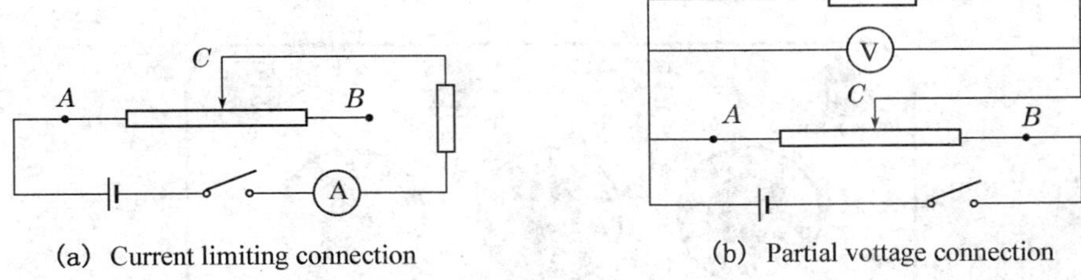

(a) Current limiting connection (b) Partial vottage connection

Fig. 3-2-12 Two connection methods of slide wire rheostat

2) Partial voltage connection: as displayed in Fig. 3-2-12 (b), the two fixed ends of the rheostat are connected with the two poles of the power supply, and the voltage can be output as soon as the the sliding end C and any fixed end are linked with the load.

When moving C, the output voltage has a continuous change from 0 to U_{AB}. It is noted that the sliding head C should be moved to position A before the switch-on of the power supply, resulting in $U_{AC}=0$ upon switching on the power supply. When point C moves to point B, the voltage U_{AC} between AC (the voltage on the load) increases. However, when point C moves to point A, the switch-on of the power supply decreases. To put it in another way, the position change of the movable head C can change the voltage between AC, and thus the voltage division can be realized. Before the experiment with the current limiting connection method, the movable head C of the rheostat should be placed at the position where there is the maximum resistance. For the experiment with the partial voltage connection method, the sliding head of the rheostat should be placed at the position where there is the minimum output voltage.

4. ZX21 Resistance Box

The resistance box is composed of several accurate fixed resistance elements, which are connected to the special change-over switch device in a certain combination. The components of the resistance box are assembled in a box, and the resistance values are marked on the panel of the box. Different resistance values can be selected by turning the knob. The resistance box is often used as the standard resistance in the laboratory.

Fig. 3-2-13 shows the panel of commonly-used rotary resistance box ZX21, and Fig. 3-2-14 depicts its internal wiring diagram. There are 6 knobs on the panel and each knob controls a set of equivalent standard resistors. Generally, there are nine equivalent standard resistors in each group. A rate marked under each knob (×0.1、×1、×10、×100、×1000、×10000) refers to the magnitude of each equivalent resistance of the group (e.g. ×100 means that the resistance of each resistor in the group is 100 Ω). The targeted resistance can be obtained by tuning on the relative knobs. The corresponding resistance value is obtained via multiplying the reading below the arrow of the knob by the corresponding magnification. Then the total resistance can be calculated by adding all the selected resistance values of each knob. As depicted in Fig. 3-2-13, the resistance value is 6935.5 Ω (when connecting "99 999.9 Ω" and "0").

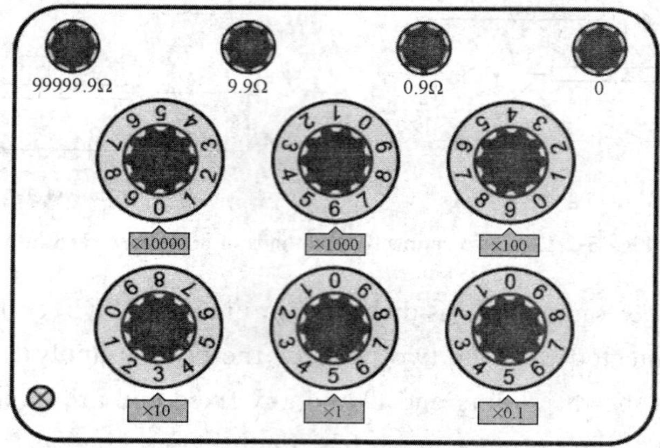

Fig. 3-2-13 ZX21 resistance box panel

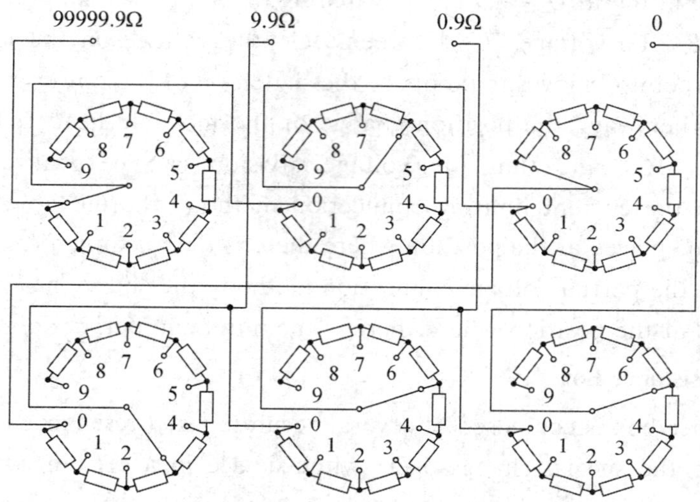

Fig. 3-2-14 ZX21 internal circuit diagram of resistance box

The four terminals of the ZX21 resistance box are marked with "0", "0.9Ω", "9.9 Ω" and "99 999.9Ω". From the internal circuit diagram, it can be seen that the resistance value adjustment range is 0.1～0.9 Ω when connecting the terminals of "0.9 Ω" and "0"; the resistance value adjustment range is 0.1～0.9 Ω for connecting the two terminals of "9.9 Ω" and "0"; the resistance adjustment range is 0.1～99 999.9 Ω for connecting the two terminals of "99 999.9 Ω" and "0". In practical use, if only the range of 0.1～0.9 Ω or 0.1～9.9 Ω is required, the wire could be connected to the terminals of "0" and "0.9 Ω" or "0" and "9.9 Ω". Therefore, the non-negligible errors of the low resistance value resulting from contact resistance and wire resistance can be avoided. The maximum current allowed for each resistance in the resistance box is different. The maximum current allowed by the resistance of the ZX21 resistance box is shown in Table 3-2-2.

Table 3-2-2　The maximum current allowed to pass through each resistance of the ZX21 resistance box

Knob magnification	×0.1	×1	×10	×100	×1000	×10 000
Allowable load current/A	1.5	0.5	0.15	0.05	0.015	0.005

The resistance box can be classified into the scales of 0.02, 0.05, 0.1, 0.2 and 0.5 based on its accuracy. The instrument error of the resistance box can usually be calculated with the following equation:

$$\Delta R_{\text{Instrument}} = (aR + bm)\%$$

Where R is the resistance value shown in the resistance box, a is the accuracy level of the resistance box, b is the coefficient related to the accuracy level, and m is the number of knobs of the resistance box used. All of these parameters in formula are marked on the name plate of the resistance box.

5. YB1601P Power Function Signal Generator

1) Panel description (Fig. 3-2-15)

(1) Power switch: pop up the power switch button to the "off" position, connect the power line, and press the power switch to connect the power.

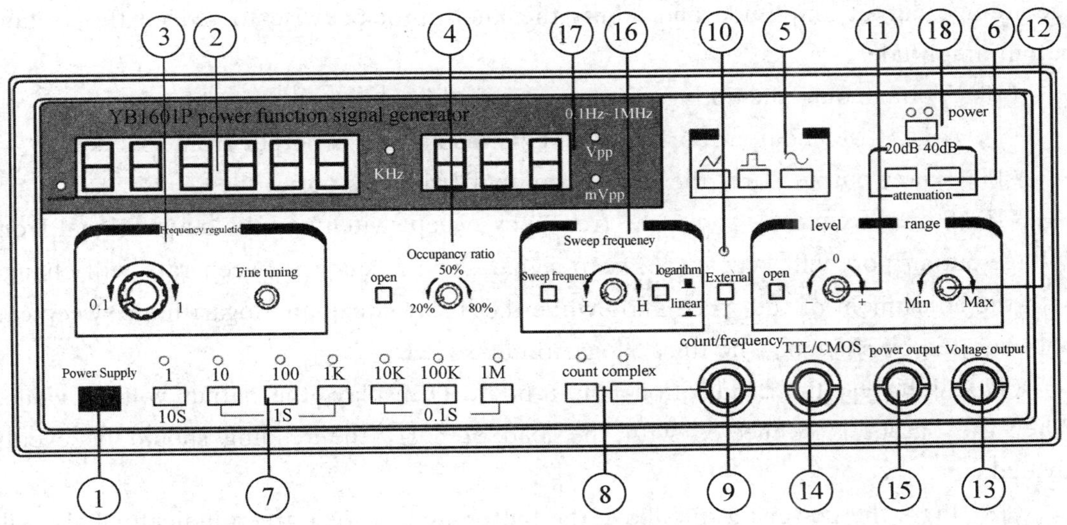

Fig. 3-2-15　YB160P power function signal generator

(2) LED display window: this window indicates the frequency of the output signal. When the "external test" switch is pressed, the frequency of the external measurement signal is displayed. If it is beyond the measuring range, the overflow indicator will be on.

(3) Frequency adjustment knob.

(4) Duty cycle: the duty cycle switch is an adjustment knob. Press the duty cycle switch, and the duty cycle indicator light will turn on. Then adjust the duty cycle knob to change the duty cycle of the waveform.

(5) Wave form: press a key of the corresponding waveform to select the desired waveform.

(6) Attenuating switch: voltage output attenuation switch, and the combinations of two switches are 20 dB, 40 dB and 60 dB.

(7) Frequency range selection switch (and frequency meter gate switch): press any one of the keys according to the required frequency.

(8) Count and reset switch: press the count key and LED display will start counting; press the reset key and LED display is cleared to zero.

(9) Count/frequency ports: they are the input ports for counting and measuring frequency.

(10) External frequency measurement switch: press the switch into the LED display window to show the frequency or count value of the external measurement signal.

(11) Level regulation: press the level adjustment switch, and the level indicator will turn on; readjust the level adjustment knob to change the DC bias level.

(12) Amplitude adjustment knob (amplitude): adjust the knob clockwise to increase the voltage output amplitude and adjust the knob anticlockwise to reduce the voltage output amplitude.

(13) Voltage out: the voltage is output from this port.

(14) TTL/CMOS output port: the TTL/CMOS signal is output from this port.

(15) Power output port: the power is output from this port.

(16) Sweep frequency: press the frequency sweep switch and the output signal from voltage output port will become a sweep signal. The frequency sweep rate will change with the adjustment of the speed knob while the linear sweep and logarithmic sweep can be generated by changing the linear/logarithmic switch.

(17) Voltage output indication: the 3-bit LED displays the output voltage value. When the output is connected with the load of 50Ω, the reading should be evenly divided.

(18) Press the power button above the button and the left green indicator light will be on, whereas when the power output is overloaded, the right red indicator light will be on.

2) How to use it

Before turning on the power switch, the operator should set the control keys as illustrated in Table 3-2-3.

Table 3-2-3 Initial key setting of YB1601P power function signal generator

POWER	eject	level	eject
Attenuation switch	eject	Sweep frequency	eject
COUNTER	eject	Duty cycle	eject

After all the control keys are set as above, turn on the power. The function signals generator default 10K sine wave and the LED display window depicts the output signal frequency.

(1) The voltage output signal is sent to the Y input port of the oscilloscope through the connection line from the voltage out port.

(2) Triangular wave, square wave and sine wave are generated as follows:

① Press the wave forms of sine wave, square wave and triangle wave, and the oscilloscope screen will display sine wave, square wave and triangular wave respectively.

② Change the frequency selection switch, and the waveform on the oscilloscope and the frequency on the LED window will clearly change.

③ When the amplitude knob is turned to the maximum clockwise, the waveform amplitude displayed on oscilloscope will be greater than or equal to $20V_{p\text{-}p}$.

④ Press the level switch and turn the level knob to the maximum clockwise, and the oscillograph waveform will move upwards; The oscillograph waveform moves downwards when turning the level knob anticlockwise. The maximum variation of the signal is over ±10 V. Note: the amplitude is limited when the signal exceeds ±10 V or ±5 V (50 Ω).

⑤ Press the attenuation switch, and the output waveform will be attenuated.

(3) Count and reset

① Press the reset key and the LED display has the zero clearing.

② Press the count key and count/frequency input, and the LED displays the start of counting.

(4) The generation of oblique wave

① Set the waveform switch to "triangle wave."

② When the duty cycle switch is pressed, the indicator light will turn on.

③ Adjust the duty cycle knob, and the triangle wave will be reshaped into an oblique wave.

(5) External measurement frequency

① Press the external test switch, and the external frequency measurement indicator will turn on.

② The external measurement signal is input by the count/frequency input.

③ Select the appropriate frequency range and the appropriate effective number from high range to low range so as to ensure measurement accuracy (Note: when there is an overflow indication, please improve the range by one gear).

(6) TTL output

① The TTL/CMOS port is connected to the y-axis input (DC input) of the oscilloscope.

② The oscilloscope will display square waves or pulse waves. The output can be used as a clock signal source of the TTL/CMOS digital circuit experiment.

(7) Scan

① Press the frequency sweep switch, and the output signal of the amplitude output port becomes the sweep signal.

② Linear/logarithmic switch: linear sweep occurs when the switch pops up with sweep mode and logarithmic sweep occurs when pressed.

③ Adjust the frequency sweep knob to change the frequency sweep rate. The sweep rate increases with the clockwise adjustment and slows down with the anti-clockwise adjustment.

(8) VCF (voltage-controlled frequency modulation): The modulation signal at 0～5 V is input through the VCF input port and the output of the amplitude output port is a voltage control signal.

(9) Frequency modulation (FM): the FM input port input voltage is the modulation signal at 10 Hz～20 kHz. Here, the amplitude port outputs a FM signal.

(10) 50 Hz sine wave: sine wave at around $2V_{p\text{-}p}$ 50 Hz is output by the AC output port.

(11) Power output: press the power button, and the upper left indicator light will turn on. The output amplitude of the amplitude potentiometer will change with the signal output from the power output port. When the output is overloaded, the right indicator light will turn on.

3.2.3 Optical Experiment Instrument

1. Telescope

Telescopes are used to observe and measure distant objects. It is composed of a long focal length objective lens and a short focal length eyepiece. A fork wire or dividing plate is installed between the eyepiece and the objective lens as the reference point for measurement. Specifically, the Kepler telescope is commonly used in the laboratory and its objective lens and eyepiece are converging lenses. The object in the distance forms an inverted reduced real image near its focal plane through the objective lens, and then the magnified and inverted virtual image is seen through the eyepiece. To use the telescope, first adjust the eyepiece so that the clear fork wire can be seen when the eye is close to the eyepiece, and then turn the focusing hand wheel to make the object image clear without the parallax on the fork wire. Parallax refers to the obvious movement of the relative position between the image and the fork wire when the observer changes the position. Given the possible impacts of parallax on the measurement accuracy, it is necessary to eliminate parallax as much as possible before the final measurements. Fig. 3-2-16 depicts the telescope for Young's modulus tester.

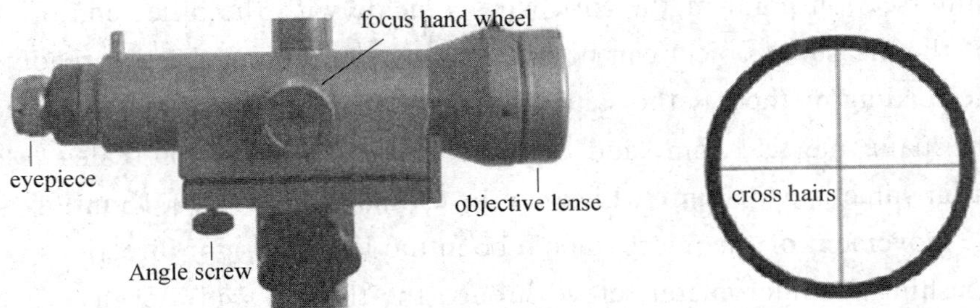

Fig. 3-2-16 Telescope in Young's modulus tester

2. Microscope

The reading microscope is a precise instrument for measuring length, which is composed of a microscope and a reading moving device, as shown in Fig. 3-2-17. The microscope tube consists of an eyepiece, objective lens and a reticle plate. Through adjusting the focusing hand wheel, the lens barrel can move up and down, so that the observer can see a clear image. Meanwhile, the lens cylinder is fixed on the micrometer screw and the micrometer drum can be rotated to push the micrometer screw forward or backward, enabling the microscope to move left and right. The moving distance can be read out from the reading device. Therefore, the reading microscope can be considered a precise instrument to measure the length based on the optical amplification principle.

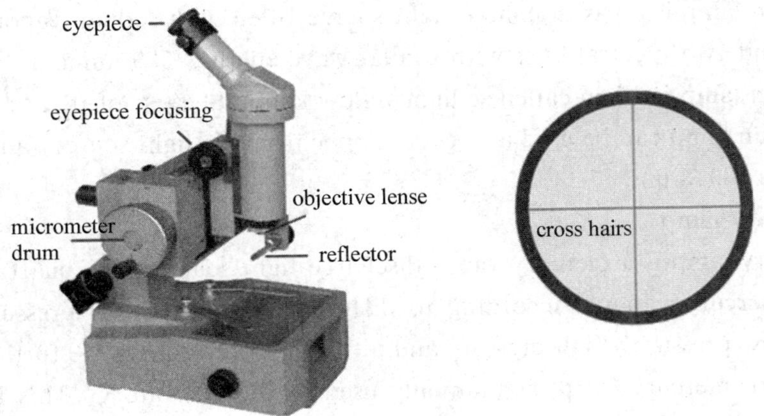

Fig. 3-2-17 Reading microscope

The method of operation is explained as follows:

1) Place the measured work piece on the worktable and press it with a spring;

2) Rotate the eyepiece to clearly see the fork wire;

3) Adjust the focusing hand wheel of the objective lens to make the object image clear on the fork wire plane with no parallax on the fork wire;

4) Rotate the drum until the intersection point of the fork wire coincides with the image of the object to be measured. Keep rotating the drum in the same direction to make the intersection point of the fork wire coincide with the other end of the image. The size of the measured object can be obtained by subtracting the two readings.

5) The reading method is the same as that of the spiral micrometer. The dividing value of the main ruler is 1 mm, and the micrometer drum has 100 scales. Meanwhile, the resolution value is 0.01 mm, which can be estimated to be 0.001 mm;

6) The movement of the microscope tube in the left and right directions is caused by the drum pushing the micrometer screw through the thread, and there is a gap between the threads. When the drum slightly rotates from one direction to the opposite direction, the screw does not move, resulting in idling. The reading errors caused by the empty transfer are referred to as return travel difference or screw idle distance. As a result, it is required to rotate the drum in one direction without coming backwards for the measurement.

3. Light Source

1) Incandescent lamp

An incandescent lamp is a kind of heat radiation light source by using electric energy to heat the tungsten filament to an incandescent state. The incandescent lamp has continuous spectrum whose composition and intensity are related to the temperature of filament. The incandescent lamp can be used as a white light source in the laboratory.

2) Sodium lamp

A sodium lamp is a gas discharge light source filled with sodium vapor in the bulb, which can emit two spectral lines with similar wavelengths—589 nm and 589.6 nm. The yellow sodium lamp is often called sodium yellow lamp. Because of its two close spectral lines, a sodium lamp can be used as a good monochromatic light source, and the average wavelength is 589.3 nm.

3) Mercury lamp

A mercury lamp is a mercury vapor discharge light source. The quartz glass tube is filled with mercury vapor. According to different mercury vapor pressures, mercury lamps with low pressure, high pressure and ultra-high pressure can be further classified. A low pressure mercury lamp is commonly used in the laboratory. This kind of lamp often works at or below one atmospheric pressure. Specifically, it usually takes 5 to 15 minutes for the mercury lamp to be stable. There are more than ten spectral lines with different wavelengths in the visible light region. Due to the strong ultraviolet radiation of the mercury lamp, the operator mustn't look directly at it. When using a sodium lamp or a mercury lamp, attention should be paid not to repeatedly switch the power supply.

Since the loss of the lamp in the starting process is far greater than that of the continuous ignition, the lamp should be turned off after the experiment is completed. After extinguishing the lamp, it should be ignited after cooling so as to extend its service life.

4) He-Ne laser

A laser is a source of stimulated radiation. Compared with ordinary light sources, a laser has the advantages of good monochromatic, strong directivity, good coherence and energy concentration. Notably, the He-Ne laser is commonly used in college physics experiments. The wavelength is 632.8 nm and the power ranges from several milliwatts to tens of milliwatts. When using the laser, attention should be paid to the following points:

(1) Beware of the electric shock because the two ends of the laser are connected with high voltage.

(2) Do not make the glow current of the laser exceed the rated current or make it lower than the threshold so as to prevent the laser flashing.

(3) The energy of the laser beam is highly concentrated. Do not look directly at the laser beam with the naked eye so as to avoid eye damage.

第4章　实验项目

4.1　拉伸法测金属丝的杨氏弹性模量

杨氏弹性模量是描述固体材料抵抗形变能力的重要物理量，它与材料自身性质有关，与材料的几何形状和所受到外力的大小无关，是选择机械构件的依据之一，也是工程技术中研究材料性质的常用参数。

测定弹性模量的方法很多，如拉伸法、振动法、弯曲法、无干涉法等。本实验提供了一种测量微小长度的方法即光杠杆法，该原理广泛应用在测量技术中，光杠杆法的装置还被许多高灵敏度的测量仪器（如冲击电流计和光点检流计）用来测量小角度的变化。在测量数据的处理上，本实验应用逐差法，这种方法在物理实验中经常用到。

实验目的

1. 学会用光杠杆法测量杨氏弹性模量。
2. 掌握光杠杆法测量微小伸长量的原理。
3. 学会用逐差法处理数据。

实验原理

固体、液体及气体在受外力作用时，形状与体积会发生或大或小的改变，这统称为形变。当外力不太大，因而引起的形变也不太大时，撤掉外力，形变就会消失，这种形变称之为弹性形变。如果加在物体上的外力过大，以致外力撤除后物体不能完全恢复原状而留下剩余形变，就称之为塑性形变。弹性形变分为长变、切变和体变三种。

最简单的弹性形变是棒状物体受外力后的伸长和缩短。设物体长 L，截面积为 S，沿长度方向施力 F 后，物体的伸长（或缩短）为 ΔL。比值 F/S 是单位面积上的作用力，称为胁强，它决定了物体的形变；比值 $\Delta L/L$ 是物体的相对伸长，称为胁变，它表示物体形变的大小。按照胡克定律，在物体的弹性限度内，胁强与胁变成正比

$$E=\frac{F}{S}/\frac{\Delta L}{L} \tag{4-1-1}$$

式中，E 为杨氏弹性模量。

实验证明，杨氏弹性模量 E 与外力 F，物体的长度 L 和截面积 S 的大小无关，而决定于棒的材料，它是表征固体性质的一个物理量。测出式(4-1-1)中的各个量，就可以得到杨氏弹性模量 E。作用在棒状物体上的力 F、棒的长度 L 和截面积 S 很容易测出，但对于棒状物体，在弹性限度内，伸长量 ΔL 甚小，用一般工具不易测准确，本实验用光杠杆法测量 ΔL。

光杠杆及望远镜尺组是用来测量长度微小变化量的实验装置。结构如图 4-1-1 所示。光杠杆下有 3 个尖足，平面镜垂直放置时与两前足共面。测量时，两前足放在固定平台沟槽内，后足支在金属丝下方的夹头上。当金属丝伸缩时，后足会随之上下移动，光杠杆镜面将向后或向前倾斜。光杠杆前方放置望远镜标尺组，用望远镜观察平面镜中标尺反射像，并读出水平叉丝所对准的数值。加置砝码使金属丝伸长 ΔL，镜面将向后倾斜 α 角，若金属丝伸长前后望远镜中标尺像的读数差为 Δn，光杠杆前后足垂直距离为 b，则有

$$\tan\alpha=\frac{\Delta L}{b}$$

$$\tan 2\alpha=\frac{\Delta n}{D}$$

当 ΔL 很小时，α 也很小，于是有近似关系

$$\alpha=\frac{\Delta L}{b}$$

$$2\alpha=\frac{\Delta n}{D}$$

消去 α，得

$$\Delta L=\frac{b}{2D}\Delta n \tag{4-1-2}$$

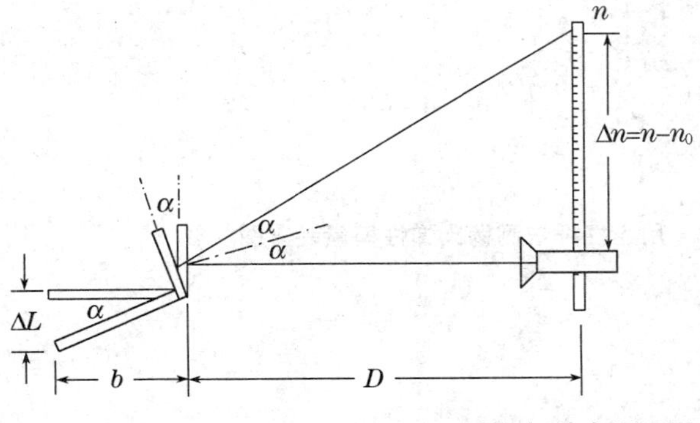

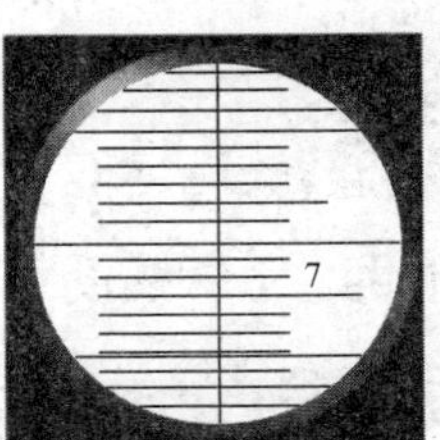

图 4-1-1 光杠杆放大原理图

ΔL 原是难以测量的微小长度,但当取 D 远大于 b 后,经光杠杆转换后的量 Δn 却是较大的量,可以从标尺上直接读得,光杠杆装置的放大倍数为 $\frac{2D}{b}$,在实验中,b 通常为 8 cm 左右,D 为 1.5~2.0 m,放大倍数可达到 35~40 倍。

由式(4-1-2)可得

$$E=\frac{F/S}{\Delta L/L}=\frac{F/\frac{1}{4}\pi d^2}{\left(\frac{b}{2D}\right)\cdot \Delta n/L}=\frac{8FLD}{\pi d^2 b\Delta n} \tag{4-1-3}$$

实验器材

YMY-1 型杨氏模量测定仪、砝码(1 kg)7 个、尺读望远镜、光杠杆、螺旋测微计、卷尺。用光杠杆法测杨氏弹性模量装置图如图 4-1-2 所示。尺读望远镜如图 4-1-3 所示,图 4-1-4 为光杠杆安装示意图。尺读望远镜和光杠杆组成测量系统。光杠杆系统则是由光杠杆镜架与尺读望远镜组成。

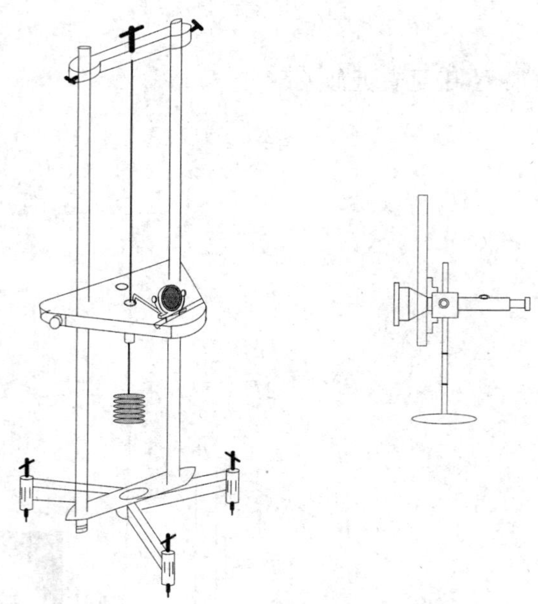

图 4-1-2 用光杠杆法测杨氏弹性模量装置图

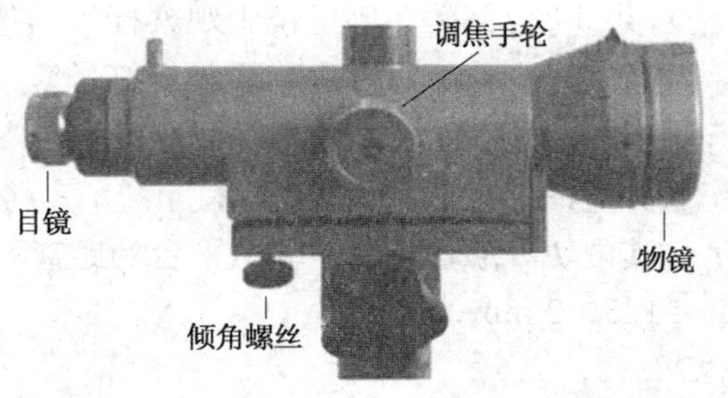

图 4-1-3　尺读望远镜

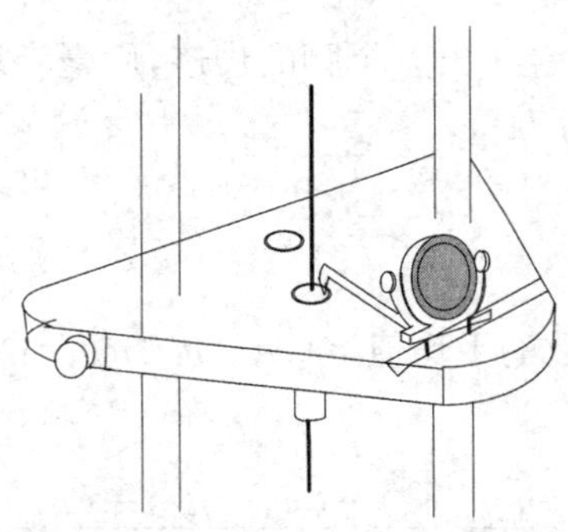

图 4-1-4　光杠杆安装示意图

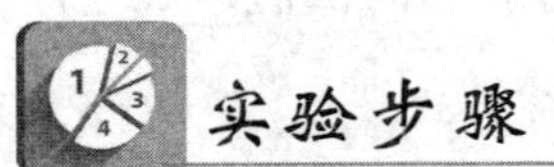

一、装置的调整

1．调节支架底部的 3 个螺丝，使平台达到水平(用水准泡检查)。光杠杆的两前足放在固定平台的小凹槽内，后足放在金属丝夹头上。

2．使望远镜与光标杆等高(如何调节)；使平面镜镜面垂直于望远镜与光杠杆连线(如何调节，怎么判断)；粗调望远镜大致水平，再调标尺零点在望远镜筒中心的高度上。

3．使望远镜对准镜面，眼睛在望远镜筒外上侧缺口向准星望去，可观测到镜面中标尺的像，若观测不到，则进行调节，从而使标尺的模糊像出现在望远镜视场中央。

4．调节望远镜目镜，使得分划板上的十字叉丝成像清晰；再调节物镜调焦手轮，使标尺刻线的像清晰，直到当眼睛在目镜前上下移动时，十字叉丝的水平线与标尺的刻度间无相对位移(即无视差)，实验操作时可放宽至视差不超过半格。

5．微调平面镜倾角(注意不让平面镜三支点移动)，使望远镜中叉丝水平线在标尺零刻线到±2 cm 左右的范围内。

二、测量

(1) 在砝码托上加一个砝码，记下十字叉丝水平线对准标尺的某一刻度 n_0'；

(2) 按顺序增加砝码(每次增加 1 个),共 5 次,每次在望远镜中观察标尺的像,并逐次记下相应的标尺刻度 n'_1、n'_2、n'_3、n'_4、n'_5(注意:经过零点后的读数要改变符号)。再加上第七个砝码,但不需记数;

(3) 按相反次序将砝码逐一取下,记录相应的标尺读数 n''_5、n''_4、n''_3、n''_2、n''_1、n''_0;

(4) 测量金属丝的长度 $L=L_0+L_1$,其中 L_0 为上卡头下端至平台的距离,用卷尺测量。L_1 为下卡头下端至平台的距离,$L_1=125\pm2$ mm;

(5) 用卷尺测平面镜到标尺的距离 D;

(6) 用螺旋测微计测金属丝平均直径 d:在金属丝的上、中、下 3 部位的不同方位上各测两次直径,将 6 个数据取平均值;

(7) 测 b,将杠杆放在纸上轻压出 3 个脚的痕迹,用毫米尺细心量出主杆尖脚至其余两尖脚迹点连线的垂直长度,即为 b。

三、数据处理

(1) 取同一负荷下标尺读数的平均值 n_0、n_1、n_2、n_3、n_4、n_5,其中 $n_i=\frac{n'_i+n''_i}{2}$,用逐差法算出平均值$\overline{\Delta n}$:

$$\overline{\Delta n}=\frac{(n_3-n_0)+(n_4-n_1)+(n_5-n_2)}{3}$$

(2) 用各数据代入式(4-1-3)计算 Y。

(3) 自列数据表,按式(4-1-3)推出计算相对不确定度的公式,求出 Y 的扩展不确定度和扩展相对不确定度,写出实验结果。

注意事项

1. 光杠杆、望远镜和标尺所构成的光学系统调节好后,在实验过程中就不可再移动,否则,所测的数据无效,实验应从头做起。

2. 不准用手触摸目镜、物镜、反射镜等光学表面,更不准用手、布类或任意纸片擦拭光学表面。

思考题

1. 光杠杆有什么优点?怎样提高光杠杆测量微小变化的灵敏度?

2. 是否可以用作图法求杨氏模量,如果以胁强为横轴胁变、胁变为纵轴作图,图形应是什么形状?

3. 做完用光杠杆测金属丝的杨氏模量实验后,请总结一下:为了能在望远镜中始终看清楚标尺刻度,镜面、尺面、镜高、望远镜高、标尺中点位置、望远镜之间相互关系应如何?与地面的关系又如何?

Chapter Ⅳ Experiments

Experiment 4.1 Measurement of Young's Modulus by Stretching Method

Young's modulus is an important physical quantity that describes the resistance of a solid material to deformation. Young's modulus is determined by the material's intrinsic properties rather than the material geometry or the external force. It is also a crucial parameter to study the properties of materials in the engineering. There are many methods to determine the Young's modulus, such as the stretching method, vibration method, bending method and non-interference method etc.. This experiment provides a method to measure the tiny extension of the iron string by the optical lever method. The optical lever is widely used to measure the slight deformation in science and industry, and it is also used for instruments with high sensitivity (such as pulsed ammeter and galvanometer) to measure small angle changes. As a common method in physics experiments, the successive difference method will be adopted to process the experiment data.

Experimental Objectives

1. To learn to how to use the optical lever method to measure Young's modulus.

2. To understand the principle of measuring small extension with the optical lever method.

3. To learn how to use the successive difference method to process the data.

Experimental Principle

Under the external force, the shape and volume of the solid, liquid and gas will change greatly or slightly. This process is called as deformation. If the external force is not large enough, the deformation will disappear with the removal of the external force. Such a deformation is called elastic deformation. If the external force on the object is so large that the object can not return to its original state even after the removal of external force, such deformation is called inelastic deformation. Elastic deformation includes longitudinal deformation, shear deformation and dilation.

The simplest form of elastic deformation is the elongation and shortening of the rod-shaped body under the external forces. Suppose an object of length L with cross-sectional area S experiencing a force along the length direction, and the object elongation (or shortening) is ΔL. The Ratio F/S is the force per unit area, known as the stress strength, which determines the object's deformation; the ratio $\Delta L/L$ is strain, the relative elongation of the object, and it represents the extent of the object deformation. According to Hooke's law, the stress strength is proportional to the strain within the elastic limit of the object, expressed as

$$E=\frac{F/S}{\Delta L/L} \tag{4-1-1}$$

Where E is Young's modulus.

The experiments' results indicate that Young's modulus E does not depend on the external force F, the length$L$ of the object and the cross-sectional area S. As a physical solid property, Young's modulus E is determined by the material of the rod. It can be obtained by measuring each quantity in Equation (4-1-1). The force F, the length L of the rod and the cross-sectional area S are easily measured. However, it is quite difficult to measure accurately the tiny elongation ΔL of the rod-shaped object within its elastic limit via ordinary tools. Therefore, optical lever is used to measure ΔL in this experiment.

Optical lever and telescopic ruler are experimental devices for measuring small variations in length, as shown in Fig. 4-1-1.

There are three sharp feet for the optical lever and the two forefeet are in the same plane of the mirror when it is placed vertically. In the measurement, the two forefeet are first placed in the fixed platform groove, and the rear foot is set on the top of the wire chucking. As the wire expands and contracts, the rear foot will move up and down and the optical lever mirror will tilt backward or forward. After setting the telescope ruler group in front of the optical lever, the operator can observe the reflection image of the ruler in the plane mirror and read the value aligned by the horizontal cross wire.

Add necessary weights to make the wire elongation ΔL, and the mirror will be tilted back by a small angle α. Suppose the coordinate difference of the scale in the telescope is Δn before and after the wire elongation, and the vertical distance between the optical lever's forefoot and the rear foot is b, the Equation will be

$$\tan\alpha=\frac{\Delta L}{b}\quad \tan 2\alpha=\frac{\Delta n}{D}$$

Because ΔL is very small, compared to b, α is also very small, there is an approximate relationship

$$\alpha=\frac{\Delta L}{b}\quad 2\alpha=\frac{\Delta n}{D}$$

After eliminating α, the following expression can be obtained

$$\Delta L=\frac{b}{2D}\Delta n \tag{4-1-2}$$

When D is given larger than b, Δn is large enough to be read directly from the scale. The magnification of the optical levers can be obtained by $2D/b$. In the experiment, b is usually about 8 cm and D is 1.5～2.0 m, thus magnification can be up to 35 to 40 times.

From Equations (4-1-1) and (4-1-2), we have

$$E=\frac{F/S}{\Delta L/L}=\frac{8FLD}{\pi d^2 b\Delta n} \tag{4-1-3}$$

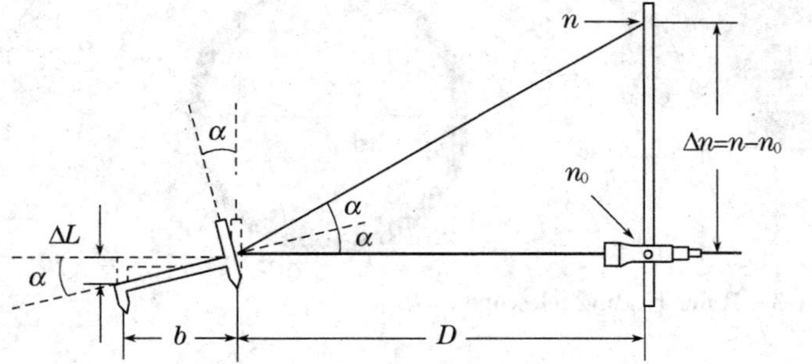

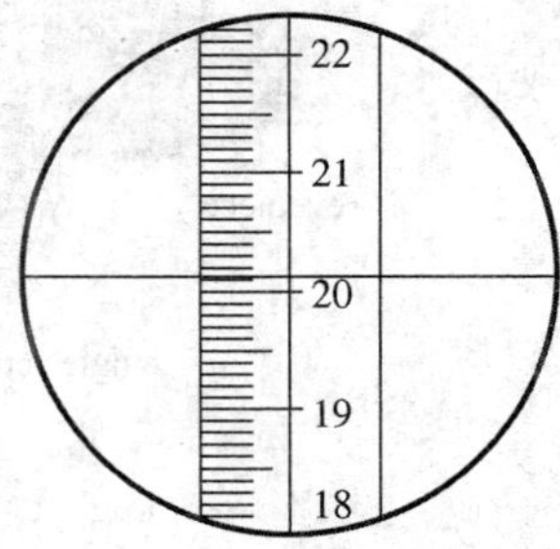

Fig. 4-1-1 The enlarged schematic for the optical lever

Experimental Instruments

YM－1 type Young's modulus instrument, 7 weights (1 kg), telescope, optical lever, micrometer screw gauge and tape.

The device for measuring Young's modulus with the optical lever method is shown in Fig. 4-1-2. The telescope device is shown in Fig. 4-1-3. The placement for the optical lever is shown in Fig. 4-1-4. A measurement system is composed of the telescope and optical lever.

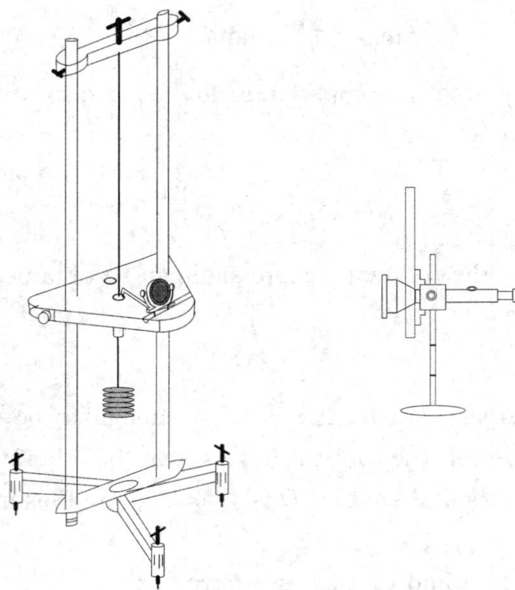

Fig. 4-1-2 Apparatus for measuring Young 's modulus of elasticity with optical lever

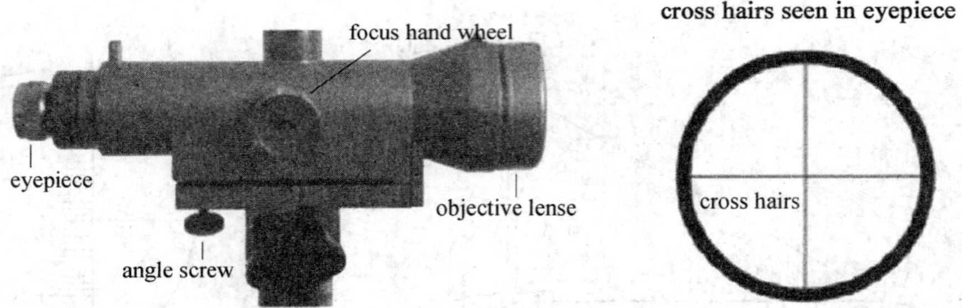

Fig. 4-1-3 Ruler reading telescope

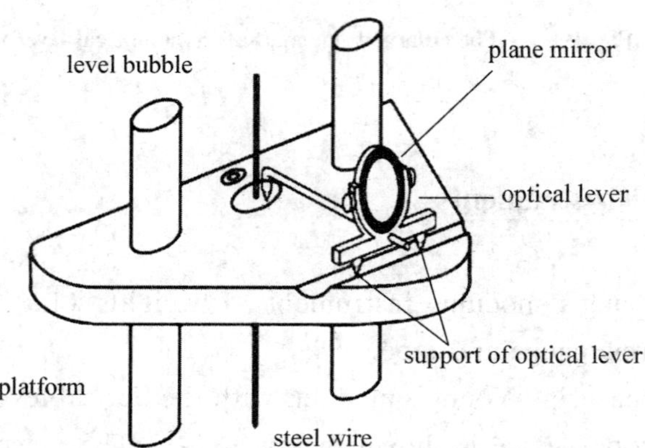

Fig. 4-1-4 Optical lever installation diagram

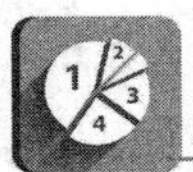

Experimental Contents and Steps

1. Device Adjustment

(1) Adjust the three screws on the bottom of instrument to level the platform (check with the level bubble). Place the two forefeet of the optical lever in the small groove in the fixed platform. Put the rear foot on the wire clamp.

(2) Set the telescope and the mirror at the same height. Make the plane mirror vertical to the line connecting the optical lever and telescope. Make the adjustment of telescope level and then set the zero point at the center of the telescope window.

(3) Point the telescope to the mirror and look into the mirror from above the telescope tube to observe the image of the ruler. If the image is not found, adjust the telescope holder until the image appears in the mirror.

(4) Adjust the telescope eyepiece until the cross image is clear. Adjust the objective lens with hand-wheel for the clear image of scale marking. When no relative displacement between the scale and the cross (namely no parallax) is found, the observer can top the up-and-down eye movement in front of the eyepiece. The parallax with less than a half grid is acceptable in the experiment.

(5) Slightly adjust the tilt angle of the telescope so that the horizontal line of the cross is within the range between -2 cm and 2 cm on the ruler image.

2. Measurement

(1) Add a weight on the weight tray and record the position n_0 of the scale.

(2)Place 5 other weights one by one and record the position as n_1, n_2, n_3, n_4 and n_5. Then, add the seventh weight, but the position record is unnecessary.

(3) Take away the weights one by one, record the position as n_5', n_4', n_3', n_2', n_1', n_0'.

(4) Measure the length of the wire $L=L_0+L_1$, where L_0 is the distance from the lower end of the upper clamp to the platform, and $L_1=(125\pm2)$ mm is the distance between the upper end of the lower clamp and the platform.

(5) Measure the distance D from the mirror to the ruler by using tape.

(6) Measure the wire diameter d using a micrometer at the upper, middle and lower parts of the wire twice to get the average diameter.

(7) Measure b. First press the lever on a paper to get three marks, and then use a millimeter to measure the vertical distance from the mark of the rear foot to the line connecting the two front feet.

3. Data Recording and Processing

(1) Calculate $\overline{\Delta n}$ with the successive difference method. $\overline{\Delta n}$ corresponds to the position change when three weights are loaded.

$$\overline{\Delta n}=\frac{(n_3-n_0)+(n_4-n_1)+(n_5-n_2)}{3}$$

(2) Calculate E by Equation (4-1-3) with the relative data.

(3) Derive the uncertainty u_E and the relative uncertainty u_{rE} of Equation (4-1-3). Then write the experimental result.

Precautions

1. Once the optical lever, telescope and ruler are well adjusted, they must not be readjusted. Otherwise, the data will become invalid and the experiment has to be reconducted.

2. It is forbidden to touch or wipe the eyepiece, objective, mirror and other optical devices with cloth or paper.

Questions

1. What are the advantages of the optical lever? How to improve the sensitivity of the optical lever to measure tiny change?

2. Can the Young's modulus be calculated by using the graphing method? What will the shape of the graph look like if the horizontal axis is the stress and the vertical axis is the strain?

4.2 电桥测电阻

电桥是采用比较法测量各种量(如电阻、电容、电感等)的一种仪器,它具有灵敏度和精确度都很高的特点,在自动控制和自动检测中得到极其广泛的应用。电桥根据用途不同有多种类型,其性能构造也各有特点。

在各种电桥中,常用的有惠斯登电桥和开尔文电桥。惠斯登电桥又称直流单臂电桥,一般用来测量 $10\sim10^6\ \Omega$ 范围的电阻;开尔文电桥又称为双臂电桥,适用于测量 $10^{-5}\sim10^2\ \Omega$ 范围的电阻。

4.2.1　用惠斯登电桥测电阻

实验目的

1. 掌握用惠斯登电桥测电阻的原理和方法。
2. 了解电桥灵敏度的概念。

实验原理

测量电阻有两个主要方法：伏安法和电桥法。采用伏安法测量电阻时，是通过测出流经电阻 R 的电流 I 和其两端的电位差 U，根据欧姆定律 $R=U/I$ 求出电阻值。这种方法，除了使用的电流表和电压表精度不高带来误差外，由于电表本身具有内阻，不论是采用内接还是外接，都不能同时准确地测出流经电阻的电流和两端电位差，因而不可避免地还存在线路本身的缺陷带来的误差。电桥法测电阻的实质是把被测电阻与标准电阻相比较，以确定其阻值。由于电阻的制造可以达到很高的精度，所以电桥法测电阻可以达到很高的精确度。

电桥分为直流电桥和交流电桥两大类。直流电桥又分为单臂电桥和双臂电桥。单臂电桥又称为惠斯登电桥，主要用途是测量中等阻值的电阻，它的基本原理线路如图 4-2-1 所示。

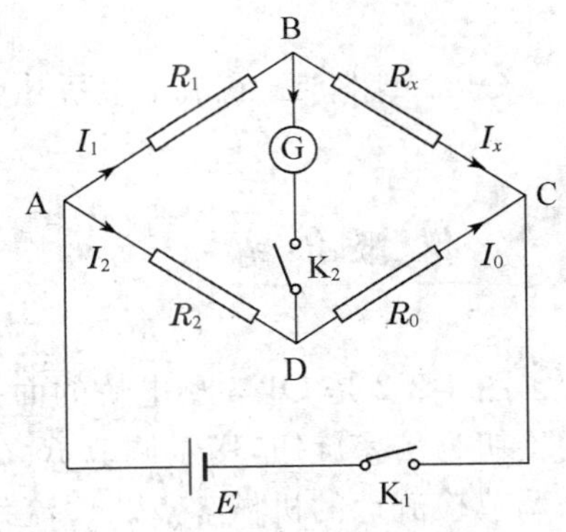

图 4-2-1　惠斯登电桥原理图

4 个电阻 R_1、R_2、R_0 和 R_x 首尾相接联成一闭合回路，每个电阻称为电桥的一个臂。A 和 C 两节点连到电源的两极上，另外的两个节点 B 和 D 间连一检流计Ⓖ。所谓“桥”就是指 BD 而言，它的作用就是将 B、D 两点的电位 U_B 和 U_D 进行比较。一般情况下，B、D 两点的电位是不相等的，检流计的指针将发生偏转。当检流计中无电流通过时，即表示 B、D 两点的电位相等，这时称为“电桥达到平衡”，此时检流计指针指零。当电桥平衡时，设流过 AB 支路的电流强度为 I_1，流过 AD 支路的电流强度为 I_2，流过 BC 支路的电流强度为 I_x，流过 DC 支路的电流强度为 I_0，则

$$I_1=I_x, I_2=I_0 \tag{4-2-1}$$

由 B、D 两点同电位可知，$U_{AB}=U_{AD}$，$U_{BC}=U_{DC}$，故 R_1、R_2、R_0 及待测电阻 R_x 应满足下列关系（平衡条件）

$$\begin{cases} I_1R_1 = I_xR_2 \\ I_1R_x = I_0R_0 \end{cases} \tag{4-2-2}$$

由式(4-2-1)和式(4-2-2)可得

$$R_x = \frac{R_1}{R_2} \times R_0 = k_r R_0 \tag{4-2-3}$$

式中,$k_r = R_1/R_2$ 为倍率;R_x 是待测电阻;R_1、R_2 是已知电阻;R_0 是测量用电阻箱。

适当选择 R_1、R_2 两个电阻的阻值,调节 R_0 的阻值,使 B、D 两点电位相等,则检流计中无电流通过,即 $I_G=0$,此时电桥平衡。实际应用时,一般先选定倍率 k_r 的数值,再调节比较臂的电阻 R_0,使检流计指针指零,即使电桥平衡。

电桥平衡后,将某一桥臂电阻 R 改变一小量 ΔR,引起检流计偏转为 ΔN 格,这时可定义电桥灵敏度为

$$S = \frac{\Delta N}{\Delta R/R} \tag{4-2-4}$$

电桥灵敏度是测量电阻的误差来源之一,通常认为指针可察觉的变化为 0.2 格,因此,可判断电桥平衡所带来的绝对误差为 $0.2R_x/S$。

实验器材

QJ23 型携带式直流单电桥,电阻面板(含 2 个电阻)。

仪器介绍

图 4-2-2 是 QJ23 型电桥的面板图。右边的 4 个旋钮是用来调节电阻 R_0 的。左上角的旋钮为倍率旋钮,用于调节 R_1/R_2。检流计安装在左下角,其上方有一个调零旋钮,用来调节检流计的机械零点。检流计下面的两个按钮开关 B、G,按下 B 接通电源,按下 G 接通检流计。B、G 按钮按下后,向顺时针方向旋转时,可以锁住。标有 R_x 的两个接线柱是接待测电阻 R_x 的。左边的 B"+"和"-"是用来连接外部电源的,本实验使用机内电源。G、"内接"和"外接"三个接线柱:当连接片连"外接"时,是采用机内检流计,当需外接检流计时,将"内接"用连接片连接,"外接"接线柱上连接外部检流计。本实验采用机内检流计。

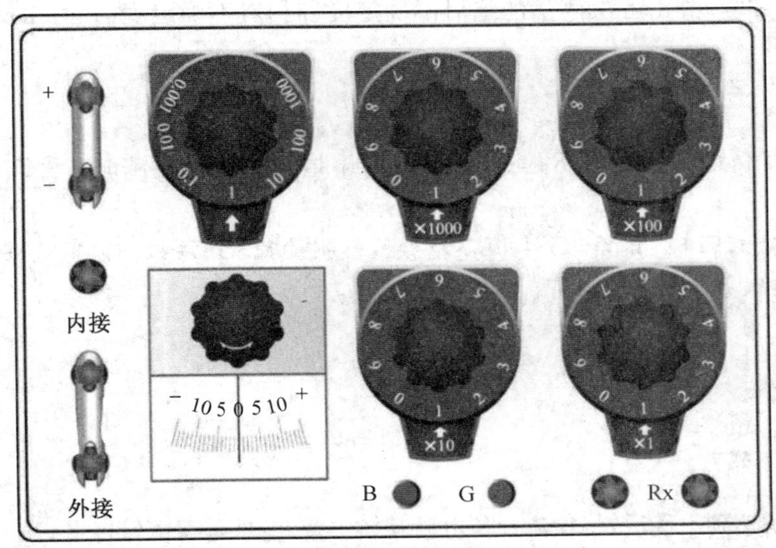

图 4-2-2 QJ23 型电桥面板图

实验内容和步骤

1. 将检流计"外接"接线柱用连接片短路，用"调零"旋钮使检流计的指针与"0"重合。

2. 把两个被测电阻 R_{x1}，R_{x2} 分别接到"R_x"两接线柱上。

根据待测电阻的标称值，选取合适的比率系数（倍率），使比较臂有效数字位数最多。具体步骤如下：

(1) 确定倍率

先将 R_0 值置 1000 Ω，倍率钮置最小值 0.001 挡。先按下"B"钮，再按下"G"钮，观察检流计指针偏转的方向。逐挡提高倍率值，直到检流计指针偏向另一方向，将倍率降低一挡。

(2) 确定 R_0 的"×1000"挡值。

R_0 的"×1000"倍率挡从 1 到 9 逐挡调节，先按下"B"钮，再按下"G"钮，观察检流计指针偏转的方向。逐挡提高"×1000"挡的值，直到指针偏向另一方向时，将"×1000"挡降低一挡。

(3) 确定 R_0 的其他各挡值。

用上述步骤 2)的方法依次调节 R_0 的"×100""×10""×1"各挡，直至检流计指针为 0 为止，读出 R_0 值。

3. 按上述方法分别测量 R_{x1}，R_{x2}，R_{x1}、R_{x2} 串联和 R_{x1}、R_{x2} 并联共 4 个阻值。将各次测得的数据 k_r 和 R_0 的值列入自拟的表格中，并计算出 R_x 值。测量电阻的最大允许误差为

$$\Delta R_x = k_r(0.2\% R_0 + 0.2 \times 1)\Omega \tag{4-2-5}$$

利用式(4-2-5)算出各次测量的绝对不确定度和相对不确定度,并写出测量的结果表达式。

注意事项

1. 为了保证测量精度,R_0 的"×1000"倍率挡至少为 1,以保证测量值 R_x 有 4 位有效数字。

2. 实验过程中,"B"和"G"按钮不准锁紧。测量时"G"按钮一按下就放开,以防检流计电流过大损坏或指针打弯。

思考题

1. 当电桥达到平衡后,将检流计与电源互换位置,电桥是否仍保持平衡?证明之。

2. 分析下列因素是否会影响电桥测电阻的误差?

(1) 电源电压不太稳定;(2) 导线电阻不能完全忽略;(3) 检流计没有调好零;(4) 检流计灵敏度不够。

3. 在电桥调节过程中,如果不论调节 R_0 为何值,检流计指针始终偏向一侧,说明什么问题?应做怎样的调整才能使电桥达到平衡?

4.2.2 用开尔文电桥测电阻

实验目的

1. 掌握用开尔文电桥测低值电阻的原理。
2. 学会用开尔文电桥测低值电阻的方法。
3. 了解测低值电阻时接线电阻和接触电阻的影响及其避免的方法。

实验原理

用惠斯登电桥测量中值电阻时,忽略了导线电阻和接触电阻(总称附加电阻)的影响,但在测量 1 Ω 以下的低值电阻时,各导线的电阻和端点的接触电阻相对被测电阻来说不可忽略,一般情况下,接线电阻和接触电阻的阻值为 $10^{-5}\sim10^{-2}$ Ω。接触电阻虽然可以用清洁接触点等措施使之减小,但终究不可能完全清除。当被测电阻仅为 $10^{-3}\sim10^{-6}$ Ω 时,其接线电阻及接触电阻值都已超过或大大超过被测电阻的阻值,这样就会造成很大误差,甚至完全无法得出测量结果。所以,用惠斯登电桥来测量低值电阻是不可能精确的,必须在测量线路上采取措施，避免接线电阻和接触电阻对低值电阻测量的影响。

为避免附加电阻的影响，本实验采用“四端钮接入法”将附加电阻部分转移到了电源回路，部分转移到比其大许多数量级的桥臂电阻的电路，组成了双臂电桥（又称为开尔文电桥），这是一种常用的测量低电阻的方法。

一、四端钮接入法

伏安法测电阻的原理图如图 4-2-3(a)所示。如果把接线电阻和接触电阻考虑在内，并将它们用普通导体电阻的符号表示，则可表示为如图 4-2-3(b)所示的等效电路。

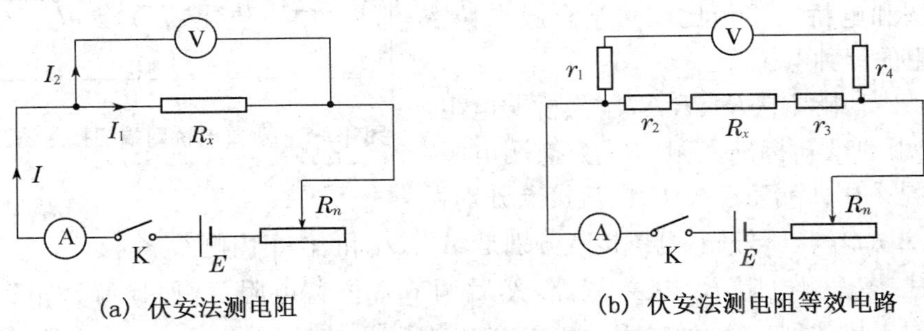

(a) 伏安法测电阻 (b) 伏安法测电阻等效电路

图 4-2-3 伏安法测量电阻电路及其等效电路

图 4-2-3(b)中，待测电阻 R_x 两侧的接触电阻和导线电阻分别以等效电阻 r_1、r_2、r_3、r_4 表示。其中 r_2、r_3 分别是连接安培表及滑线变阻器用的两根导线与被测电阻两端接头处的接触电阻及导线本身的接线电阻，r_1、r_4 是伏特表和安培表、滑线变阻器接头处的接触电阻和接线电阻。通过安培表的电流 I 在接头处分为 I_1、I_2 两支，I_1 流经安培表和 R_x 间的接触电阻再流入 R_x，I_2 流经安培表和伏特表接头处的接触电阻再流入毫伏表。因此，r_2、r_3 应算作与 R_x 串联；r_1、r_4 应算作与毫伏表串联。通常伏特表内阻较大，r_1 和 r_4 对测量的影响不大，而 r_2 和 r_3 与 R_x 串联在一起，被测电阻实际应为 $r_2+R_x+r_3$，若 r_2 和 r_3 数值与 R_x 为同一数量级，或超过 R_x，显然不能用此电路来测量 R_x。

若将测量电路改为如图 4-2-4 所示的电路，即将待测低电阻 R_x 两侧的接点分为两个电流接点 C_1-C_2 和两个电压接点 P-P，C_1-C_2 在 P-P 的外侧。显然伏特表测量的是 P-P 之间一段低电阻两端的电压，这就消除了 r_2 和 r_3 对 R_x 测量的影响。这种测量低电阻或低电阻两端电压的方法就称为“四端钮接入法”，它广泛应用于各种测量领域中。例如，为了研究高温超导体在发生正常超导转变时的零电阻现象和迈斯纳效应，必须测定临界温度 T_c，正是用通常的四端钮接入法，通过测量超导样品电阻随温度的变化而确定的。低值标准电阻正是为了减小接触电阻和接线电阻而设有 4 个端钮。

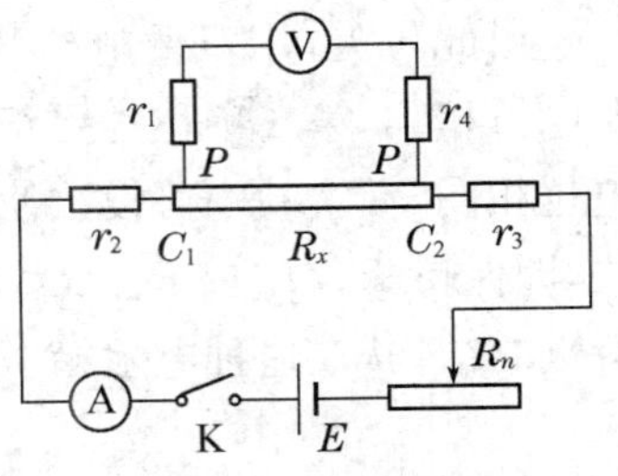

图 4-2-4 电阻的四端钮接法

二、用开尔文电桥测量导体的电阻率

用惠斯登电桥测量电阻,测出的 R_x 值中,实际上含有接线电阻和接触电阻(统称为 R_j)的成分(一般为 $10^{-3}\sim10^{-4}$ 数量级),通常可以不考虑 R_j 的影响,而当被测电阻达到较小值(如几十欧姆以下)时,R_j 所占的比重明显。因此,需要从测量电路的设计上来考虑。双臂电桥正是把四端钮接入法和电桥的平衡比较法结合起来精密测量低电阻的一种电桥。

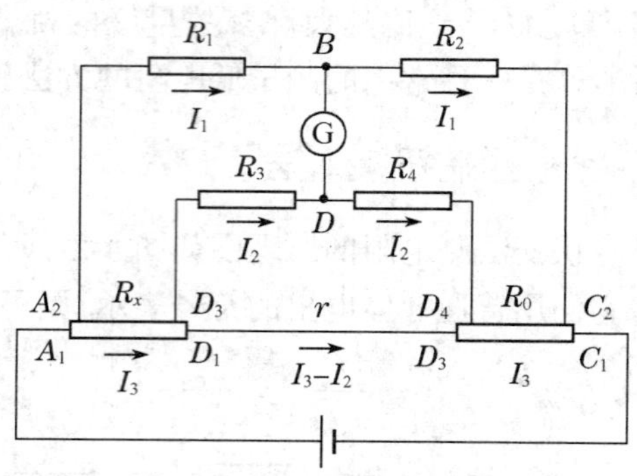

图 4-2-5　双臂电桥四端钮接入法测低电阻

如图 4-2-5 中,R_1、R_2、R_3、R_4 为桥臂电阻。R_0 为比较用的已知标准电阻,R_x 为被测电阻。R_0 和 R_x 是采用四端钮接线法,电流接点分别为 C_1、D_2 和 A_1、D_1 位于外侧;电压接点分别是 A_2、D_3 和,位于内侧。

测量时,将被测电阻 R_x 接入线路,然后调节各桥臂电阻值,使检流计指针为零,即 $I_G=0$,这时电桥达到平衡,通过 R_1 和 R_2 的电流相等,图中以 I_1 表示;通过 R_3 和 R_4 的电流相等,以 I_2 表示;通过 R_x 和 R_0 的电流也相等,以 I_3 表示。因为 B、D 两点的电位相等,故根据基尔霍夫定律有

$$\begin{aligned}I_1R_1&=I_3R_x+I_2R_3\\I_1R_2&=I_3R_s+I_2R_4\\I_2(R_3+R_4)&=(I_3-I_2)r\end{aligned}\tag{4-2-6}$$

联立求解,可得到

$$R_x=\frac{R_1}{R_2}R_0+\frac{rR_4}{R_3+R_4+r}\left(\frac{R_1}{R_2}-\frac{R_3}{R_4}\right)\tag{4-2-7}$$

由此可见,用双臂电桥测电阻,R_x 的结果由等式右边的两项来决定,其中第一项与单臂电桥相同,第二项称为更正项。为了更方便测量和计算,使双臂电桥求 R_x 的公式与单臂电桥相同,在实验中可设法使更正项尽可能为零。在双臂电桥测量时,通常可采用同步调节法,即保持 $R_1/R_2=R_3/R_4$,使得更正项能接近零。在实际的使用中,通常将电桥做成一种特殊的结构,在这种电阻箱中转臂的任一位置上都保持 $R_1=R_2$,$R_3=R_4$,这时上式变为

$$R_x=\frac{R_1}{R_2}R_0=k_rR_0\tag{4-2-8}$$

也就是说,当电桥平衡时就可以消除附加电阻 r 的影响。因此双臂电桥平衡时,式(4-2-7)成立,或者说,式(4-2-7)是双臂电桥的平衡条件,根据式(4-2-7)可以得出低电阻 R_x。

但是在这里必须指出,在实际的双臂电桥中,很难做到 R_1/R_2 与 R_3/R_4 完全相等,所以 R_x 和 R_0 电流接点间的导线应使用较粗的、导电性良好的导线,以使 r 值尽可能小,这样,即使 R_1/R_2 与 R_3/R_4 两项不严格相等,但由于 r 值很小,更正项仍能趋近于零。

一般说，用具有滑线盘的双臂电桥测量电阻时，其最大允许误差 ΔR_x，在准确度等级 $\alpha=0.2, 0.5, 1, 2$ 时，测量最大允许误差

$$\Delta R_x = \alpha\% \cdot R_{max} \tag{4-2-9}$$

式中，R_{max} 是具有滑线盘的电桥的量程。本实验所用的双臂电桥 $\alpha=0.2$，$R_{max}=k_r\times 0.11\ \Omega$。

一段导体的电阻与该导体材料的物理性质和它的几何形状有关。实验指出，导体的电阻与其长度 L 成正比，与其横截面面积 A 成反比，即：

$$R_x = \frac{\rho L}{A} \tag{4-2-10}$$

式中，比例系数 ρ 称为导体的电阻率，它的大小表示导电材料的性质，可按下式求出

$$\rho = \frac{RA}{L} = \frac{R_x \pi d^2}{4L} \tag{4-2-11}$$

电阻率的相对不确定度

$$u_{r\rho} \approx \sqrt{\left(\frac{u_{Rx}}{R_x}\right)^2 + 4\left(\frac{u_d}{\bar{d}}\right)^2} \tag{4-2-12}$$

式中，$\bar{d}$ 为金属圆棒直径的平均值。

实验器材

QJ42 型直流双臂电桥，连接架，铜、铝棒各一根，螺旋测微计，游标卡尺，导线若干。

仪器介绍

一、QJ42 型携带式直流面板

箱式双臂电桥的形式多样，本实验采用 QJ42 型携带式直流双臂电桥，图 4-2-6 为其面板配置图。各部分名称如下：① 检流计，其上有机械调零器；② 电位端接线柱（P_1、P_2）；③ 电流端接线柱（C_1、C_2）；④ 倍率开关；⑤ 电源选择开关；⑥ 外接电源接线柱；⑦ 标尺；⑧ 读数盘 R_b；⑨ 检流计按钮开关；⑩ 电源按钮开关。

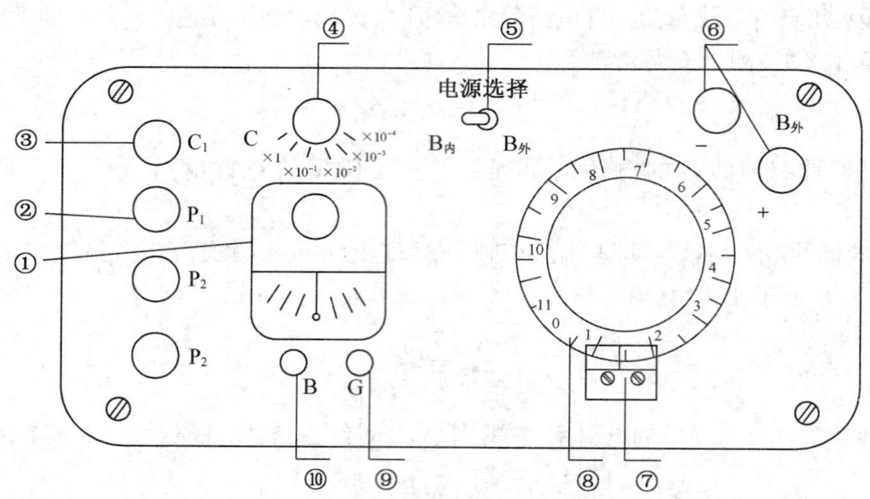

图 4-2-6 QJ42 型携带式直流双臂电桥

二、使用方法

1. 将检流计指针调到“0”位置。

2. 将被测电阻 R_x 的四端接到双臂电桥的相应四个接线柱上。

3. 估计被测电阻值,并将倍率开关旋到相应的位置上。

4. 当测量电阻时,应先按“B”后按“G”按钮,并调节读数盘 R_0,使检流计重新回到“0”位。断开时应先放“G”后放“B”按钮。注意:一般情况下,“B”按钮应间歇使用。此时电桥已处平衡,而被测电阻 R_x 为:R_x=(倍率开关的示值)×(读数盘的示值)欧。

5. 使用完毕,应把倍率开关旋到“G 短路”位置上。

实验内容和步骤

1. 将待测的一段金属圆棒夹到连接架上,接成图 4-2-7 所示的一个“四端电阻”,将 C_1、P_1、P_2、C_2 依次接到双臂电桥的 C_1、P_1、P_2、C_2 四个接线柱上。

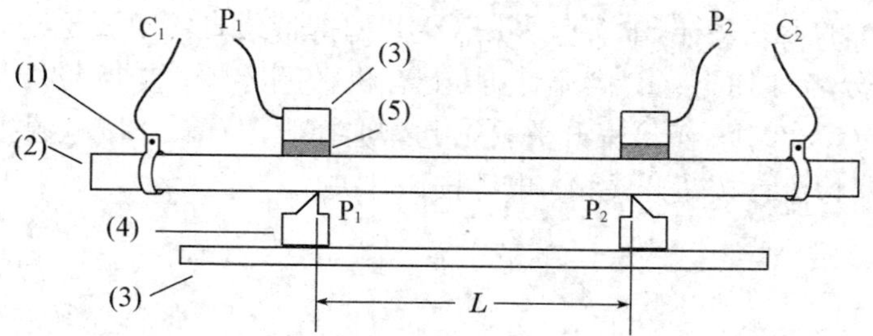

图 4-2-7 待测材料接成“四端电阻”示意图

2. 用 QJ42 型携带式直流双臂电桥的使用方法测量该金属圆棒的电阻 R_x。

3. 用螺旋测微计测出圆形导体的直径，在不同的部位测 5 次，取平均值。用游标卡尺测量 P_1、P_2（两夹具尖刀口内侧）间导体的长度，测量一次。

4. 自拟数据表，由式(4-2-11)算出 $\bar{\rho}$ 值，由式(4-2-12)算出待测量 ρ 的相对不确定度 $u_{r\rho}$，再计算其绝对不确定度 $u_\rho=\bar{\rho}u_{r\rho}$，并写出结果表达式。

5. 对于其他待测的金属圆棒，重复上述过程测得结果。实验完毕后关掉"B_1"开关。

思考题

1. 为什么双臂电桥能够大大减小接线电阻和接触电阻对测量结果的影响？

2. 根据测量结果说明，为了减小电阻率 ρ 的测量误差，在被测量 R_x、d 和 l 三个直接测的量中，应特别注意哪个物理量的测量。为什么？

3. 如果低电阻的电流接头和电压接头互相接错，这样做有什么问题？

Experiment 4.2　Measuring the Resistance with Electric Bridges

Circuit bridge is the instrument to measure various quantities (such as resistance, capacitance, inductance, etc.) by using a comparative method. With the advantages of high sensitivity and accuracy, electric bridge has a wide application in automatic control and automatic detection. There are various types of bridge-circuits with special performances and structures to serve different functions.

Among all the bridge-circuits, Wheatstone bridge and Kelvin bridge are the commonly-used types. Wheatstone bridge, also known as DC Wheatstone bridge, is generally used to measure resistance with the range of $1\sim10^6\ \Omega$; Kelvin bridge, also known as double bridge, is suitable for measuring resistance with the range of $10^{-5}\sim10^2\ \Omega$.

4.2.1　Measure the Resistance with the Wheatstone Bridge

Experimental Objectives

1. To master the principle and method of measuring resistance with Wheatstone bridge.

2. To understand the concept of electric bridge's sensitivity.

Experimental Principle

To measure the resistance with *I-U* method, it is necessary to measure the current I running through the resistor R and the potential difference U. Then the resistance can be calculated according to Ohm's law $R=U/I$.

Apart from the errors caused by the inaccuracy of the ammeter and voltmeter, the intrinsic resistances of the ammeter and the voltmeter may also lead to the inaccuracy of the current and the potential difference between the two ends, with both internal connection and external connection with the resistor. The basic rule of measuring resistance via electric ridge method is to compare the measured resistance with the standard resistance so as to determine its value. A resistor can be made with a great accuracy, and thus the electric bridge method can assist to achieve high accuracy in the resistance measurement.

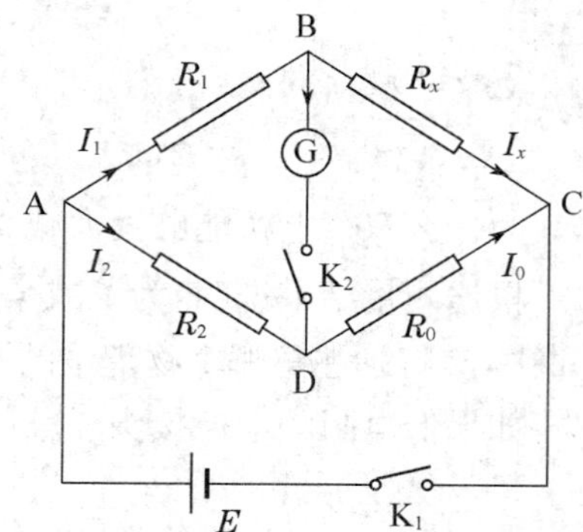

Fig. 4-2-1 The circuit of wheatstone bridge

The main purpose of Wheatstone bridge is to measure the medium resistance. The schematic circuit for it is shown in Fig. 4-2-1.

Three adjustable resistors with resistences R_0, R_1, R_2 as the resistance are connected with resistor R_x of unknown resistance to form a closed loop, junctions A and C are connected across a DC source, and the other two junctions B and D are connected to a galvanometer Ⓖ, which can detect very small current. If the potential at B equals that at D, the galvanometer indicator will not deflect. Therefore, when the galvanometer reads zero, there is no current passing through the galvanometer, which is referred to as "bridge-balance". Denote the current through the AB branch is I_1, the current through the branch AD is I_2, the current through the branch BC is I_x, and the current through the branch DC is I_0. When the bridge is balanced, we have

$$I_1=I_x,\quad I_2=I_0 \tag{4-2-1}$$

Junctions B and D have the same potential, thus $U_{AB}=U_{AD}$ and $U_{BC}=U_{DC}$, and R_0, R_1, R_2 and R_x satisfy (balance conditions)

$$\begin{cases} I_1R_1=I_2R_2 \\ I_xR_x=I_0R_0 \end{cases} \tag{4-2-2}$$

From Equation (4-2-2), we obtain

$$R_x=\frac{R_1}{R_2}R_0=k_rR_0 \tag{4-2-3}$$

Where $k_r=R_1/R_2$ is called the magnification.

Adjust the resistance of resistors R_0, R_1, R_2 appropriately, so that junctions B and D have equal potential. And there is no current in the galvanometer, namely $I_g=0$, and then the bridge is balanced. In practical applications, the magnification k_r is generally selected first, and then the resistance R_0 of the comparison arm is adjusted so as to make the galvanometer pointer zero for the balance of bridge.

After the bridge is balanced, change the resistance R_0 of a bridge arm into a small amount of ΔR, resulting in ΔN of the galvanometer deflection, and then the sensitivity of bridge circuit can be defined as

$$S=\frac{\Delta N}{\Delta R/R} \tag{4-2-4}$$

Bridge's sensitivity is one of the sources of error in measuring the resistance. It is usually considered that the perceptible change of the indicator is 0.2 grid. Therefore, the absolute error of the bridge balance is $0.2R_x/S$.

Experimental Instruments

QJ23 portable DC Wheatstone bridge, Resistance box (including two resistances), and two resistors with unknown resistance.

Fig. 4-2-2 shows the panel of the QJ23 wheatstone bridge. The four knobs on the right are used to adjust the resistance R_0. The knob on the upper left corner is to adjust ratio k_r, i.e. R_1/R_2. The built-in galvanometer is installed in the lower left corner, and there is a knob on it for mechanical zero setting. Button switches "B" and "G" correspond to "On" and "Off" of the galvanometer in the circuit shown in Fig. 4-2-2. When buttons "B" and "G" are pressed and rotated in the clockwise direction, they can be locked. The two terminals labeled R_x are used to connect the resistor R_x. The terminals "+" and "−" are shorted with a sheet metal in order to use the built-in source. The built-in galvanometer is used for this experiment, the other two terminals are also shorted as shown in Fig. 4-2-2 so that the label "external" on the panel can be seen.

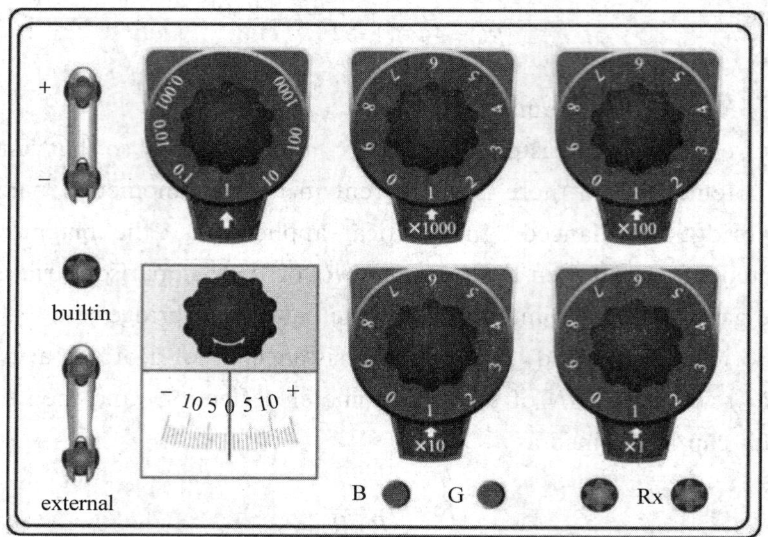

Fig. 4-2-2 QJ23 Wheatstone bridge

Experimental Contents and Steps

1. Adjust the calibrating knob to align the galvanometer pointer to 0.

2. Connect the two resistors R_{x1}, R_{x2} to the "R_x" terminals. Choose the appropriate ratio k_r, so that the resistor has the most significant digits. Specific steps are as follows:

(1) Determine the ratio k_r

First set R_0 to 1000 Ω and the ratio knob to 0.001.

Press buttons "B" and "G". If the galvanometer indicator deflects abruptly, release button "G" immediately.

Increase the ratio k_r step by step and repeat the operation until the indicator of the galvanometer turn to the opposite side, then turn the ratio to a lower one.

(2) Determine R_0 value on the "× 1000"

Adjust the knob of "× 1000" from 1 to 9. First press the "B" button, and then press the "G" button. Observe the galvanometer indicator's deflection direction. Increase the value of "×1000" gradually until the indicator turns to the other direction. Then gear the "× 1000" knob to a lower block.

(3) Determine the other block values of R_0

Adjust the "×100", "×10", "×1" knobs sequentially with the method of step (2) until the galvanometer indicator points at zero and record the value of R_0.

3. Measure the resistances of R_{x1} and R_{x2}, the total resistance of R_{x1} and R_{x2} in series, and the total resistance of R_{x1} and R_{x2} in parallel with the above method. Fill the

values of the measured data k_r and R_0 into the table and calculate the value of R_x. The maximum allowable error of the measured resistance is

$$\Delta R_x = k_r(0.2\% R_0 + 0.2 \times 1\ \Omega) \tag{4-2-5}$$

Calculate the absolute uncertainty and relative uncertainty of each measurement with Equation (4-2-5), and write down the result expression of the measurement.

Precautions

1. In order to ensure the accuracy of measurement, the R_0 "×1000" magnification block should be at least 1, so that the measured value R_x has four significant digits.

2. During the experiment, the "B" and "G" buttons are not allowed to be locked, and the "G" button should be released immediately after it is pressed so as to prevent the galvanometer from being damaged due to excessive current or the indicator bending.

Questions

1. Under the balance of the bridge, if the positions of galvanometer and power are exchanged positions, will the bridge still be in balance?

2. Will the following factors influence the error of resistance measurement in the bridge circuit?

(1) The voltage of the power supply is not stable;

(2) The resistance of the wire cannot be completely ignored;

(3) The galvanometer is not properly tuned to zero;

(4) The sensitivity of the galvanometer is not high enough.

3. In adjusting the bridge circuit, if the galvanometer indicator is always one-sided biased whatever the value of R_0 is set, what are the possible reasons behind? What kind of measures could be taken to make the bridge circuit balanced?

4.2.2 Measure the Resistance with Kelvin Bridges

Experimental Objectives

1. To master the principle of measuring the resistance with Kelvin bridge.

2. To learn how to measure the resistance with Kelvin bridge.

3. To understand the impact of wire resistance and contact resistance when measuring the low resistance and how to avoid it.

Experimental Principle

When measuring the medium resistance with the Wheatstone bridge, the influence of the wire resistance and the contact resistance (referred to as additional resistance) is ignored. However, when measuring low resistance below 1 Ω, the resistance of each conductor line and the contact resistance of the end point are not negligible. Generally, the resistance of the connection resistance and contact resistance are about $10^{-5}\sim10^{-2}$ Ω. The contact resistance can be reduced by cleaning the contact point and other techniques, yet it cannot be completely removed. When the measured resistance is only $10^{-3}\sim10^{-6}$ Ω, the values of wiring resistance and contact resistance have exceeded the resistance of measured resistance. As a result, large errors will be generated or even the final measurement result cannot be obtained. Therefore, it is impossible to measure the low-value resistance precisely with the wheatstone bridge, and it is a must to take necessary actions to prevent the negative impact of the wiring resistance and the contact resistance on the measurement of low-value resistor.

In order to prevent the impact of additional resistance, this experiment applies the "four-terminal access method" to transfer the partial additional resistance to the circuit, and to transfer partly the additional resistance to the arm resistance with resistor that is much larger than the measured low-value resistor. In this way, the double-arm bridge (also known as the Kelvin bridge) can be formed and commonly used to measure the low-value resistor.

1. Four-Terminal Access Method

The schematic diagram of resistance measurement with the volt-ampere method is shown in Fig. 4-2-3(a). If the wiring resistance and contact resistance are taken into account and are represented by symbols of common conductor resistance, they can be expressed as an equivalent circuit as shown in Fig. 4-2-3(b).

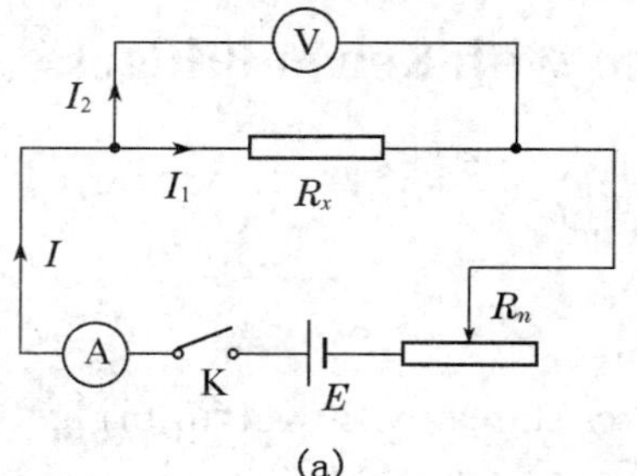

(a)

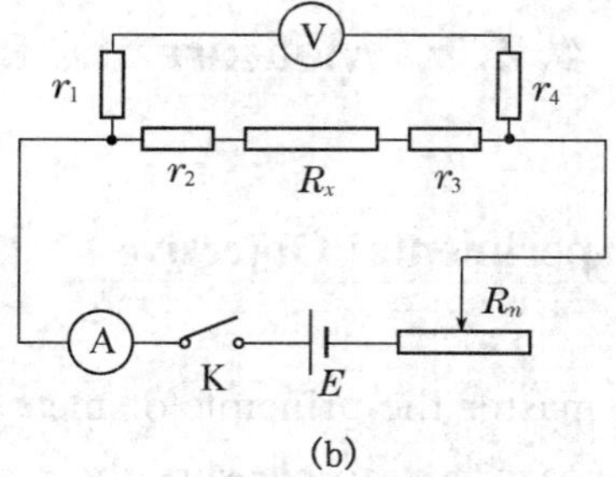

(b)

Fig. 4-2-3

In Fig. 4-2-3(b), the contact resistance and the wire resistance on both sides of the resistor R_x to be measured are represented by the equivalent resistance r_1, r_2, r_3, and r_4. r_2 refers to the contact resistances between the two ends of the measured resistance and the two wires connecting the ammeter to the sliding rheostat while r_3 is the wiring resistance respectively. r_1 and r_4 stand for the contact resistance between the voltmeter, the ammeter and the junction of the sliding rheostat, and the wiring resistance respectively. The current I through the ammeter is divided into I_1 and I_2 at the joint. I_1 flows through the contact resistor between the ammeter and R_x and then flows into R_x. I_2 flows through the junction resistance between the junctions of the ammeter and the voltmeter and then flows into the voltmeter. Therefore, r_2 and r_3 should be seen as the serial resistors of R_x, while r_1 and r_4 should be seen as the serial resistors of voltmeter. Generally, the voltmeter has a large internal resistance while r_1 and r_4 have less effect on the measurement. Since r_2 and r_3 are connected in series with R_x, the measured resistance should be $r_2+R_x+r_3$. If the values of r_2 and r_3 are of the same magnitude or larger than R_x, it is obvious that this circuit can not be used to measure R_x.

The circuit to be measured can be changed as shown in Fig. 4-2-4. The contacts on both sides of the low resistance R_x to be measured are divided into two current contacts C_1-C_2 and two voltage contacts P_1-P_2, and C_1-C_2 are on the outside of P_1-P_2. Obviously, the voltmeter measures the voltage across the low resistance between P_1 and P_2, which eliminates the effect of r_2 and r_3 on the R_x measurement. Such a method of measuring the low resistance at both ends of the voltage is called "four-terminal access method". This method has been widely used in a variety of measurement areas. For example, its application in the measurement of zero resistance and the study of Meissner effect. On these occasions, the critical temperature T_c of the high-temperature superconductor in the normal superconductivity transition needs to be measured. It is with the four-terminal access method that the critical temperature T_c can be determined by measuring the resistance change of the superconducting samples with temperature. The four end buttons of the low-value standard resistor are especially designed to reduce the contact resistance and the wiring resistance.

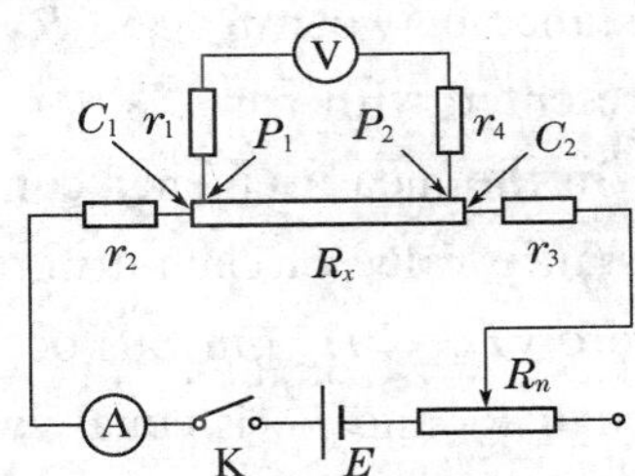

Fig. 4-2-4 Four-terminal connection of resistors

2. Measure the Resistivity of a Conductor with a Kelvin Bridge

The resistance R_x measured by the Wheatstone bridge contains the wire resistance and the contact resistance (collectively referred to as R_j). Given that the magnitude order of R_j is around 10^{-3} to 10^{-4}, the influence of R_j can usually be ignored. When the measured resistance has a small value (e.g. tens of ohms or less), the proportion of R_j is obvious. Therefore, great attentions should be taken to the circuit design. With the combination of the four-terminal connection and the balance of the bridge, the double-arm bridge can accurately measure the low resistance.

In Fig. 4-2-5, R_1, R_2, R_3, and R_4 are the bridge arm resistance, R_0 is the standard resistor with a specific resistance for comparison, and R_x is the unknown resistance to be measured. R_0 and R_x are connected by using four-terminal method. The current contact positions of C_1-D_2 and A_1-D_1 are located at the outer end of the resistors, while the voltage contact positions of C_2-D_4 and A_2-D_3 are located on the inner end of the resistors. The wire resistance and contact resistance between R_0 and R_x, and voltage external resistor of R_0 and R_x are together represented with r.

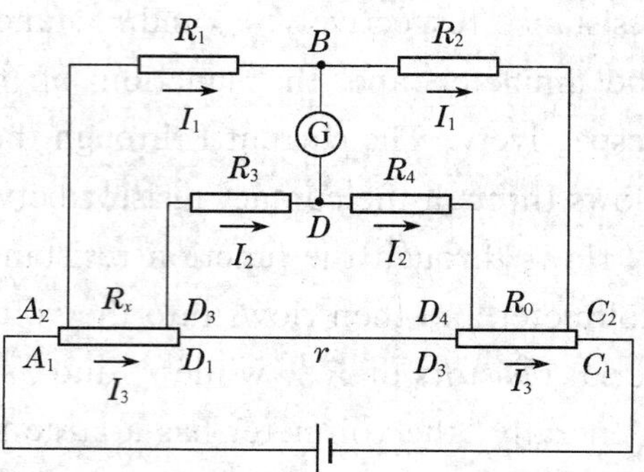

Fig. 4-2-5 The Circuit diagram of Kelvin Bridge

In the measurement, connect the resistor R_x into the circuit, and then adjust the resistance value of each arm of the bridge circuit until the galvanometer indicator points to zero ($I_G=0$). On this occasion, the bridge reaches equilibrium, the current passing through R_1 and R_2 is equal, which is indicated by I_1. The equal current through the R_3 and R_4 is recorded as I_2, and the equal current through the R_x and R_0 is represented as I_3, thus the current passing through r is (I_3-I_2). Given the equal potentials at B and D points, so based on Kirchhoff's law, we have

$$\begin{cases} I_1R_1=I_3R_x+I_2R_3 \\ I_1R_2=I_3R_0+I_2R_4 \\ I_2(R_3+R_4)=(I_3-I_2)r \end{cases} \tag{4-2-6}$$

By solving these equations together, we obtain

$$R_x=\frac{R_1}{R_2}R_0+\frac{rR_x}{R_3+R_4+r}\left(\frac{R_1}{R_2}-\frac{R_3}{R_4}\right) \tag{4-2-7}$$

In the measurement of the resistance with a double-arm bridge circuit, the R_x is determined by the two terms on the right sides of the above equations, where the first term is the same as the single-arm electric bridge and the second term is called the correction term.

For more convenient measurement and calculation, the formula for R_x with the double-arm bridge circuit can be converted into the same one in Wheatstone bridge, which means adjusting the correction term in the experiment close to zero as much as possible. In practical measurement with double-arm bridge circuit, the synchronous adjustment method is often adopted to maintain $R_1/R_2=R_3/R_4$, so that the correction term is close to zero. In actual experiment, the bridge circuit is usually made into a special structure with $R_1=R_2$, $R_3=R_4$, then the above formula is converted as

$$R_x=\frac{R_1}{R_2}R_0=k_rR_0 \tag{4-2-8}$$

In other words, when the bridge circuit is balanced, the influence of the additional resistance r can be eliminated. Therefore, when the double-arms bridge circuit is balanced, Equation (4-2-8) is valid, or Equation (4-2-8) is the balance condition of the double-arm bridge circuit. Thus, we can obtain the low resistance R_x via this Equation.

However, it is difficult to make R_1/R_2 and R_3/R_4 exactly same in the actual double-arm bridge circuit, so the wire between R_x and R_0 should be thicker and better conductive to make the value of r as small as possible. In this way, even if R_1/R_2 and R_3/R_4 are not strictly same, the small value of r can make the correction term approach zero.

In measuring the resistance by using a double-arm bridge circuit with a sliding rheostat, the maximum allowable error is defined as ΔR_x. When the accuracy level is 0.2, 0.5, and 1, 2, the maximum allowable error is measured as the following equation:

$$\Delta R_x=\alpha\% \cdot R_{max} \tag{4-2-9}$$

Where R_{max} is the range of the bridge circuit with the sliding rheostat, and the parameters of the double-arm bridge circuit used in this experiment are as follows: $\alpha=0.2$ and $R_{max}=kr\times 0.11\ \Omega$.

The resistance of a conductor is related to its physical properties and geometry. The experiment's result indicates that a conductor's resistance is proportional to its length L, and inversely proportional to its cross-sectional area A, shown as

$$R_x=\rho\frac{L}{S} \tag{4-2-10}$$

Here ρ is the resistivity of the conductor, which can be obtained by the following formula

$$\rho=\frac{R_xS}{L}=\frac{\pi R_xd^2}{4L} \tag{4-2-11}$$

The relative uncertainty of resistivity is

$$u_{r\rho}=\sqrt{\left(\frac{u_{Rx}}{\overline{R}_x}\right)^2+4\left(\frac{u_d}{\overline{d}}\right)^2+\left(\frac{u_L}{\overline{L}}\right)^2} \tag{4-2-12}$$

Where $\overline{d}$ is the average diameter of the metal rod.

Experimental Instruments

QJ44 type DC double-arm bridge circuit, a connection frame, a copper rod, an aluminum rod, a micrometer screw gauge, a vernier caliper and some wires.

The Introduction of Instruments

1. QJ42 Portable DC Panel

There are different types of the box-type double-arm bridge circuit, and the QJ42 portable DC double-arm bridge is adopted in this experiment. Fig. 4-2-6 shows its panel configuration diagram. Each of the items is illustrated as follows: ① galvanometer, it has a mechanical 0 adjuster ② potential terminal (P_1, P_2); ③ current terminal (C_1, C_2); ④ magnification switch; ⑤ power selector switch; ⑥ external power supply terminal; ⑦ ruler; ⑧ reading plate R_0; ⑨ galvanometer button switch; ⑩ power switch.

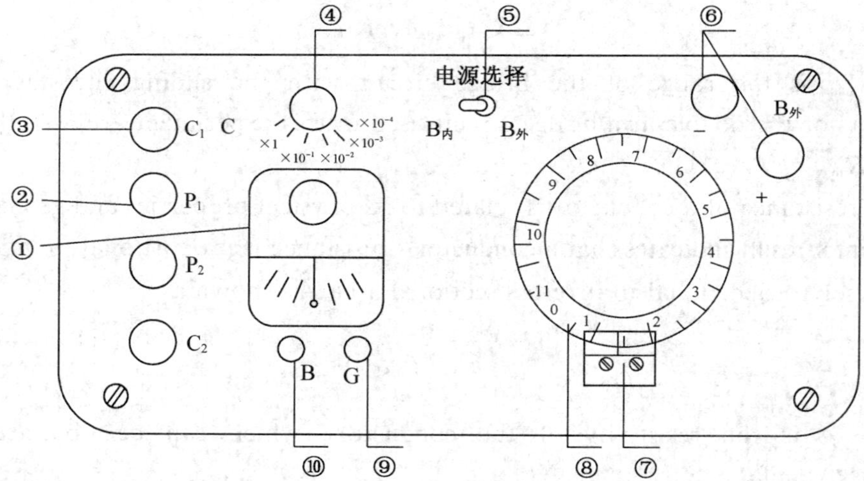

Fig. 4-2-6 QJ42 portable DC double-arm bridge

2. Operation Instructions

(1) Set the galvanometerindicator to the "0" position.

(2) Connect the four ends of the measured resistor R_x to the corresponding four terminals of the double-arm bridge.

(3) Estimate the resistance of R_x, and turn the magnification switch to the corresponding position.

(4) For the measurement, first press the "B" button and then press the "G" button. Next adjust the reading plate R_0 until the galvanometer is back to the "0" position. When disconnecting it, release the "G" button first and then the "B" button. The "B" button should be used intermittently. When the bridge is balanced, the measured resistance can be obtained as $R_x=$ (the indicated value of magnification switch) $\times$ (the value shown in the reading plate).

(5) After the experiment, the magnification switch should be turned to the "G short circuit" position.

Experimental Contents and Steps

1. Connect a piece of metal rod to the connecting frame with the four-terminal method shown as Fig. 4-2-7. C_1, P_1, P_2, and C_2 terminals are connected to the C_1, P_1, P_2, C_2 of the double-arm bridge respectively.

2. Measure the resistance R_x of the metal rod using the QJ44 type portable double-arm bridge circuit.

3. Measure the diameter of the conductor with a micrometer screw caliper. Measure it at 5 different positions and take the average value. Measure the length of the rod between P_1 and P_2 (inside the two tips) once with a Vernier caliper.

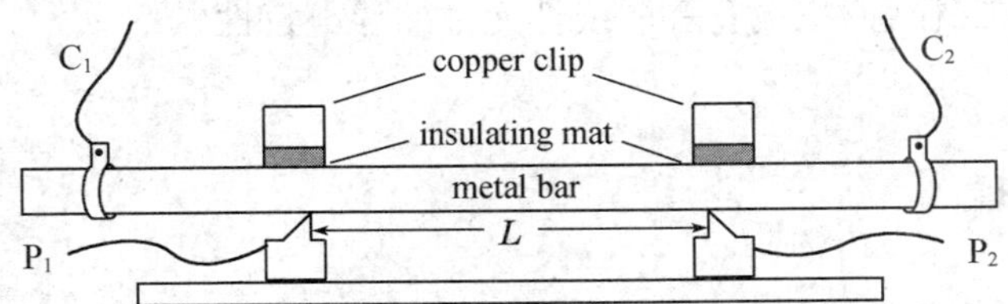

Fig. 4-2-7　A schematic diagram of the four-terminal resistance of the material to be tested

4. Compile the data tables. Calculate the value of $\bar{\rho}$ from Equation (4-2-11), calculate the relative uncertainty $u_{r\rho}$ of ρ to be measured by Equation (4-2-12). Write down the final result after calculating its absolute uncertainty $u_\rho=\bar{\rho}u_{r\rho}$.

5. Repeat the above process for other metal rods to be tested. Turned off "B_1" switch after the experiment.

Questions

1. Why can a double-arm bridge circuit greatly reduce the impacts of wiring resistance and contact resistance on the measurement results?

2. According to the measurement results, in order to reduce the measurement error of resistivity ρ, which parameter should be particularly paid attention to among the three directly measured quantities of R_x, d and l? Explain the reason.

3. What problems would be aroused if the current contact and voltage contact of low-resistance are connected incorrectly?

4.3 线性电阻和非线性电阻的伏安特性曲线

给一个元件施加以直流电,用电压表测出该元件两端的电压,用电流表测出通过元件的电流。通常以电压为横坐标、电流为纵坐标,画出该元件的电流和电压关系曲线,称为该元件的伏安特性曲线。这种研究元件特性的方法称为伏安法。伏安特性曲线为直线的元件称为线性元件,如金属电阻;伏安特性曲线为非直线的元件称为非线性元件,如二极管、三极管等。伏安法的主要用途是测量研究线性和非线性元件的电特性。非线性电阻总是与一定的物理过程相联系,如发热、发光和能级跃迁等。

实验目的

1. 测绘电学元件的伏安特性曲线,学习用图线表示实验结果。
2. 了解半导体二极管、三极管等非线性元件的导电特性。

实验原理

一般金属导体的电阻是线性电阻,它与外加电压的大小和方向无关,其伏安特性是一条直线。图 4-3-1 是某金属电阻的伏安特性曲线示意图,直线通过一、三象限,电阻始终为一定值,等于直线斜率的倒数,即 $R=\dfrac{U}{I}$。

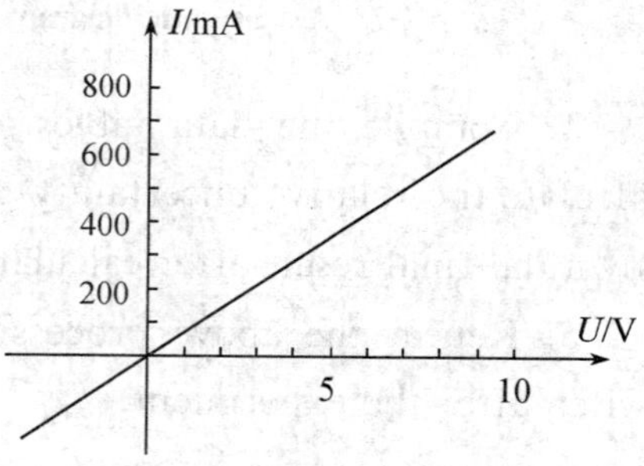

图 4-3-1 线性电阻的伏安特性曲线

常用的晶体二极管是非线性电阻,其电阻不仅与外加电压的大小有关,而且还与方向有关。为了了解晶体二极管的导电特性,下面对它的结构与电学性能作一简单介绍。

晶体二极管又叫半导体二极管。纯净半导体又称本征半导体,其导电性能介于导体和绝缘体之间,如果在纯净的半导体中适当地掺入极微量的杂质,则半导体的导电能力就会有上百万倍的增加。根据掺入杂质的不同,半导体可分为两种类型:某些杂质(如磷、砷等)加到半导体中会产生许多带负电的电子，这种半导体叫电子型(或 N 型)半导体；另一些杂质(如硼)加到半导体中会产生许多缺少电子的空穴，这种半导体叫空穴型(或 P 型)半导体。

晶体二极管是由两种具有不同导电性能的 P 型和 N 型半导体结合，在两种半导体交界面处形成的 PN 结所构成的，PN 结具有单向导电的性能。它有正、负两个电极，正极由 P 型半导体引出，负极由 N 型半导体引出，如图 4-3-2(a)所示。在电路图中二极管用图 4-3-2(b)所示的符号表示。PN 结的形成和导电性能可作如下解释：

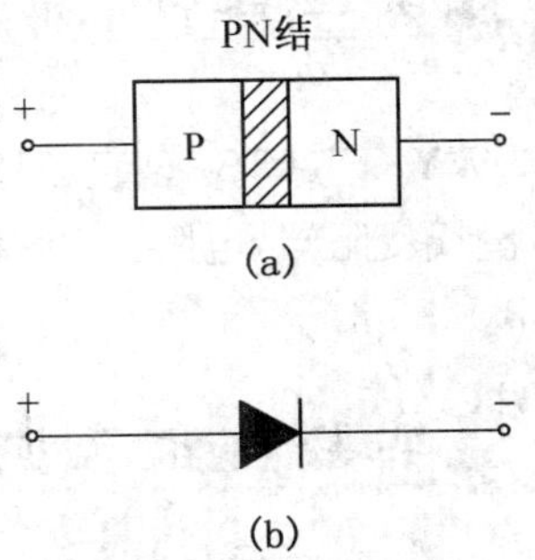

图 4-3-2 晶体二极管的 PN 结和表示符号

如图 4-3-3 (a)所示，由于 P 区中空穴的浓度比 N 区大，空穴便由 P 区向 N 区扩散。同样，由于 N 区的电子浓度比 P 区大，电子便由 N 区向 P 区扩散。结果在 P 型和 N 型半导体交界面的两侧附近，形成了带正、负电的薄层区，称为 PN 结。这个带电薄层内的正、负电荷产生了一个电场，它对载流子(空穴和电子)的作用力恰好与载流子扩散运动的方向相反，使扩散受到内电场的阻力作用，所以这个带电薄层又称为阻挡层。当扩散作用与内电场作用相等时，P 区的空穴和 N 区的电子不再减少，阻挡层也不再增厚，达到动态平衡，这时二极管中没有电流。

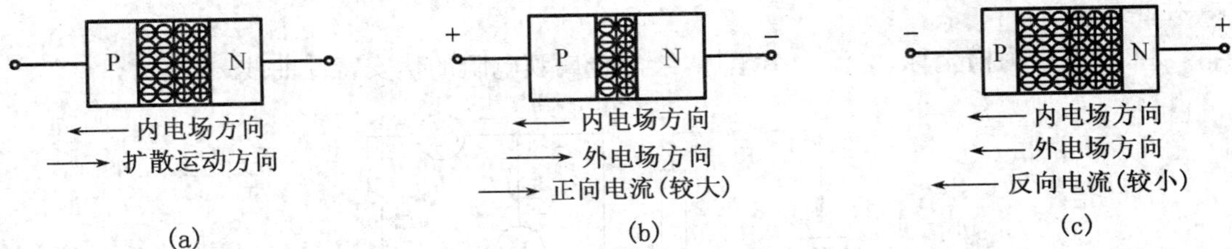

图 4-3-3 PN 结的形成和单向导电特性

如图 4-3-3 (b)所示，当 PN 结加上正向电压(P 区接正、N 区接负)时，外电场与内电场方向相反，因而削弱了内电场，使阻挡层变薄。这样，载流子就能顺利地通过 PN 结，形成比较大的电流，所以 PN 结的正向电阻较小。

如图 4-3-3(c)所示，当 PN 结加上反向电压(P 区接负，N 区接正)时，外加电场与内电场方向相同，因而加强了内电场作用，使阻挡层变厚。这样，只有极少数载流子能够通过 PN 结，形成很小的反向漏电流，所以 PN 结的反向电阻很大。

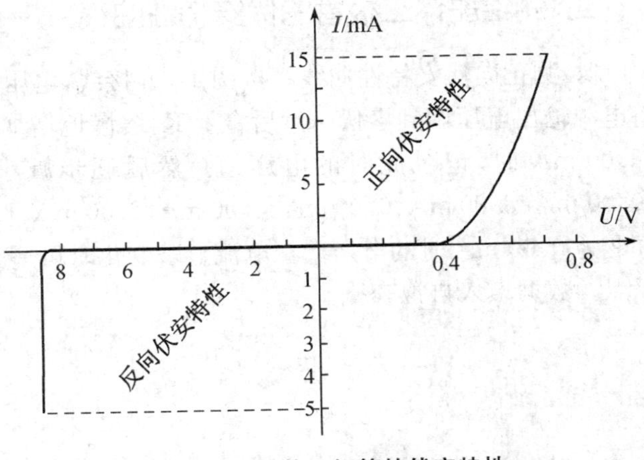

图 4-3-4 晶体二极管的伏安特性

图 4-3-4 是某种晶体二极管的伏安特性曲线，第一象限是正向的，第三象限是反向的。从图中可看出，电流和电压不是线性关系，各点

的电阻都不相同。凡具有这种性质的电阻,就称为非线性电阻。

实验器材

WYK－303B_2 型直流稳压电源,RLS-A_5 数显式直流电流表,RLS-V_6 数显式直流电压表,滑线变阻器,电阻,二极管,导线若干。

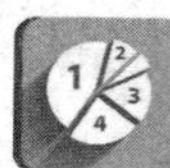

实验内容及步骤

一、测绘金属膜电阻的伏安特性曲线

1. 按图 4-3-5 所示电路图接好线路。要求 $R \gg R_A$(R_A 是毫安表内阻),选择 $R=1\ \text{k}\Omega$。电流表量程选择 20 mA,电压表量程选择 20 V,滑线变阻器滑动端置中部位置。

2. 首先调节稳压电源,使其面板上的电压表指示在 12 V 左右,然后调节滑线变阻器,使线路中电压指示依次为 0.00 V,1.00 V,2.00 V,…,10.00 V,并读出相应的电流值,数据填入自拟表格。

3. 以电压为横坐标,电流为纵坐标,绘出金属膜电阻的伏安特性曲线。

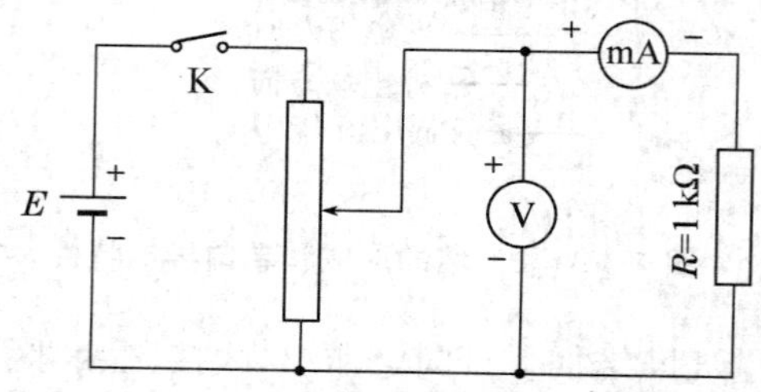

图 4-3-5 测电阻伏安特性的电路

二、测绘晶体二极管的伏安特性曲线

1. 测正向伏安特性曲线。按图 4-3-6 接线,电压表量程选择 2 V,电流表量程选20 mA。将电源输出电压调到零伏。然后合上 K,缓慢地增加电压,观察电流的变化,当电流升高到 15.00 mA 时,记录此时的电压值。然后逐步减小电压,分别记录电流为 12.00 mA、10.00 mA、8.00 mA、6.00 mA、4.00 mA、2.00 mA、1.00 mA、0.50 mA、0.5 mA 以及电流恰好变为 0.00 mA 时的电压。在电流显著变化时可适当再增加几个测量点,如隔 0.5 mA 测一点。数据填入自拟表格。

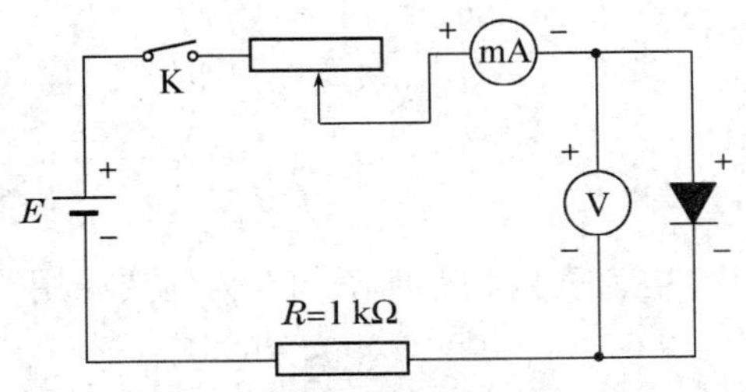

图 4-3-6　晶体二极管正向伏安特性测量电路

2. 测反向伏安特性曲线。按图 4-3-7 接线，电压表量程选择 20 mA，电压表量程选择 20 V。电路图中 1 kΩ 电阻起着保护二极管的作用。合上 K，逐步增大电压，观察电流的变化，当电流升高到 5 mA 时，停止增大电压，记录此时的电压值。然后逐步减小电压，分别记录电流为 4 mA、3 mA、2 mA、1 mA、0.5 mA、0.1 mA、0.05 mA 以及电流恰好变为 0.00 mA 时所对应的电压。数据填入自拟表格。

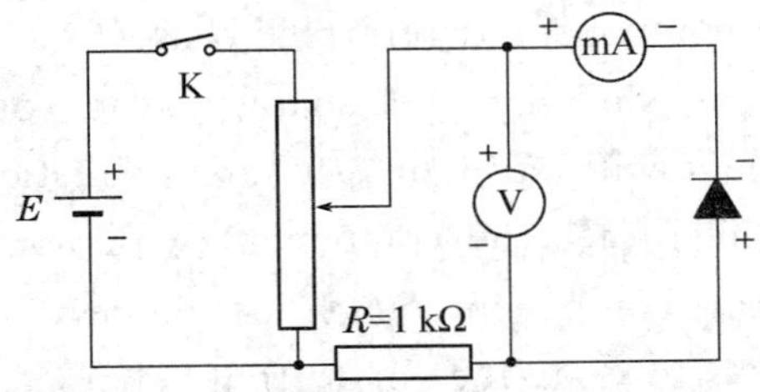

图 4-3-7　测晶体二极管反向伏安特性电路

注意事项

1. 每次电源开启前，必须检查电源输出电压调节旋钮是否已经调至最小（逆时针旋到底），如确已将输出电压调至零伏，方可合上开关 K。

2. 测量二极管伏安特性时必须注意二极管的极性，不能接反。

3. 本实验用 3DG182 型三极管的 b、e 极作为二极管。其主要参数是：允许通过的最大正向电流为 15 mA；最大反向电流不超过 5 mA。实验时，不能超出这一范围，否则易损坏晶体管。

4. 测正向伏安特性时，如果电压加到 750 mV 时，电流值仍小于 2 mA，则说明 PN 结已损坏或线路中接触不良，应请教师检查原因。

5. 测反向伏安特性时，若稍加电压（例如 2V）即有电流出现，说明 PN 结已损坏，需更换二极管。

思考题

在图 4-3-6 和图 4-3-7 中,电流表接法有何不同? 为何要采用这样的接法?

Experiment 4.3 Investigating Voltage-Current Characteristic of Linear Resistors and Nonlinear Resistors

For a DC driven electronic component, a voltmeter can be used to measure its voltage, and an ammeter can be used to measure its current. Utilizing the voltage and the current to plot their relationship in the coordinate, the obtained curve between current and voltage of the component is referred to as *U-I* characteristic curve of the component. The method of studying the characteristics of the element is called voltammeter. The component with straight *U-I* characteristic curve is called a linear element, such as a metal resistor. The component with nonlinear *U-I* characteristic curve is called a nonlinear component, such as diodes, triodes, etc. The main application of voltammeter is to study the conductivity characteristics of linear and non-linear components. Nonlinear resistors are always associated with some physical processes such as heat, light, energy level transition and so on.

Experimental Objectives

1. To draw the *U-I* characteristic curves of electronic components, and learn to process the experimental data in graphs.

2. To understand the conductivity characteristics of non-linear components such as diodes and triodes.

Experimental Principle

General metal conductor resistance is a linear resistor, which will not be changed with the magnitude and direction of the applied voltage, and its *U-I* characteristic curve is a straight line. Fig. 4-3-1 is a *U-I* characteristic curve of a metal resistance, where the straight line is located in the first and third quadrants. The resistance is always a constant, which is equal to the reciprocal of the straight slope, namely $R=U/I$.

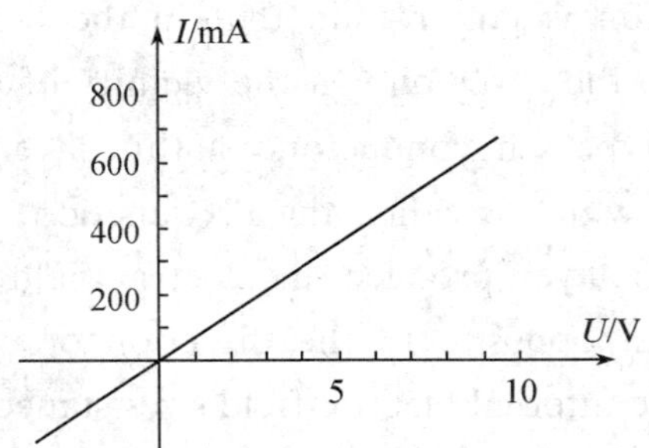

Fig. 4-3-1　Volt-ampere characteristic curve

Transistor (diode, triode and so on) is a non-linear resistor, and its resistance not only depends on the magnitude of the applied voltage but also on the current direction. In order to understand the conductive properties of a diode, this section will give a brief introduction to its structure and its electronic property.

Diode is also known as semiconductor diode. Pure semiconductor, called as intrinsic semiconductor, has conductivity between the conductor and the insulator. If a tiny amount of impurities is incorporated in a pure semiconductor, the semiconductor's conductivity will increase by a million times. According to the doped impurities, the semiconductor can be classified into two types: the electronic semiconductor and the hole semiconductor. Some impurities (such as phosphorus, arsenic, etc.) doped in the semiconductor will produce a lot of negatively charged electrons, and thus the semiconductor with these impurities is referred to as electronic (or N-type) semiconductor. Other impurities (such as boron) doped in the semiconductor will produce many positive-charged holes, and thus such kind of semiconductor is called a hole type (or called P-type) semiconductor.

The diode is the combination of the P-type and N-type semiconductors with different conductivities, and it is composed of a PN junction formed at the interface of these two semiconductors. With the one-way conductive performance, the PN junction has positive and negative electrodes. The positive electrode is contacted on P-type semiconductor side while the negative electrode is contacted on the N-type semiconductor side, as shown in Fig. 4-3-2 (a). In the circuit diagram, the diodes are represented by the symbols shown in Fig. 4-3-2(b).

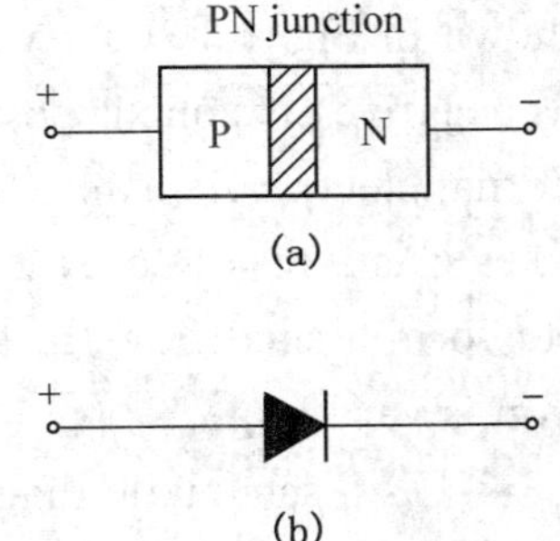

Fig. 4-3-2　Diagram of semiconductor diode

PN junction's formation and conductivity: As shown in Fig. 4-3-3 (a), the concentration of the holes in the P region is larger than that in the N region, and thus the holes will diffuse from the P region to the N region. Similarly, since the electron concentration in the N region is larger than that in the P region, the electrons will diffuse from the N region to the P region. In the vicinity of both sides of the interface between the P-type and N-type semiconductors, a thin layer region with positive and negative charges is formed, which is called the PN junction. The positive and negative charges in this charged thin layer produce an electric field, whose force on carriers (holes and electrons) is just opposite to the direction of carrier diffusion, thus the diffusion is suppressed by the internal electric field's resistance. As a result, this charged layer is also known as the barrier layer.

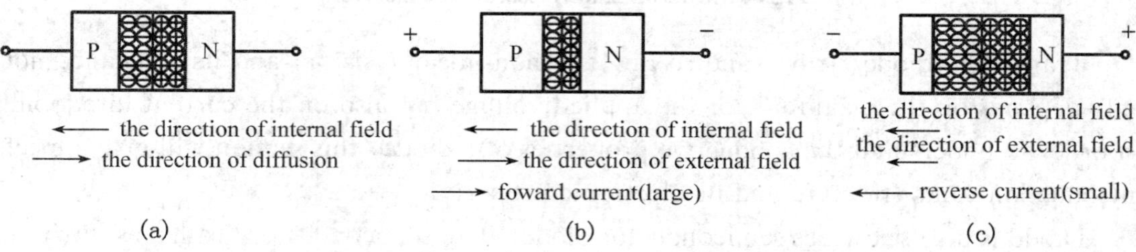

Fig. 4-3-3 PN junction's formation and conductivity

When the diffusion and the internal electric field force offsets, the holes in the P region and the electrons in the N region are no longer reduced. The thickness of the barrier layer is no longer increased and thus a dynamic balance is achieved. At this moment, there is no current in the diode.

As shown in Fig. 4-3-3(b), when the PN junction is biased with the forward voltage (P zone is positive, N zone is negative), the external electric field and the internal electric field are in the opposite direction, weakening the internal electric field and thinning the barrier layer. In this way, the carrier can smoothly go through the PN junction and form a relatively large current, and thus the forward resistance of the PN junction is small.

As shown in Fig. 4-3-3(c), when the PN junction is biased with the reverse voltage (P zone is negative, N zone is positive), the applied electric field is in the same direction as the internal electric field, strengthening the internal electric field to make the barrier layer thicker. As a result, only a very small number of carriers can pass through the PN junction to form a small reverse leakage current, and thus the reverse resistance of the PN junction is very large.

Fig. 4-3-4 is a schematic diagram of the *U-I* behavior of a diode. The curve in the first quadrant is positive, and the part in the third quadrant is inverted. As can be seen from the figure, the current and voltage is not linearly related and the resistance at each point is not the same. Resistance with these features is called nonlinear resistance.

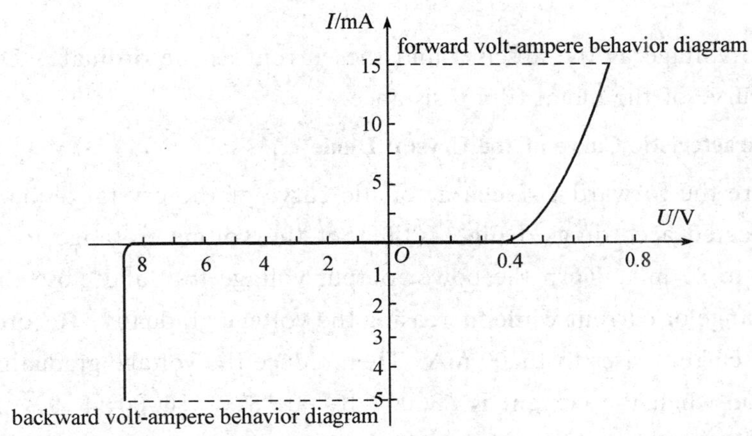

Fig. 4-3-4　Schematic diagram of the volt-ampere behavior of a crystal diode

Experimental Instruments

WYK-303B$_2$ DC power supply, RLS-A$_5$ digital display DC ammeter, RLS-V$_6$ digital display DC voltmeter, sliding rheostat, resistance, diode and wires.

Experimental Contents and Steps

1. *U-I* Characteristic Curve of Metal Film Resistors

Connect the lines according to Fig. 4-3-5. $R \gg R_A$ (R_A is the resistance of millimeter) is required. Set $R = 1\ \text{k}\Omega$. Set the ammeter range to 20 mA and voltmeter range to 20 V. The sliding end of the sliding rheostat is placed in the middle position.

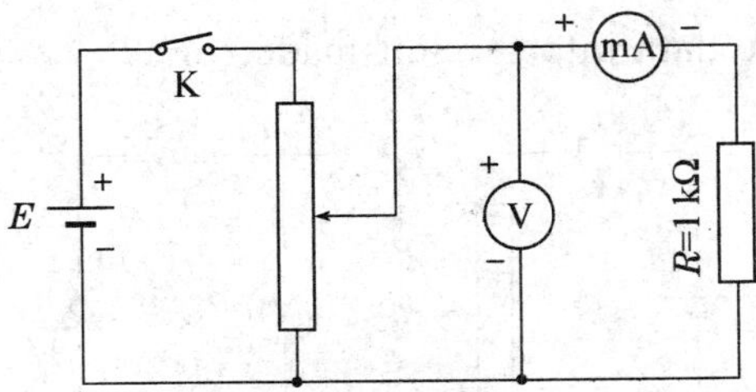

Fig. 4-3-5 The circuit for testing a 1 kΩ resistor

(1) First adjust the power supply until the voltage on the panel is around 12 V. Next adjust the sliding rheostat and set the voltage as 0.00 V, 1.00 V, 2.00 V, ⋯ and 10.00 V respectively. Record the corresponding current value and fill the data into the self-made

table.

(2) Set the voltage as the abscissa and the current as the ordinate. Draw the *U-I* characteristic curve of the metal film resistance.

2. *U-I* Characteristic Curve of the Crystal Diode

(1) Measure the forward *U-I* characteristic curve of the crystal diode. The circuit should be connected according to Fig. 4-3-6. Set the voltmeter range to 2 V and the ammeter range to 20 mA. Turn the power output voltage to 0 and close the switch K. Observe the change of current while increasing the voltage gradually. Record the voltage value when the current rises to 15.00 mA. Then reduce the voltage gradually and record the voltage value when the current is 12.00, 10.00, 8.00, 6.00, 5.00, 4.00, 3.00, 2.00, 1.00, 0.50, 0.05 and 0.00 mA. When the current changes significantly, more measurement points can be added, for example, measuring the interval of every 0.5 mA. All the collected data are filled into the self-made table.

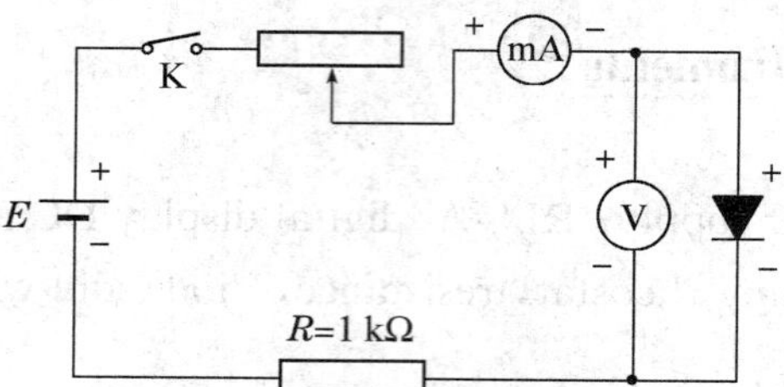

Fig. 4-3-6 The circuit for testing a diode

(2) Plot the reverse *U-I* characteristic curve of the crystal diode. The circuit should be connected according to Fig. 4-3-7. Set the ammeter range to 20 mA and the voltmeter range to 20 V. The 1 kΩ resistor in the circuit protects the diode from being damaged. After closing K, observe the change of current while increasing the voltage gradually. Record the voltage value when the current rises to 5.00 mA. Then reduce the voltage gradually and record the voltage value when the recording current is 4, 3, 2, 1, 0.5, 0.05 and 0.00 mA. Fill the data into the self-made form.

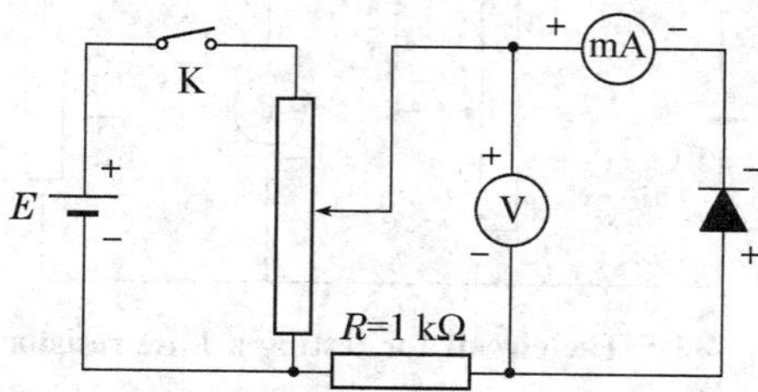

Fig. 4-3-7 The circuit for the crystal diode

Precautions

1. Before switching on the power, check whether the power supply adjustment knob has been adjusted to the minimum (counterclockwise rotation to the end). Only if the output voltage has been adjusted to zero volts can the K switch be closed.

2. To measure the *U-I* characteristic of the diode, the polarity of the diode must not be reversed.

3. The b and e poles of 3DG182 transistor are adopted as a diode in this experiment. The maximum forward current allowed is 15 mA while the maximum reverse current is within 5 mA. The current must not go beyond this range, otherwise the transistor will be damaged.

4. In measuring the forward *U-I* characteristic curve of the diode, if the current value is still smaller than 2 mA even when the voltage is increased to 750 mV, it means that the PN junction has been damaged or there is a poor contact in the line. Ask for the teacher's assistance.

5. When measuring the reverse *U-I* characteristic curve of the diode, if there is a current even at very small voltage (such as 2 V), it means that the PN junction has been damaged and should be replaced.

What are the differences in ammeter connections in Fig. 4-3-5 and Fig. 4-3-6? Why should such a connection be adopted?

4.4 示波器原理与使用

在模拟电路和数字电路实验中，需要使用若干仪器、仪表观察实验现象和结果。常用的电子测量仪器有万用表、逻辑笔、普通示波器、存储示波器、逻辑分析仪等。其中，示波器是一种用途广泛的电子仪器，它可以直接观察电信号的波形，测量电压的幅度、周期(频率)等参数。用双踪示波器还可测量两个电压之间的时间差或相位差。配合各种传感器，它可用来观测非电学量(如压力、温度、磁感应强度、光强等)随时间的变化过程。因此，是科学研究和工程技术中重要的常用电子仪器，广泛应用于无线电测量中。

实验目的

1. 了解示波器的主要组成部分及其工作原理。
2. 掌握示波器和信号发生器的基本使用方法。
3. 掌握用示波器观察电信号的波形和李萨如图形的方法。
4. 学习用示波器测量电信号的幅度、周期(频率)的方法。

实验原理

一、示波器的结构及工作原理

示波器的基本结构主要包括示波管、扫描电路、同步触发电路、X 轴和 Y 轴放大器、电源等部分,如图 4-4-1 所示。

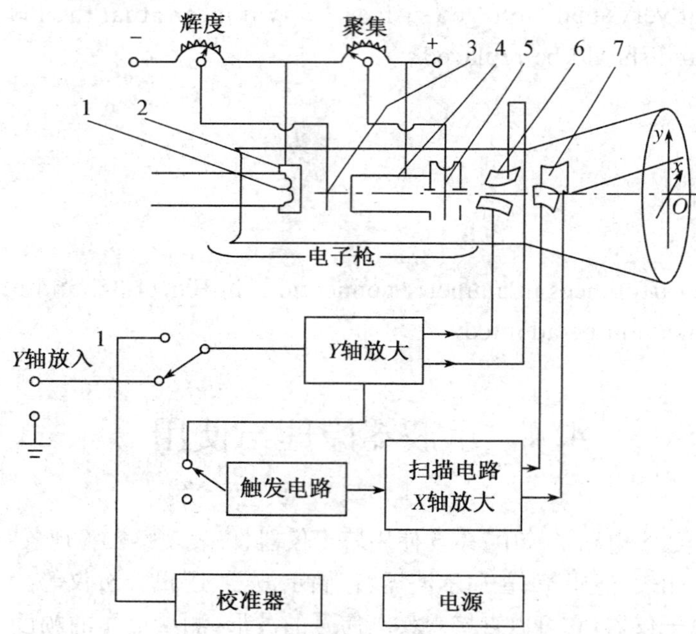

图 4-4-1　示波器的结构示意图

1—灯丝;2—阴极;3—控制栅极;4—第一阳极;
5—第二阳极;6—垂直偏转板;7—水平偏转板。

1. 示波管

示波管是示波器的"心脏",主要由安装在高真空玻璃管中的电子枪、偏转系统和荧光屏 3 部分组成。

(1) 电子枪(图 4-4-7)用来发射一束强度可调且能聚焦的高速电子流,它由灯丝 1、阴极 2、控制栅极 3、第一阳极 4 和第二阳极 5 等 5 部分组成。灯丝通电后加热阴极,使阴极发射电子。控制栅极的电势比阴极低,对阴极发出的电子起排斥作用,只有初速度较大的电子才能穿过栅极的小孔射向荧光屏,而初速度较小电子则被电场排斥回阴极。栅极用来控制阴极发射的电子数,从而改变屏上的光斑亮度。示波器面板上的"亮度"旋钮用来调节栅极电势以控制光斑亮度。阳极电势比阴极电势高很多,由于阳极对电子的加速作用,使电子获得足够的能量射向荧光屏而激发荧光屏上的荧光物质。第二阳极电势比第一阳极电势高,当第一阳极与第二阳极之间的电势差调节合适时,具有静电透镜作用,其电场使电子射线聚焦,在屏上形成明亮、清晰的小圆点光斑。面板上的"聚焦"旋钮是用来调节第一阳极电势的,所以第一阳极又称为聚焦阳极,第二阳极称为加速阳极。有些示波器面板上还有"辅助聚焦"旋钮,是用来调节第二阳极电势的。

(2) 偏转系统是由垂直(Y)偏转板 6 和水平(X)偏转板 7 组成。偏转板用来控制电子束运动,在偏转板上加适当电压,电子束通过时运动方向发生偏转,在荧光屏上产生的光点位置随之改变。

(3) 荧光屏由示波管末端玻璃屏的内表面上涂了一层荧光粉所构成的。荧光屏受电子轰击后发光而形成光点,光点的亮度取决于电子束的电子数量,大小由电子束的粗细决定。它们分别由"亮度"和"聚焦"旋钮来调节。不同材料的荧光粉发光的颜色不同,发光过程延续的时间也不同。如果电子束光斑需要长时间停留在屏上不动时,宜将光斑亮度减弱,以免使荧光屏局部过热,造成永久损伤。

2. 扫描电路及示波器显示波形原理

当偏转板加上一定的电压后,电子束将受到电场的作用而偏转,光点在荧光屏上移动的距离与偏转板上所加的电压成正比。若在 Y 偏转板上加正弦电压 $U_y=U_{ym}\sin\omega t$,X 偏转板不加电压,荧光屏上光点只是做上下方向的正弦振动,振动频率较快时,看起来是一条垂直线,只有同时在 X 偏转板上加入一个与时间成正比的锯齿形电压 $U_x=kt$ 时,才能将亮点沿 Y 方向的振动展开,从而在屏幕上显示信号电压 U_y 和时间 t 的关系曲线,其示波原理如图 4-4-2 所示。如果光点沿 X 轴正向匀速地移动了 U_y 的一个周期之后,迅速反跳到原来开始的位置上,再重复 X 轴正向匀速运动,则光点的正弦运动轨迹就和前一次的运动轨迹重合起来。每一个周期都重复同样的运动,光点的轨迹就能保持固定位置。重复频率较大时,可在屏上看见连续不动的一个周期函数曲线(波形)。光点沿 X 轴正向的匀速运动及反跳的周期过程,称为扫描。获得扫描的方法,是在 X 轴偏转板上加一个周期与时间成正比的电压,称扫描电压或锯齿波电压,由示波器内的扫描电路产生,锯齿波的周期 T_x(或频率 $f_x=1/T_x$)可以由电路连续调节。当扫描电压的周期 T_x 是信号电压周期 T_y 的 n 倍时,即 $T_x=nT_y$ 或 $f_y=nf_x$,屏上将显示出 n 个周期的波形。

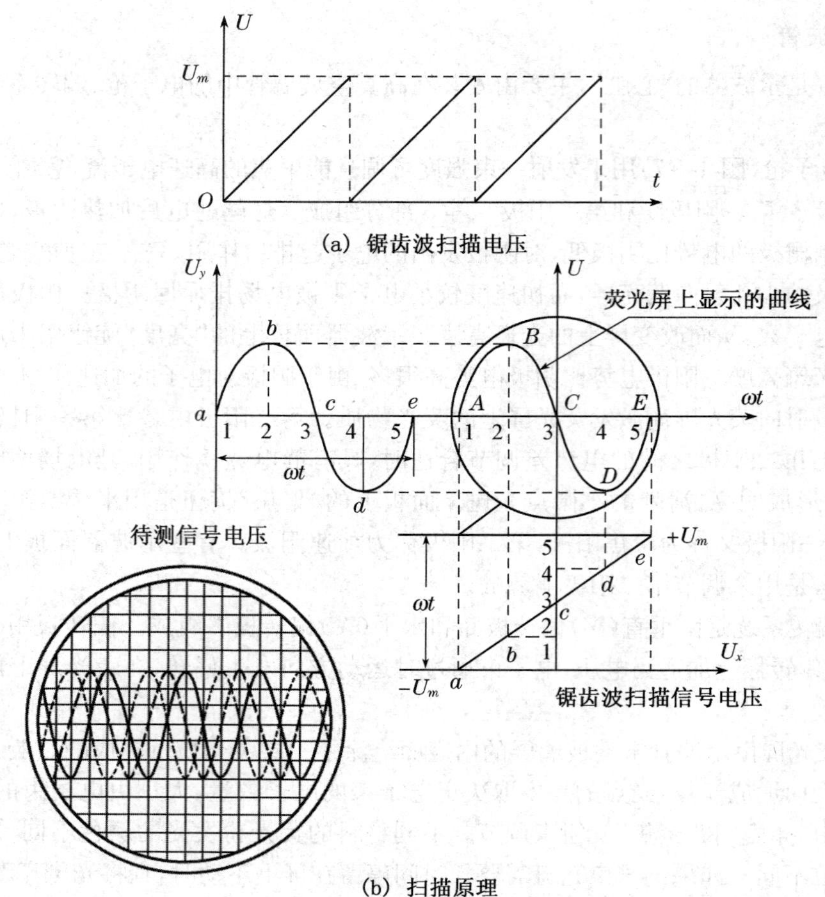

(b) 扫描原理

图 4-4-2 示波器的扫描原理

3. 整步

由于信号电压和扫描电压来自两个独立的信号源,它们的频率难以调节成准确的整数倍关系,屏上的波形发生横向移动,如图 4-4-3 所示,造成观察困难。克服的办法是,用 Y 轴的信号频率去控制扫描发生器的频率,使扫描频率与信号频率准确相等或成整数倍关系。电路的这种控制作用称为"整步"或"同步",通常由放大后的 Y 轴电压控制扫描电压的产生时刻,这一过程称为触发扫描。

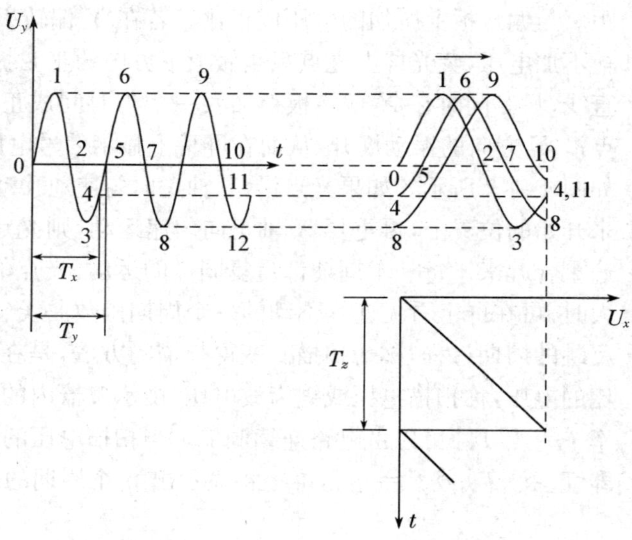

图 4-4-3 $T_x=7T_y/8$ 时显示的波形

整步电路从垂直放大电路中取出部分待测信号,输入到扫描发生器,迫使锯齿波与待测信号同步,此称为“内同步”;若同步电路信号是从仪器外部输入,则称“外同步”;如果同步信号从电源变压器获得,则称为“电源同步”。

4. 放大器

一般示波器垂直和水平偏转板的灵敏度不高,当加在偏转板上的电压较小时,电子束不能发生足够的偏转,光点位移很小。为了便于观测,需要预先把小的输入电压经放大后再送到偏转板上,为此设置垂直和水平放大器。示波器的垂直偏转因素是指光迹在荧光屏 Y 方向偏转一格时对应的被测电压的峰-峰值(电压峰-峰值与有效值的关系为:$U_{P-P}=2\sqrt{2}U_{ef}$),其单位为 mV/格或 V/格(一格为荧光屏上一格的长度,通常为 1 cm)。如某示波器的垂直偏转因素为 10 mV/格,即当 Y 轴输入电压的峰—峰值为 30 mV 时,光迹在 Y 方向偏转 3 格。

水平放大器将扫描电压放大后送到 X 偏转板,以保证扫描线有足够的宽度。水平偏转因素是指光迹在 X 方向偏转一格对应的扫描时间,其单位为 s/格、ms/格或 μs/格。此外,水平放大器亦可直接放大外来信号,这时示波器可作 X-Y 显示用。

5. 电源

用以供给示波管及各部分电子线路所需的各种交直流电源。

二、示波器的基本应用

1. 观察交流电压信号波形

将待检测交流电压信号输入示波器的 CH1 或 CH2 通道,X 轴处于扫描状态,适当调节“V/DIV”,“SEC/DIV”及“电平”等旋钮,即可在荧光屏上显示出稳定的电压波形。

2. 测量交流信号的幅度

通过比较法可定量测出信号电压的大小。一般示波器内部都有校准信号,将校准信号和待测信号分别输入示波器,并保持 Y 轴增幅(V/格)不变,根据两个信号在 Y 轴的偏转量和校准信号的幅度,即可求出待测信号的幅度。对于 YB4328 双踪示波器,经校准后,待测电压的幅度可直接由信号在 Y 轴的偏转量和“V/DIV”的取值计算。

3. 测量交流信号的频率(周期)

(1) 利用扫描频率求信号频率

由扫描原理可知,只有当输入信号的频率为扫描频率的整数倍时,波形才是稳定的,利用这一关系可以求得未知频率。通过示波器的“SEC/DIV”旋钮位置能够直接得到扫描频率(周期),这种方法实际上也是一种比较法。

(2) 利用李萨如图形求信号频率

当 X 轴输入扫描电压时,示波器显示 Y 轴输入电压信号的瞬变过程。当 X 和 Y 轴均输入正弦电压信号,荧光屏上光迹的运动是两个相互垂直谐振动的合成。当两个正弦电压的频率成简单整数比时,合成轨迹为一稳定的曲线,称为李萨如图形(表 4-4-1),其合成过程如图 4-4-4 所示。

表 4-4-1　李萨如图形

图形							
N_x	2	2	2	4	6	6	4
N_y	2	4	6	6	4	8	2
$f_y : f_x$	1 : 1	1 : 2	1 : 3	2 : 3	3 : 2	3 : 4	2 : 1

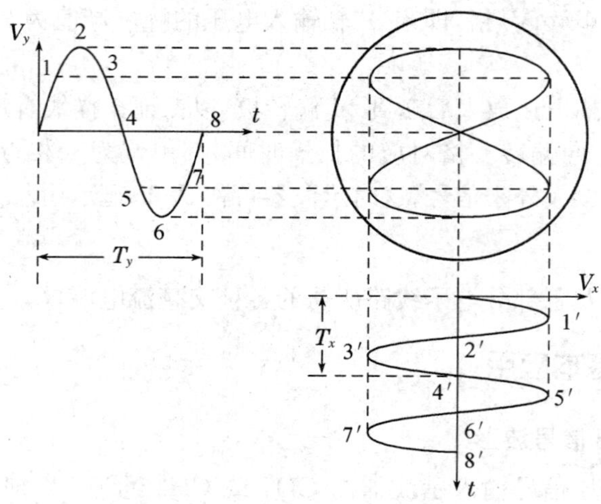

图 4-4-4　两个相互垂直的正弦振动的合成

利用李萨如图形可以比较两个电压的频率。当李萨如图形稳定后,对图形作水平和竖直割线(两条割线均应与图形有最多的相交点),若设水平割线与图形的交点数为 N_x、竖直割线与图形的交点数为 N_y,则 X、Y 轴上的电压频率 f_x、f_y 与 N_x、N_y 有如下关系

$$\frac{f_x}{f_y}=\frac{N_y}{N_x} \tag{4-4-1}$$

因此,只要知道 f_x 或 f_y 的其中一个,就可以求出另一个。

实验器材

YB4328 型双踪示波器或 LDS20205 型数字示波器,信号发生器,整流滤波板,导线若干。

实验内容和步骤

一、示波器使用前的检查及基本练习

1. 熟悉面板上有关控制键的作用。

2. 开机前，先把“水平位移”和“垂直位移”置于中间，“亮度”和“聚焦”置于最小（即逆时针旋到底），触发源置于CH1，垂直方式置于CH1或CH2，扫描速率（Sec/Div）置1～2 ms，可用于观察校准信号。耦合方式先置于GND，平时置于AC（交流输入）。

3. 接上电源，闭合开关，预热约3 min后，调节“亮度”“聚焦”“垂直位移”和“水平位移”等旋钮，使屏幕上出现一个亮度适中、清晰，但不同步的校准方波信号，而后将“电平”旋钮调节至指示灯亮，即可得到稳定的方波波形。这些说明示波器工作正常，并处于待测状态。

二、正弦电压波形观察

步骤1　按基本练习，调节好旋钮。

把待测信号（由低频信号发生器输出500 Hz电压）接到CH1（或CH2），垂直方式选择置CH1（或CH2），耦合方式置“AC”。调节CH1（或CH2）的灵敏度（Volts/Div），使波形在垂直方向的振幅合适。

步骤2　选择与待测信号频率相应的扫描速率（Sec/Div），并调节“扫描微调”，使波形初步稳定。再将“电平”旋至指示灯亮，得到稳定的波形。

步骤3　改变扫描周期，使 T_x 分别等于 T_y、$2T_y$ 和 $3T_y$，使屏上分别出现含1，2，3个周期的波形。

步骤4　改变输入信号的频率（为50 Hz和5 kHz），重复步骤2、步骤3。

三、整流滤波波形观察

1. 按基本练习调节好各旋钮。

2. 从整流、滤波实验板上依次输出“交流”“整流（半波、全波）”“滤波”等到CH1输入，观察整流、滤波波形。观察整流和滤波波形时，耦合方式置“DC”（直流输入）。

四、电压测量、时间测量

1. 直流电压测量

（1）Y 轴（CH1或CH2）输入方式耦合方式置GND，并调节触发电平，使屏幕上出现一条扫描基线。按被测信号的幅度和频率将 Y 轴灵敏度“Volts/Div”挡级和扫描速率“Sec/Div”置于适当位置，然后调节“垂直位移”，使扫描基线位于荧光屏的某一特定基准位置。

（2）将输入耦合方式置“DC”，将滤波输出接入CH1（或CH2）输入，调“电平”旋钮使波形稳定。

(3) 测量直流分量(图 4-4-5),计算方法如下:

若"Volts/Div"挡级的标称值为 K(V/cm)(1 格=1 cm),峰到扫描线的距离为 B cm,则被测信号波峰值 $U_1=KB$(V);谷到扫描线的距离为 C cm,则被测信号波谷值 $U_2=KC$(V)。被测信号直流分量为:

$$U=\frac{U_1+U_2}{2} \tag{4-4-2}$$

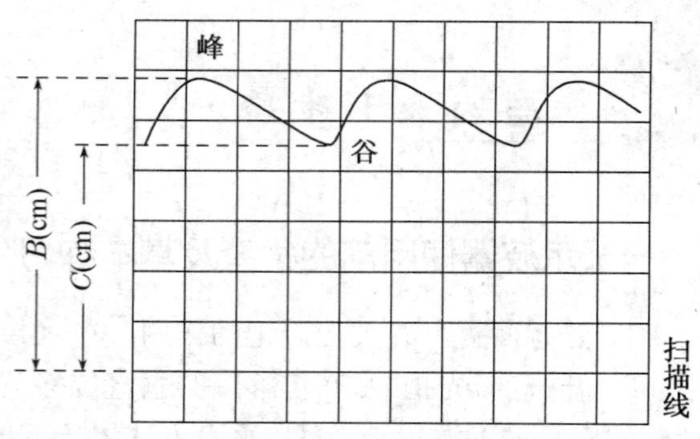

图 4-4-5　测量直流电压

2. 交流电压测量

(1) Y 轴(CH1 或 CH2)耦合方式置于"AC",Y 轴灵敏度"Volts/Div",扫描速率"Sec/Div"置适当位置。整流、滤波板的"交流"输出接入 Y 轴输入端,调"扫描微调""触发电平",使波形稳定。

(2) 根据屏上的坐标刻度读出信号波形的峰-峰值 D cm(图 4-4-6)。如"Volts/Div"挡级标称值为 K(V/cm)(设 1 格=1 cm),则被测信号的峰-峰值为:

$$U_{\mathrm{P-P}}=KD \tag{4-4-3}$$

而交流电压的有效值为

$$U_{\mathrm{ef}}=\frac{\sqrt{2}}{4}U_{\mathrm{P-P}} \tag{4-4-4}$$

3. 时间测量

将"整流""滤波"板的"交流"输出接到 Y 轴(CH1 或 CH2)输入,选择适当的"Volts/Div"和"Sec/Div"档值,调"扫描微调"至校准位置(即顺时针旋到底),调"电平"使波形稳定,则屏上被测信号波形两特定点之间的时间间隔(图 4-4-6)可按下列方法计算。

若扫描速率"Sec/Div"档极标称值为 K(ms/cm)(设 1 格=1 cm),两点之间的水平距离为 B(cm),则两点时间间隔值 t 为

$$t=KB(\mathrm{ms}) \tag{4-4-5}$$

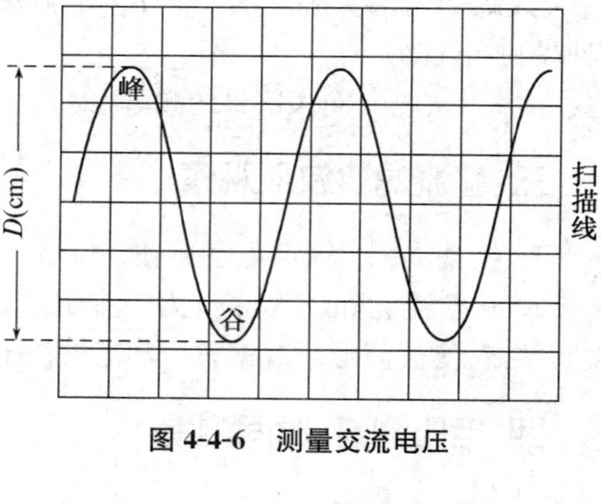

图 4-4-6　测量交流电压

要测量交流信号的周期,测出两相邻峰或谷的水平距离 B,代入式(4-4-5)就可得出。

五、观察李萨如图形、测信号频率

1. 按基本要求调好各旋钮。
2. 将"Sec/Div"左旋至"X—Y"方式。

3. 用示波器内部的“试验信号”(其输出端在示波器背面)作为待测信号输入到CH1,而把低频信号发生器输出的交变信号输入到CH2。

4. 调节低频信号发生器输出的信号频率,使图形分别呈 $N_y/N_x=1:1,1:2,1:3,2:3,2:1$,记录各自的 f_y 值,再根据式(4-4-1)计算出 N_y/N_x 为不同值时的试验信号频率 f_x,并求出 f_x 的平均值。

注意事项

1. 荧光屏上的光点亮度不能太强,而且不能让光点长时间停留在荧光屏的某一点,尽量将亮度调暗些,可以看得清为准,从而避免损坏荧光屏;

2. 在实验过程中如果暂不使用示波器,可将亮度旋钮逆时针旋至尽头,截至电子束的发射,使光点消失。不要经常通断示波器的电源,以免缩短示波管的使用寿命;

3. 示波器的所有开关及旋钮均有一定的转动范围,决不可用力过大,以免损坏仪器。

4. 在对所观察的信号进行定量测量时,一定要把有关微调旋钮顺时针旋到底,以关闭微调。

思考题

1. 当 Y 轴输入端有信号,但屏上只有一条垂直亮线是什么原因?如何调节才能使波形沿 X 轴展开?

2. 用示波器观察周期为0.2 ms的信号电压,若在屏上呈现5个周期的稳定波形,扫描电压的周期应等于多少?

3. 若示波器观察的波形不断向右移动,说明扫描频率偏高还是偏低?

4. 在用李萨如图形测频率时,如果 X 与 Y 轴正弦信号频率相等,但荧光屏上的图形还在不停转动,为什么?

Experiment 4.4 The Principle of the Oscilloscope and Its Operation

In the experiments of analog and digital circuits, a number of instruments are required to observe the experiment phenomena. The common instruments include the multimeter, logic pen, general oscilloscope, storage monitor, logic analyzer, and etc. As a widely applied electronic instrument, the oscilloscope can be used to observe the waveform of the electrical signal directly and measure the amplitude of voltage, cycle (frequency) and other parameters. The dual trace oscilloscope can also be applied to measure the time difference and phase difference between the two voltages. Together with a variety of sensors, the oscilloscope can be used to observe the changing process of non-electrical quantities (such as pressure, temperature, magnetic flux density, light intensity, etc.) over time. Therefore, the oscilloscope is an important electronic instrument for scientific research and engineering technology.

Experimental Objectives

1. To understand the main components of the oscilloscope and its working principle.

2. To master the basic usage of the oscilloscope and signal generator.

3. To grasp the method of observing the waveform of electric signals and Lissajous figures with oscilloscopes.

4. To learn the method of measuring the amplitude and period (frequency) of an electrical signal with an oscilloscope.

Experimental Principle

1. The Structure and Working Principle of an Oscilloscope

The basic structure of the oscilloscope includes the oscilloscope tube, a scanning circuit, a synchronous trigger circuit, X and Y-axis amplifiers, a power supply and other parts, as shown in Fig. 4-4-1.

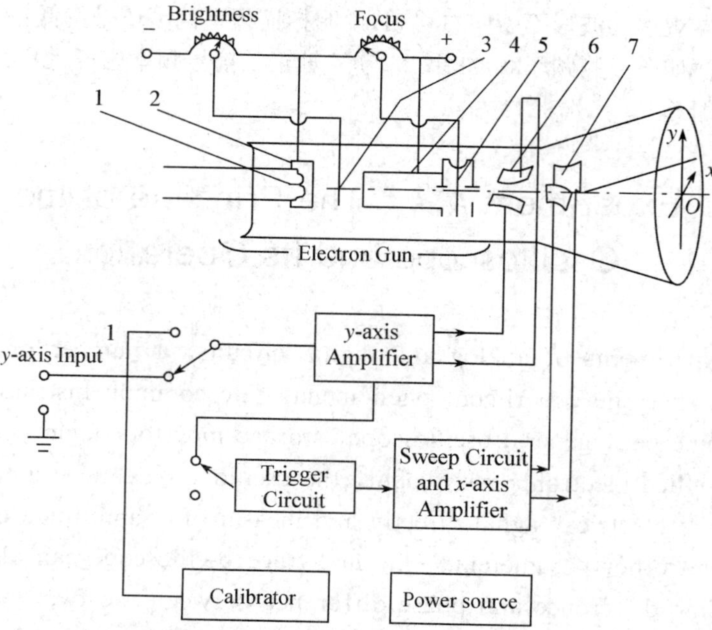

Fig. 4-4-1 The schematic structure of the oscilloscope

(1) Oscilloscope tube

As the heart of the oscilloscope, the oscilloscope tube mainly consists of three parts: the electron gun installed in high vacuum glass tube, the electron beam deflection system and the phosphor screen.

① The electron gun emits a high-speed electron flow with adjustable intensity and focusing ability, which consists of five parts: filament "1", cathode "2", gate "3", first anode "4" and second anode "5". When the filament is heating the cathode, electrons can be emitted. The potential of the gate is lower than that of the cathode and has a repulsive effect on the electrons coming from the cathode. Only the electrons with large initial velocities can pass through the small hole of gate and arrive at the screen, while the electrons with low initial velocities will be repelled by the electric field and go back to the cathode. The gate is used to control the number of electrons coming from the cathode, changing the brightness of the spot on the screen. On the oscilloscope panel, the "brightness" knob is used to adjust the gate potential to control the brightness of the spot. The potential of the anode is much higher than that of the cathode. Because of the accelerating effect of the anode on the electrons, the electrons with high energy collide with the phosphor molecules on the screen to emit fluorescence. The second anode potential is higher than the first anode potential. When the potential difference between the first anode and second anode is adjusted properly, it has the function of electrostatic lens. The electric field causes the electron beam to be focused so as to form a bright and clear dot spot on the screen. The "focusing knob" on the panel is used to adjust the first anode potential, so the first anode is also called the focusing anode, while the second anode is called as accelerating anode. Some oscilloscope panels also have an "auxiliary focus knob", which is used to adjust the second anode potential.

② The electron beam deflection system consists of vertical (Y) deflector plate "6" and horizontal (X) deflector plate "7". The deflector plate is used to control the movement of the electron beam. If the deflector plate is applied an appropriate voltage, the movement direction of the electron beam will deflect while passing and the spot position on the screen will change accordingly.

③ The phosphor screen is glass with its inner surface coated by a layer of phosphor material. The phosphor screen will glow to form a light spot when it is bombarded by electrons. Its brightness depends on the electron number of the electron beam, and its size is determined by the thickness of the electron beam. They are adjusted by the "brightness" and "focus" knobs respectively. Different phosphor materials have different colors, and their lifetimes also differ. If the electron beam spot needs to stay on the screen stationarily for a long time, it is advisable to weaken the spot brightness, so as not to overheat the screen, resulting in permanent damage.

(2) Scanning circuit and the principle of oscilloscope displaying waveform

When the deflector plate is applied with a certain voltage, the electron beam will deflect under the electric field, and the distance of the spot moving on the screen is proportional to the voltage applied to the deflector plate. If the Y deflector plate is applied with the sinusoidal voltage $U_y = U_{ym} \sin \omega t$ and no voltage exerts on the X deflector, the spotlight on the screen goes up and down in the form of the sine vibration. The fast vibration frequency makes the trace appear to be a vertical line. Only when a time-proportional serrated voltage $U_x = kt$ is applied to the X deflector board at the same time, the bright spot vibrate along the Y direction so that the relation curve of the signal voltage U_y vs the time t can be formed on the screen. The principle of wave displaying is shown in Fig. 4-4-2. If the light spot can quickly return to the original position after moving evenly along the positive X-axis for a uniform U_y period and repeat the even motion along the positive X-axis, the sinusoidal motion trajectory of the spot will coincide with the previous trajectory. Repeating the same movement in each period, the spot trajectory will maintain a fixed position. Moving with a higher repetition rate, a continuous curve of the periodic function (waveform) can be formed

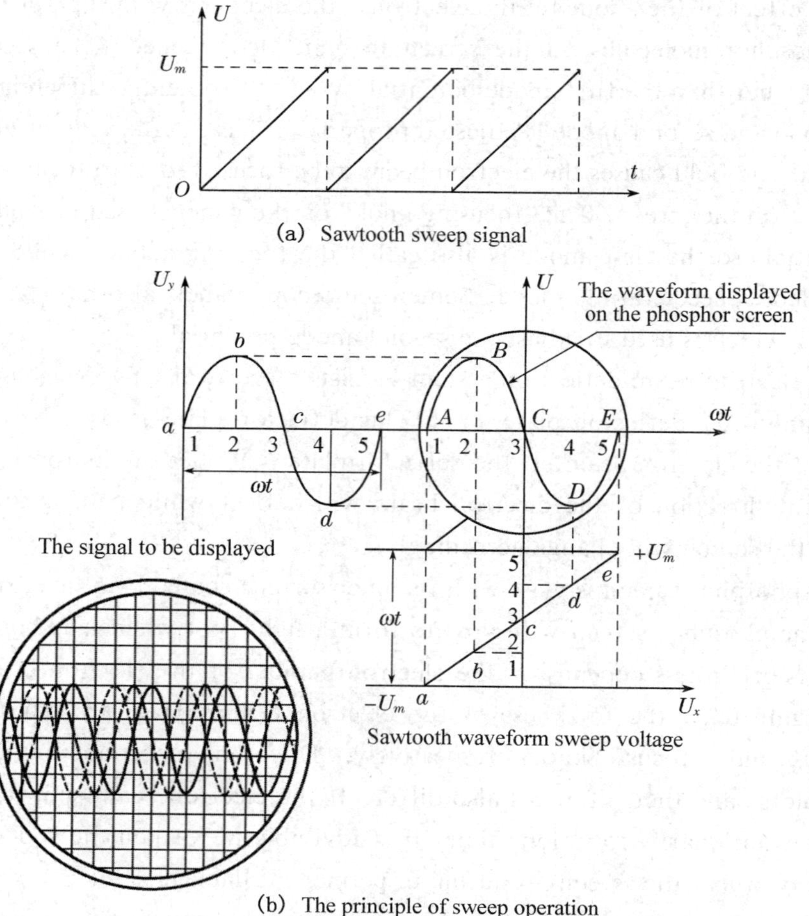

Fig. 4-4-2 The principle of displaying a voltage signal on the screen

on the screen. The periodic process, during which the light spot moves evenly along the positive X-axis and returns to the original position, is called scanning. To obtain scanning, a voltage with its period proportional to the time has to be applied on the X-axis deflection plate. Such a voltage is referred to as the scanning voltage or serrated wave voltage. Given that the serrated wave is generated in the oscilloscope scan circuit, its period T_x(or frequency $f_x=1/T_x$) can be continuously adjusted by the circuit. When the period of the scan voltage T_x is n times of the period T_y of the signal voltage, namely $T_x=nT_y$ or $f_y=nf_x$, the screen will display n cycles of the waveform.

(3) Synchronization

Since the signal voltage and the scanning voltage come from two independent signal sources, it is difficult to adjust their frequencies to an exact integer multiple. As a result, the horizontal move of their waveforms on the screen, as shown in Fig. 4-4-3, makes it difficult for the observer to distinguish. To overcome this problem, the Y-axis signal frequency should be used to control the frequency of the scan generator, so that the scanning frequency and signal frequency have an exactly equal or an integer multiple relationships. Such as effect of the circuit control is called "synchronization". It is the amplified Y-axis voltage that often controls the generation time of the scanning voltage. This whole process is called triggering scanning.

The whole step circuit takes out part of the signal to be tested from the vertical amplification circuit and inputs it to the scan generator to force after taken out from the vertical amplification circuit, part of the signal to be tested is input to the scan generator, which makes the serrated wave to synchronize with the signal to be measured. Such a process is called "internal synchronization". The process that the synchronous circuit signal is input from the outside part of the instrument is called "outer synchronization". When the synchronization signal is obtained from the power transformer, this process is called "power synchronization".

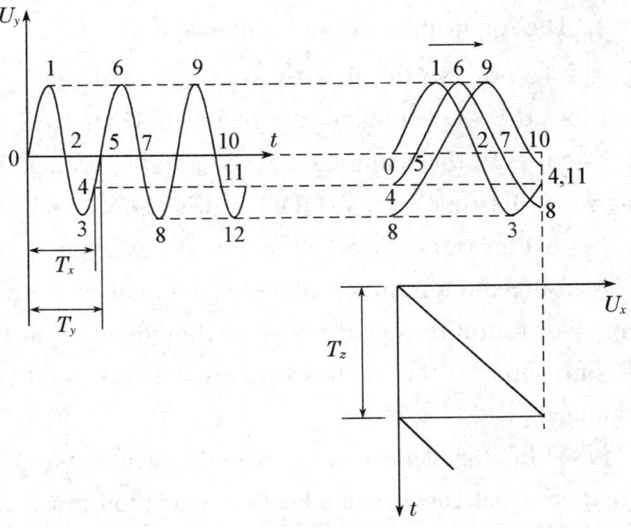

Fig. 4-4-3 The waveform displayed as $T_x=7T_7/8$

(4) Amplifier

In general, the vertical and horizontal deflector plates don't have the high sensitivity. The low voltage applied to the deflector plate cannot force the electron beams deflect enough, resulting in the small displacement of the light spots. To achieve a better observation, a small input voltage needs to be magnified and then sent to the deflector board, and thus the vertical and horizontal amplifiers have to be set in advance. The vertical deflection factor of the oscilloscope refers to the corresponding peak-to-peak value of the measured voltage when the light track deflects one grid in the Y direction of the screen (the relationship between the peak value of the voltage and the effective value is $2U_{P-P}=\sqrt{2}\,U_{ef}$), the unit of which includes mV/div or V/div (div is the length of a grid on the screen, usually 1 cm.). For example, if the vertical deflector factor of an oscilloscope is 10 mV/div, it means that the voltage peak-to-peak value is 30 mV and the track is deflected by 3 grids in the Y direction.

The horizontal amplifier amplifies the scan voltage and sends it to the X deflector to ensure the scan line to be wide enough. The horizontal deflector factor refers to the scanning time when the light track deflects one grid in the X direction, the unit of which includes s/div, ms/div or or μs/div. In addition, the horizontal amplifier can also directly amplify the external signal, and then the oscilloscope can be used for X-Y display.

(5) Power Source

The power source provides the oscilloscope and various parts of the electronic circuit with AC or DC.

2. The Application of the Oscilloscope

(1) To observe the signal wave of AC voltages.

Input the AC voltage signal to the CH1 or CH2 channel of the oscilloscope. When the X-axis is in the scanning status, a stable voltage waveform can be displayed on the screen by adjusting the "V/DIV", "SEC/DIV" and "Voltage waveform" knobs.

(2) To measure the amplitude of the AC signal.

Measure the amplitude of voltage signal with the comparison method. Generally, there is a calibration signal in the oscilloscope. Input the calibration signal and the signal to be measured to the oscilloscope respectively, and keep the Y-axis sensitivity (V/div) unchanged.

Based on the deflection of the two signals on the Y-axis and the amplitude of the calibration signal, the magnitude of the signal to be measured can be calculated. When YB4328 dual trace oscilloscope is calibrated, the amplitude of voltage to be measured can be directly calculated from the deflection of the signals on the y-axis and the value of "V/DIV".

(3) To measure the frequency (cycle) of the AC signal.

① Using the scanning frequency to calculate the signal frequency. According to the principle of scanning, the waveform is stable only when the frequency of the input signals is an integer times of the scanning frequency. Such a relationship can be utilized to calculate the unknown frequency. The scanning frequency (cycle) can be directly obtained by choosing the "SEC/DIV" knob position of the oscilloscope, which is actually a kind of comparison in essence.

② Using Lissajous figures to find the signal frequency. When the scanning voltage is input to the X-axis, the oscilloscope displays the transient process of Y-axis inputs voltage signal. When the sinusoidal voltage signal is input to the X and Y axis, the motion of the light track on the screen is a superposition of two perpendicular harmonic vibrations. When the frequency of two sinusoidal voltages is in a ratio of an integer, the superposition trajectory is a stable curve, which is called a Lissajous figure (Table 4-4-1). The detailed superposition process is shown in Fig. 4-4-4.

Table 4-4-1　Lissajous figures

Figure							
N_x	2	2	2	4	6	6	4
N_y	2	4	6	6	4	8	2
$f_y : f_x$	1 : 1	1 : 2	1 : 3	2 : 3	3 : 2	3 : 4	2 : 1

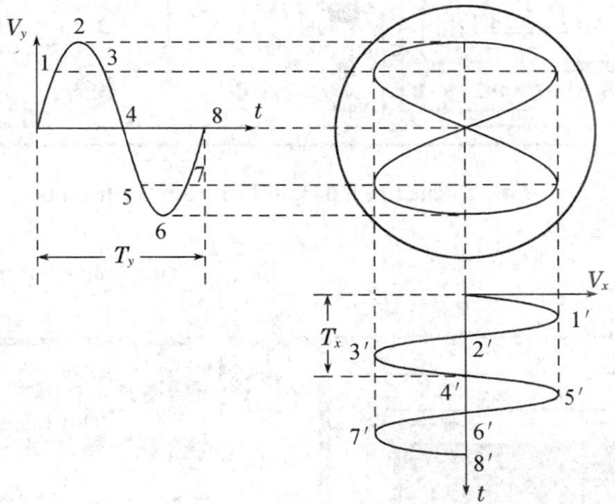

Fig. 4-4-4　Image curve from two mutually perpendicular signals

Lissajous figures can be used to compare the frequency of two voltages. When Lissajous figures are stable, the horizontal and vertical lines can be drawn on the graphics (two lines should have the most intersection points with the Lissajous figure). Suppose the number of intersection points between the horizontal line and the Lissajous figure as N_x, and the number of intersection points between the vertical secant and the Lissajous figure as N_y, the relationship between the frequency f_x and f_y of voltage on the x and y axes and N_x and N_y can be deduced as follows:

$$\frac{f_x}{f_y}=\frac{N_y}{N_x} \tag{4-4-1}$$

Therefore, if either of f_x or f_y is known, the other one can be obtained.

Experimental Instruments

YB4328 dual trace oscilloscope or LDS20205 digital oscilloscope, signal generator, rectifier filter panel and wires. YB4328 dual trace oscilloscope panel is shown in Fig. 4-4-5. LDS20205 digital oscilloscope panel is shown in Fig 4-4-6.

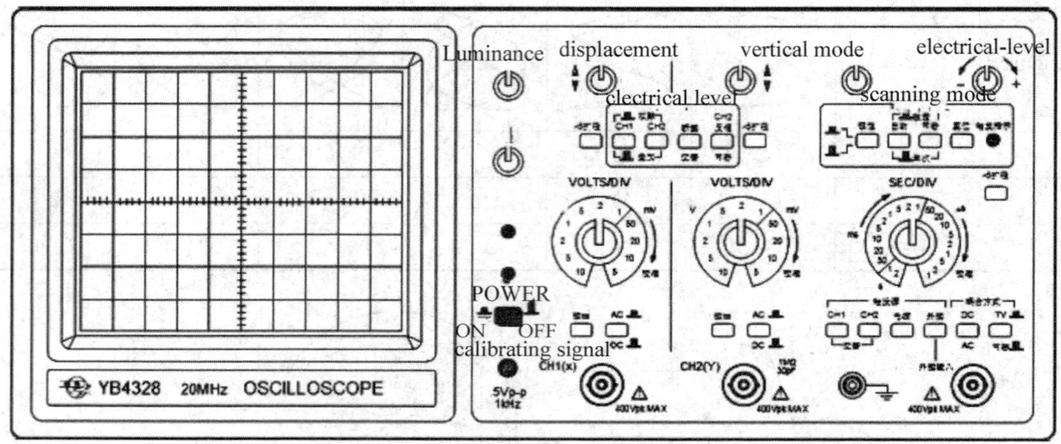

Fig. 4-4-5 Panel of YB4328 dual trace oscilloscope

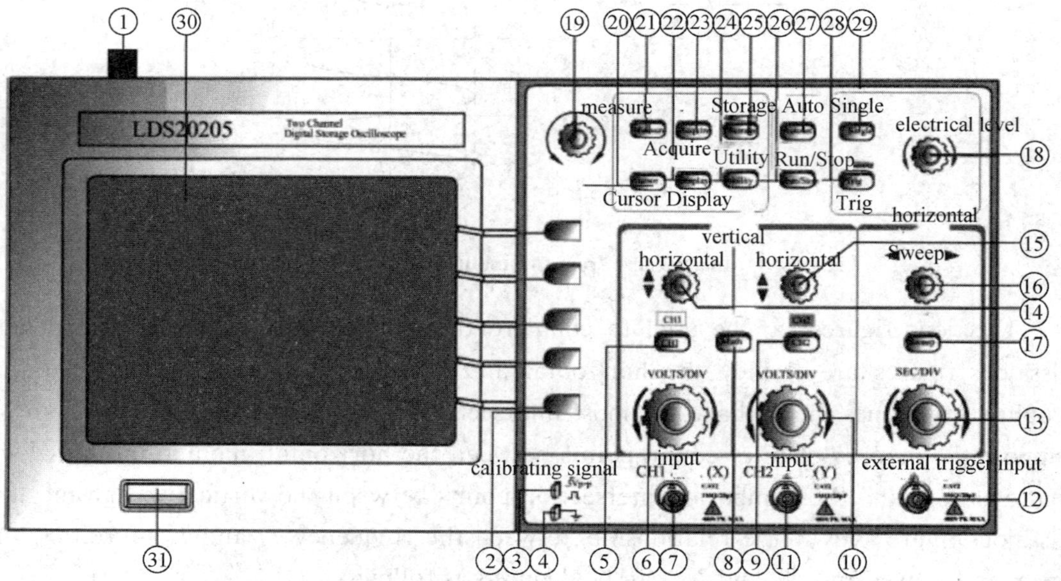

Fig. 4-4-6 Panel of LDS20205 digital oscilloscope

Experimental Contents and Steps

1. Inspection of Oscilloscope Before Use and Its Basic Practice

(1) Be acquainted with the functions of the panel.

(2) Before turning on the oscilloscope, set the "vertical displacement" and "vertical displacement" knobs in the middle, and set the "brightness" and "focus" knobs at the minimum (rotate them counterclockwise to the end). The trigger source is placed at CH1, and the vertical mode is placed at CH1 or CH2. With the scanning frequency (Sec/Div) at (1~2) ms, the calibration signal can be observed. The coupling is firstly placed at GND, which is usually placed at AC (AC input).

(3) Turn on the power source, and switch on the oscilloscope. After preheating for about 3 min, adjust the knobs of "brightness", "focus", "vertical displacement" and "horizontal displacement" until the screen shows an unsynchronized calibration square wave signal with a moderate brightness and clarity. Then, adjust the knob of "electrical balance" until the indicator is on to get a stable square wave waveform, which shows that the oscilloscope works well and is available for experiment.

2. The Observation of Sinusoidal Voltage Waveform

(1) Adjust the knob according to the basic requirement. Connect the signal to be tested (output by the 500 Hz low-frequency signal generator) to CH1 (or CH2), and select CH1 (or CH2) in the vertical mode and "AC" in the coupling mode. Adjust the sensitivity (Volts/Div) of CH1 (or CH2) so as to obtain the appropriate amplitude of the waveform in the vertical direction.

(2) Select the scanning frequency (Sec/Div) corresponding to the frequency of the signal to be measured, and adjust the "sweeping fine tuning" until the waveform is initially stable. Turn the knob of the "electrical balance" until the indicator light is on and a stable waveform is obtained.

(3) Change the scan cycle to make T_x equals T_y, $2T_y$, and $3T_y$, resulting in the waveforms with 1, 2, and 3-cycle on the screen respectively.

(4) Change the frequency of the input signal (which is 50 Hz and 5 kHz) and repeat steps (2) and (3).

3. The Observation of Rectification Filter Waveform

(1) Adjust the knobs according to the basic requirements.

(2) Output "AC," "Rectifier (half-wave, full-wave)" and "filter" from the rectifier and filter board to CH1 in turn. The coupling is placed at "DC" (DC input) for observing the waveform of rectification and filter.

4. The Measurement of Voltage and Time

(1) DC voltage measurement

① The input coupling mode of the input to Y-axis (CH1 or CH2) is set at GND. Adjust the trigger balance until a scan baseline appears on the screen. Adjust the sensitivity "Volts/Div" and scanning frequency "Sec/Div" of Y-axis to the appropriate position according to the magnitude and frequency of the signal to be tested, and then adjust the "vertical displacement" to set the scan baseline to be at a specific reference position on the screen.

② Set the input coupling mode to "DC", and connect the filter output to CH1 (or CH2), and then adjust the "electrical balance" knob to stabilize the waveform.

③ Measure the DC component (Fig. 4-4-7), and the calculation method is as follows:

If the value of "Volts/Div" is K (V/cm) (1 div=1 cm), and the distance from the peak to the scan line is B (cm), the peak value of the measured signal is $U_1=KB$ (V); if the distance from valley to scan line is C (cm), the valley value of measured signal is $U_2=KC$. The DC component of the signal to be tested is

$$U=\frac{U_1+U_2}{2} \tag{4-4-2}$$

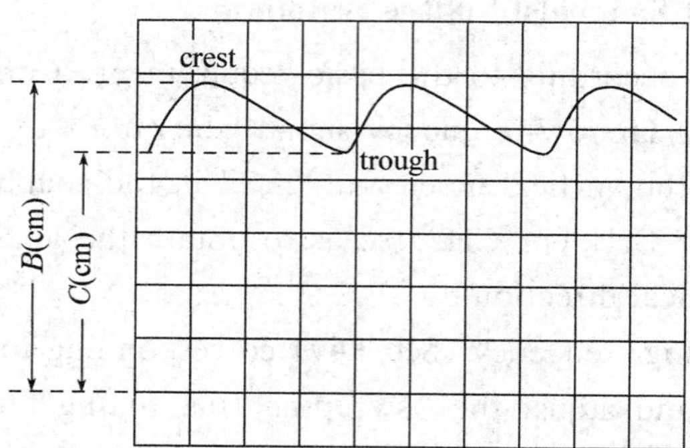

Fig. 4-4-7 Measurement of the DC voltage

(2) The measurement of AC voltage

① The Y axis (CH1 or CH2) coupling mode is set to "AC". Y-axis sensitivity "Volts/Div" and the scanning frequency "Sec/Div" are set to appropriate positions. Connect the rectifier and filter panel's AC output to the Y-axis input, and adjust "sweeping fine tuning" and "electrical balance" knobs so as to make the waveform stable.

② Read the peak-to-peak value D (cm) of the signal waveform according to the coordinate scale on the screen, as shown in Fig. 4-4-8. If the value of "Volts/Div" is K (V/cm), the peak-to-peak value of the measured signal is

$$U_{P-P}=KD(\mathrm{V}) \tag{4-4-3}$$

The effective value of AC voltage is

$$U_{\mathrm{eff}}=\frac{\sqrt{2}}{4}U_{P-P} \tag{4-4-4}$$

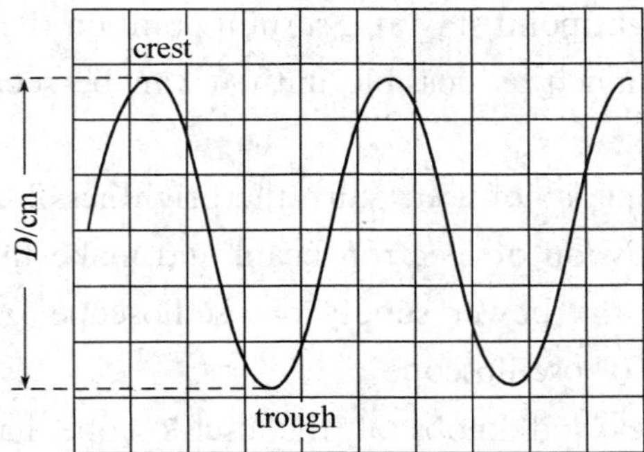

Fig. 4-4-8 Measurement of AC voltage

(3) The measurement of time

Connect the "Rectifier" and "Filtering" panel's AC output to the Y-axis (CH1 or CH2) input, and select the appropriate values of "Volts/Div" and "Sec/Div". Adjust the "sweeping fine tuning" to the calibration position (turn clockwisely to the end) and set the "electrical balance" to stabilize the waveform. In this way, the time interval between the two specific points of the on-screen signal waveform (it can be seen in Fig. 4-4-8) can be calculated as follows:

If the value scanning frequency "Sec/Div" stall is K (ms/cm) (assuming 1div= 1 cm), and the horizontal distance between two points is B (cm), the time interval t between the two points is

$$t=KB(\mathrm{ms}) \tag{4-4-5}$$

To measure the period of the AC signal, it is necessary to measure the horizontal distance B between two adjacent peaks or troughs. The time interval t can be calculated by putting the distance value into Equation (4-4-5).

5. Observe Lissajous Figures and Measure the Signal Frequency

(1) Adjust the knobs according to the basic requirements.

(2) Turn counter-clockwisely the "Sec/Div" knob to the mode of "X-Y".

(3) Use the "test signal" (whose output end is on the back of the oscilloscope) inside the oscilloscope as the signal to be tested and input it to CH1, and input the alternating signal output by the low-frequency signal generator to CH2.

(4) Adjust the frequency of the signal output from the low-frequency signal generator, and the graph is displayed as $N_y/N_x=1:1$, $1:2$, $1:3$, $2:3$, and $2:1$, and

record their respective f_y value. Calculate the frequency f_x of the test signal when N_y/N_x is different according to Equation (4-4-1), and then obtain the average value of f_x.

Precautions

1. The brightness of the light point on the fluorescent screen cannot be too strong. Don't let the light point stay at a certain point on the screen for a long time. Dim the brightness as much as possible until it can be seen clearly, so as not to damage the screen.

2. If the oscilloscope is not used, turn the brightness knob counterclockwise to the end to stop the emission of electron beam and make the light spot disappear. Don't turn on and off the power supply of oscilloscope frequently, so as not to shorten the service life of oscilloscope.

3. All the switches and knobs of the oscilloscope have a certain range of rotation, and the force cannot be too strong, so as to avoid damaging the instrument.

4. In the quantitative measurement of the signal, be sure to close the fine-tuning knob by turning it clockwisely to the end.

Questions

1. The Y axis input has a signal, yet the screen has only one vertical bright line, explain the possible reasons. How can the input be adjusted to make the waveform spread along the X axis?

2. In observing a signal of period 0.2 ms with the oscilloscope, if the screen displays a persistent waveform of five cycles, what is the period of the sweep voltage?

3. If the waveform observed by the oscilloscope continues to move to the right, is the scanning frequency high or low?

4. In measuring frequency with the Lissajous figure, if the sinusoidal signal frequencies of the X and Y-axis are equal, yet the graphics on the screen are still non-stop rotating, what are the possible reasons for this phenomenon?

4.5 分光计的调整和使用

分光计是一种测量光线偏转角的仪器,实际上就是一种精密的测角仪。由于不少物理量如折射率、波长等往往可以用光线的偏折来量度,因此分光计是光学实验中的一种基本仪器。在分光计的载物台上放置色散棱镜或衍射光栅，它就成为一台简易的光谱仪器；在分光计上装上光电探测器，还可以对光的偏振现象进行定量的研究。为了保证测量的精确，分光计在使用前必须调整。分光计的调整方法对一般光学仪器的调整也有一定的通用性，因此学习分光计的调整方法也是使用光学仪器的一种基本训练。

实验目的

1. 了解分光计的结构，学会正确的调节和使用方法。
2. 掌握角游标原理和用分光计测角度的方法。
3. 用反射法测定三棱镜顶角。

实验原理

一、分光计的结构

分光计是用来准确测量角度的仪器。它由五部分组成，即支架、载物台、阿贝式自准直望远镜、可变狭缝的平行光管、游标刻度盘。KF-JJY1 型分光计结构如图 4-5-1 所示。

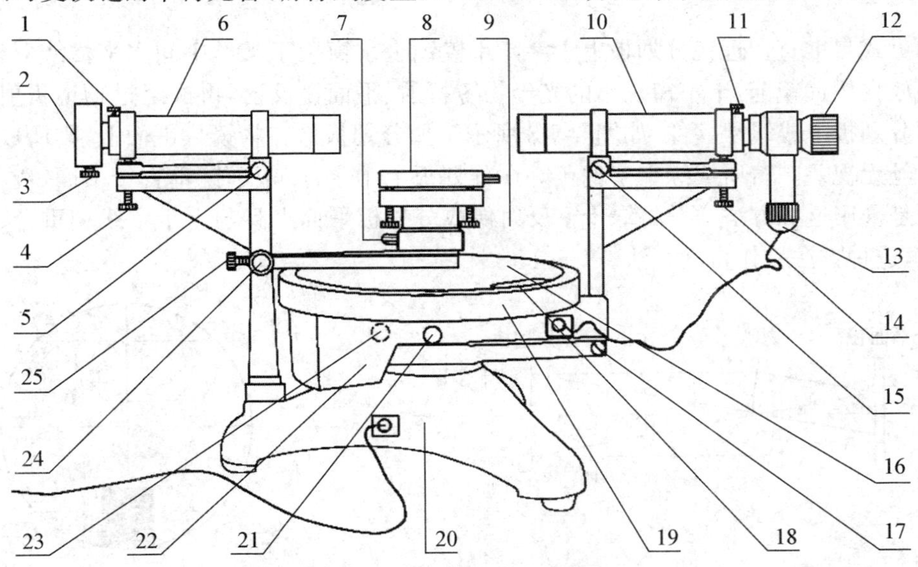

图 4-5-1 KF-JJY1 型分光计结构图

1—平行光管狭缝锁紧螺钉；2—平行光管狭缝装置；3—平行光管狭缝调节螺钉；4—平行光管倾斜度调节螺钉；5—平行光管水平方向调节螺钉；6—平行光管，7—载物台锁紧螺钉；8—载物台；9—载物台调平螺钉；10—望远镜；11—望远镜目镜锁紧螺钉；12—望远镜目镜调焦螺旋；13—小电珠；14—望远镜倾斜度调节螺钉；15—望远镜水平方向调节螺钉(背面)；16—游标盘；17—转座水平方向微调螺钉(背面)；18—游标；19—刻度盘；20—底座；21—转座与刻度盘锁定螺钉；22—转座；23—望远镜止动螺钉(背面)；24—游标盘微动螺钉；25—游标盘止动螺钉。

1. 自准直望远镜

装有“阿贝”目镜的望远镜称阿贝式自准望远镜,其结构如图 4-5-2(a)所示。它用以观察平行光进行的方向。与普通望远镜相类似,它由物镜与目镜组成。改变物镜至目镜的距离,可以使不同距离远处的物体成像清晰。望远镜调焦于无穷远时,则可使从无穷远处来的平行光成像最清晰。

为了测量,物镜与目镜之间装有分划板,分划板上有叉丝,如图 4-5-2(b)所示。目镜与分划板,物镜与分划板的距离均可调节,分划板应位于目镜焦平面上。目镜是由场镜和接目镜组成的,KF-JJY1 型分光计的目镜采用了阿贝目镜,在目镜与叉丝之间装了一个带透光小十字窗的全反射小三棱镜,小灯发出的光经小三棱镜反射后将叉丝的照亮。从目镜望去分划板上一部分被小三棱镜遮住,故只能看到叉丝的其他部分。

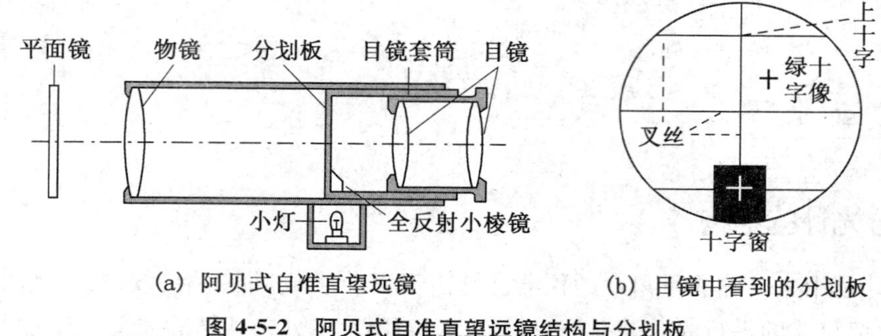

(a) 阿贝式自准直望远镜　　(b) 目镜中看到的分划板

图 4-5-2　阿贝式自准直望远镜结构与分划板

阿贝式自准直望远镜分划板上十字与小棱镜十字窗关于叉丝中间水平横线对称,将一平面镜放在望远镜前,十字窗发出的光经物镜折射、平面镜反射,再次经过物镜折射进入望远镜在分划板上成绿十字像,如图 4-5-3 所示。如分划板恰在物镜焦平面上,则所成绿十字像与叉丝无视差。若所成绿十字像正好在分划板上十字,则望远镜垂直于平面镜。望远镜调节就是采用这种方法——将绿十字反射像成分划板平面上且与上十字线相重合、与叉丝无视差,这种方法称作自准直法。

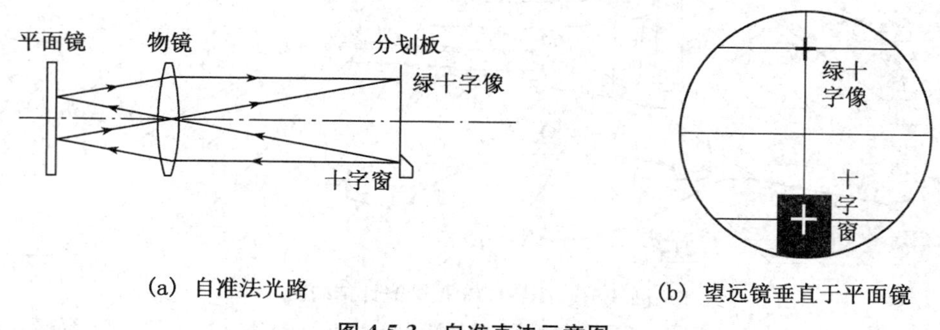

(a) 自准法光路　　(b) 望远镜垂直于平面镜

图 4-5-3　自准直法示意图

2. 平行光管

平行光管的结构如图 4-5-4 所示。平行光管的一端装有可变宽度的狭缝体,前后移动狭缝体,使其位于平行光管凸透镜的焦面上,这时就能使照在狭缝上的光线经凸透镜后成为平行光线。平行光管的倾角也可由座架上的螺丝来调节。

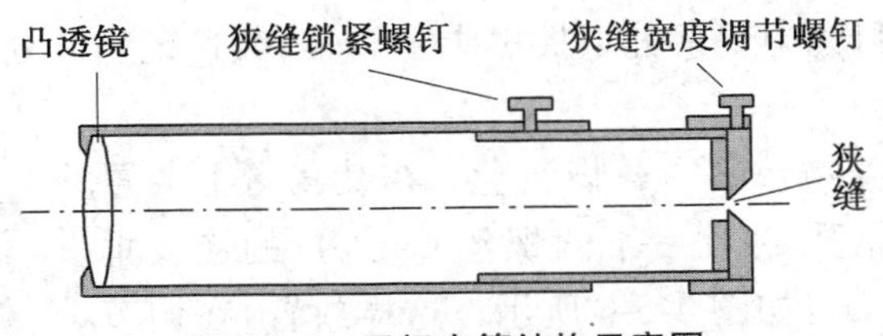

图 4-5-4　平行光管结构示意图

3. 读数盘

在底座的中央固定一中心轴，度盘和游标盘套在中心轴上，可以绕中心轴旋转，度盘下端有一推力轴承支撑，使旋转轻便灵活。度盘上刻有 720 等分的刻线，每一格的分度值为 30′。为了消除度盘的偏心误差，采用两个相差 180°的游标 A、B 进行读数，游标如图 4-5-5 所示。游标上的 30 格与刻度盘上的 29 格角度相等，故游标的最小分度值为 1′。读数时应先看游标零刻度线所指的位置。

读数时读出度盘零线和游标盘零线之间的整 30′的值，设为 α；再找出游标上与度盘上恰好重合的刻线，读出不到 30′的部分，设为 β；两数值之和（$\alpha+\beta$）即为读数值。图 4-5-5 的读数为 205°11′。

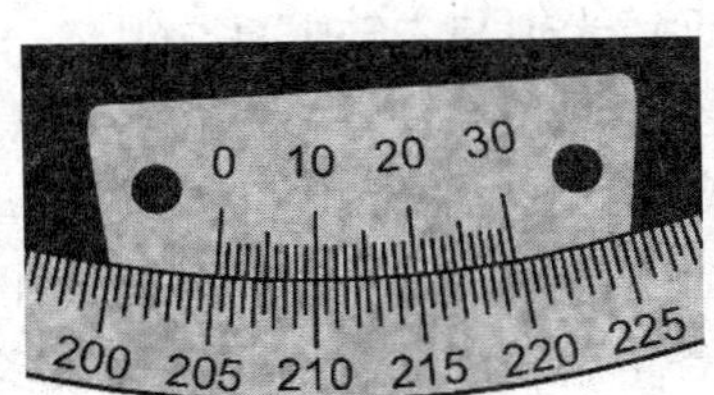

图 4-5-5　分光计的度盘和角游标

实验中，望远镜转动角度与两游标转动的角度相等。测量望远镜转动角度可以测量望远镜转动前后两游标零刻线的角坐标，共 4 个值，求两游标的平均得到结果：

$$\theta=\frac{1}{2}(\theta^{A}+\theta^{B})=\frac{1}{2}[(\phi_2^{A}-\phi_1^{A})+(\phi_2^{B}-\phi_1^{B})] \tag{4-5-1}$$

式中，θ 为望远镜转动的角度；ϕ_1^{A}、ϕ_1^{B} 为起始 A、B 游标的角坐标；ϕ_2^{A}、ϕ_2^{B} 为转过后 A、B 游标的角坐标。

4. 载物台

载物台可绕主轴回转和沿轴升降，上面可放光学元件，如三棱镜、光栅等。有三只调节螺钉 G_1、G_2 和 G_3 可改变小平台倾斜度，见图 4-5-6。载物台也有锁紧螺丝固定位置。

二、分光计的调节

分光计的调节要求是：平行光管聚焦于无穷远使之能发出平行光；望远镜聚焦于无穷远使之能接收平行光；平行光管和望远镜的光轴分别与分光计的主轴垂直。

调节前，先旋紧度盘锁紧螺丝。调节时，先用目视法粗调，再分下列四步细调。

1. 用自准直法调节望远镜,使其聚焦于无穷远

(1) 接通电源,从目镜中观察分划板黑十字线,调节目镜与分划板之间的距离,直至看清黑十字线为止。

(2) 调载物台,使载物台贴紧载物底盘。在载物台上放置一平面反射镜,使其镜面与载物台下两螺丝 G_2、G_3、的连线垂直,并通过另一个螺丝 G_1,如图 4-5-6 所示。这样放置,在调节平面镜的倾斜度时,只需调 G_2 或 G_3 就行,而与 G_1 无关。

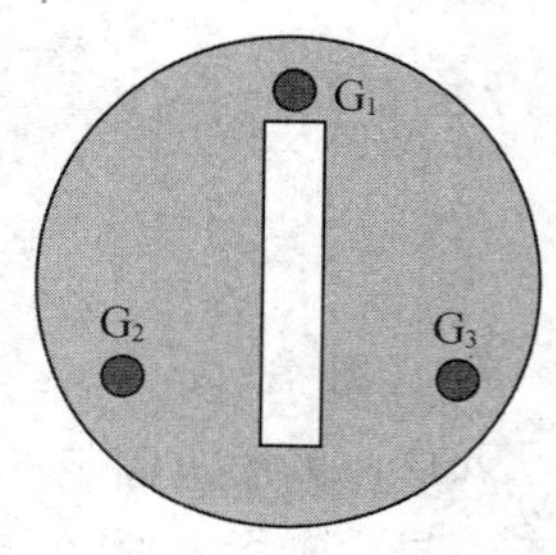

图 4-5-6 平面镜在载物台的放置方法

(3) 找绿十字反射像。调节物镜,使绿十字与黑十字叉丝无视差(自准直调焦),并将绿十字像移动至上十字线。

(4) 放正分划板十字线的方位,使载物台连同反射镜相对于望远镜转动时,绿十字像的移动方向与叉丝水平线相平行。上述步骤调好后,锁紧目镜筒。

2. 调节望远镜光轴与分光计主轴垂直

(1) 转载物台,使平面镜绕轴转 180°,使平面镜的另一镜面正对着望远镜。从目镜中观察是否有绿十字像出现,若没有,调螺丝 14(此时不要调 G_1 或 G_2),使绿十字像出现在视场中,调节使平面镜二面反射回来的绿十字像都出现在视场中。

(2) 用"各半调整法"使望远镜光轴与主轴垂直。观察一面反射回来的绿十字像是否在上横线上。如不在,调平台螺丝 G_2 或 G_3,使绿十字像与上黑十字线靠拢一半,再调望远镜,使绿十字与在上横线上(在上横线上任一位置均可)。此时,望远镜光轴与平面镜面垂直(但不等于望远镜光轴已和分光计主轴垂直)。再将平面镜转 180°,观察另一面绿十字像是否在上横线上。若不在,以上述"各半调整法"进行调节,反复多次(一般 3 次即可),直至两个面反射回来的绿十字像均在上横线上,此时,望远镜光轴与分光计主轴垂直,将望远镜水平位置固定。

注意:在以后整个实验过程中,望远镜的各调节螺丝均不能再动。

3. 调节平行光管,使其聚焦于无穷远

(1) 用光照平行光管的狭缝,把调节好的望远镜转到对准平行光管的位置,从目镜中观察狭缝像,同时调节狭缝与透镜间的距离,直至清晰地看清狭缝像,且与叉丝无视差为止。此时狭缝恰好在平行光管物镜的焦平面上,经物镜出射的光为平行光。

(2) 微转狭缝体,使狭缝像与竖十字线平行,然后锁紧。

(3) 调节狭缝像宽度,使之与黑十字线宽度大体相同。

4. 调节平行光管光轴与分光计主轴垂直

调节平行光管的水平度,使狭缝像的长度被分划板的中间水平线平分(调节时,可先使狭缝像偏离十字线的竖线,以使在视场中能见到全部狭缝像),如图 4-5-7 所示,调好后,将平行光管的水平位置固定。

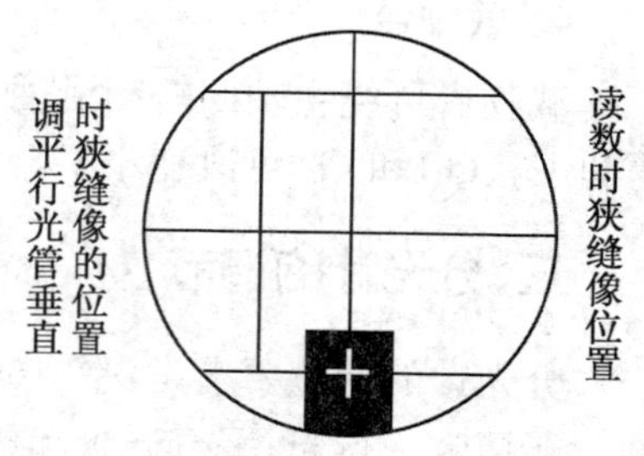

图 4-5-7 狭缝像长度被下横线平分

三、三棱镜主截面的调整

分光计调整完毕后，在载物台上放待测三棱镜。进行测量前，三棱镜的主截面必须与分光计的主轴垂直，这样望远镜才与三棱镜的侧面垂直。此时，望远镜已垂直于分光计主轴，不能再调节望远镜。具体的调节方法是：

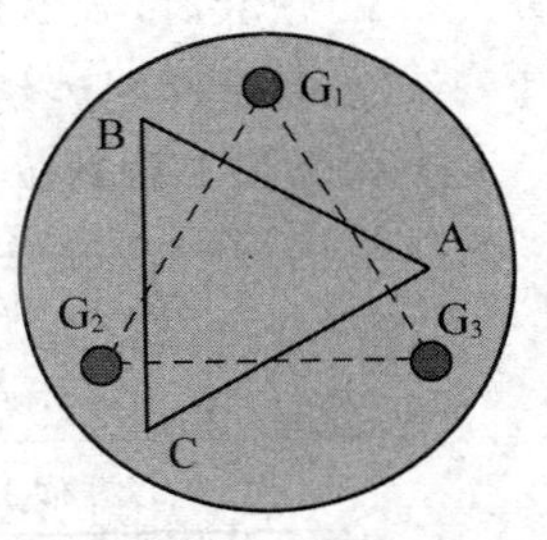

图 4-5-8 三棱镜在载物台上的放置

1. 在载物台上按图 4-5-8 所示的位置放置三棱镜，其中 AB 垂直于 G_1G_2，BC 垂直于 G_2G_3，AC 垂直于 G_1G_3。这样放置的好处是，调节 AB 面与垂直于望远镜时，只要调 G_2；调节 AC 面与垂直于望远镜时，只调 G_3；因调节会同时影响到 AB 面和 AC 面，所以不调 G_1。

2. 将望远镜分别转到垂直于 AB 面和 AC 面的位置，并分别将绿十字像调至叉丝上横线。只用 G_2 调 AB 面的绿十字；只用 G_3 调 AC 面的绿十字。

实验器材

KF-JJY1 型分光计，平面反射镜，玻璃三棱镜。

实验内容

测三棱镜顶角 A，转动载物台，使棱镜顶角正对已调好的平行光管，如图 4-5-9 所示。由平行光管发出的一束平行光（平行于主截面）入射到棱镜的顶角上，光线 1 经 AB 面反射，光线 2 经 AC 面反射（BC 为毛面）。可以证明，二反射光线的夹角 θ 与棱镜顶角 A 有如下关系：

$$A=\theta/2 \tag{4-5-2}$$

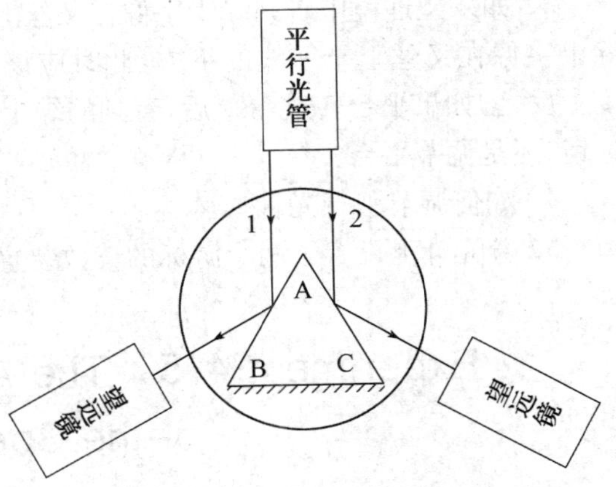

图 4-5-9 测量三棱镜顶角原理示意图

转动望远镜，从 AB 面的反射光中找到狭缝的像，微调望远镜，使狭缝像与十字线的竖线重合，记录两游标的读数 ϕ_1^A、ϕ_1^B，用相同方法找到另一反射光的角坐标 ϕ_2^A、ϕ_2^B，则得顶角为：

$$A=\frac{1}{2}\theta=\frac{1}{4}(\theta^A+\theta^B)=\frac{1}{4}[(\phi_2^A-\phi_1^A)+(\phi_2^B-\phi_1^B)] \tag{4-5-3}$$

稍微转动平台的位置，重复测量三次，求其顶角的平均值。

注意事项

1. 读数时,狭缝像应对准叉丝竖线;

2. 在计算望远镜转过的角度时,要注意望远镜是否转过刻度盘的零值点。例如望远镜从位置Ⅰ转到位置Ⅱ,游标A未经过零点,游标B经过了零点,读数如下

望远镜位置	Ⅰ	Ⅱ
游标A	$175°45'(\phi_1^A)$	$295°43'(\phi_2^A)$
游标B	$355°46'(\phi_1^B)$	$115°43'(\phi_2^B)$

游标A未经过零点,望远镜转过的角度为:

$$\theta^A=\phi_2^A-\phi_1^A=295°43'-175°45'=119°58'$$

游标B经过零点,望远镜转过的角度为:

$$\theta^B=\phi_2^B-\phi_1^B=(360°-355°46')+115°43'=119°57'$$

思考题

1. 了解分光计各主要部件的功能,熟悉分光计上各调节螺丝的作用和调节方法,及分光计正确使用时需满足的要求。

2. 望远镜聚焦无穷远如何调节,怎样判断是否调好?如何判断平行光管是否聚焦无穷远?

3. 如果望远镜中看到绿十字像在叉丝上十字线的上方,而当平面镜转过180°后看到的绿十字像在叉丝上十字线的下方,此时应该调节望远镜的倾斜度,还是应调节平台的倾斜度?反之,如果平台转过180°后,看到的绿十字像仍然在叉丝上十字线上方,这时应调节望远镜,还是调节平台?

4. 如何调节平行光管?

5. 分光计上设置了两个游标的读数装置,其目的是什么?

Experiment 4.5 The Alignment and Operation of the Spectrometer

The spectrometer is a sophisticated goniometer to measure the light deflection angle. The spectrometer is a basic instrument in optical experiments since many parameters such as refractive index and wavelength can be identified by measuring the deflection angle of light. The spectrum structure can be observed by simply placing a dispersive prism or diffraction grating on the stage of the spectrometer. The addition of a photodetector to the spectrometer allows a quantitative study of the polarization of light. The pre-use adjustment of the spectrometer must be conducted in order to ensure the accuracy of measurement. Given the method of the spectrometer alignment is valid for other optical instruments, learning how to conduct the spectrometer alignment is a basic training for the use of optical instruments.

Experimental Objectives

1. To understand the structure of the spectrometer, and learn how to correctly align and how to use the spectrometer.

2. To grasp the principle of angular vernier calliper and the method of measuring angle by a spectrometer;

3. To master how to determine the vertex angle of the prism with the reflection method.

Experimental Principle

1. The Structure of the Spectrometer

A spectrometer is an instrument used to accurately measure the angle. It consists of five parts: bracket, stage, Abbe type self-collimation telescope, collimator with variable slit and Vernier caliper dial. The structure of KF-JJY1 type spectrometer is shown in Fig. 4-5-1.

(1) Self-collimation telescope

The telescope equipped with the Abbe eyepiece is called Abbe self-collimation telescope, shown in Fig. 4-5-2(a). It is often used to observe the direction of incident directional light. Similar to other telescopes, it consists of an objective and an eyepiece. Changing the distance between the objective lens and the eyepiece, the clear image of objects at different distances can be obtained. When the telescope focuses on infinity, the directional light from infinity can be clearly imaged.

For the convenience of experiment, a transparent plate with reticle is located between the objective lens and the eyepiece, shown as Fig. 4-5-2(b). The distance between eyepiece and reticle, and the distance between objective and reticle can be adjusted. The reticle should be located in the eyepiece focal plane. The eyepiece is composed of a field lens and an eyepiece. The eyepiece of KF-JJY1 type spectrometer is an abbe eyepiece. Between eyepiece and reticle it is equipped with a small transparent total reflection prism, whose small cross window can allow light to pass through. The

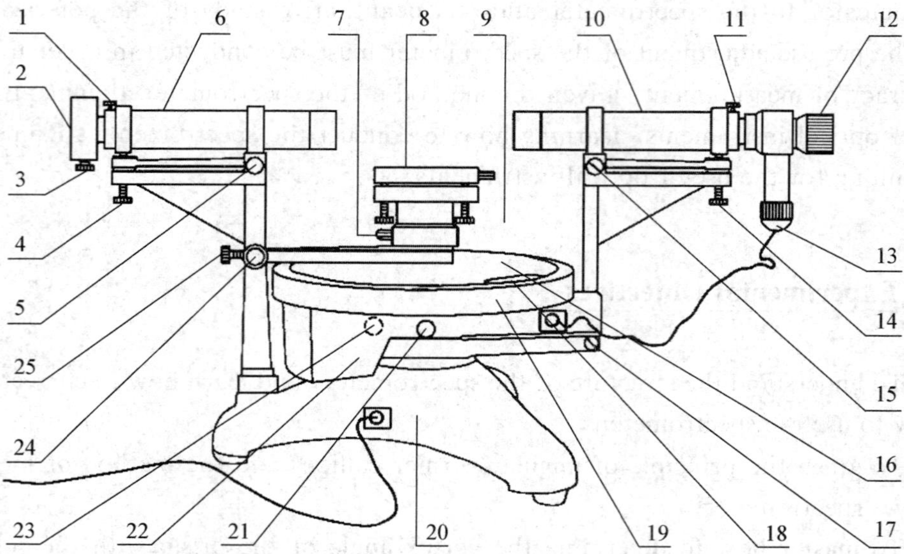

Fig. 4-5-1 KF-JJY1 type spectrometer structure chart

1 - Locking screw of the slit in the directional light tube. 2 - Slit in the parallel light tube. 3 - Slit adjustment screw in the directional light tube. 4 - Tilt adjustment screw of the directional light tube. 5 - Horizontal direction adjustment screw of the directional light tube. 6 - Parallel light tube. 7 - Locking screw of stage. 8 - Stage. 9 - Screw for leveling Stage. 10 - Telescope. 11 - Screw for locking eyepiece in the telescope. 12 - Screw for focusing eyepiece in the telescope. 13 - Light. 14 - Screw for adjusting telescope tilt. 15 - Screw for horizontally adjusting telescope (Back). 16 - Vernier calliper dial. 17 - Fine - tuning screw for holder rotation (Back). 18 - Vernier calliper. 19 - Main angle calliper coordinate. 20 - Holder. 21 - Screws for locking transposition and dial. 22 - Swivel mount. 23 - Screw for locking telescope (Back). 24 - Screw for fretting Vernier plate. 25 - Screw for locking Vernier calliper.

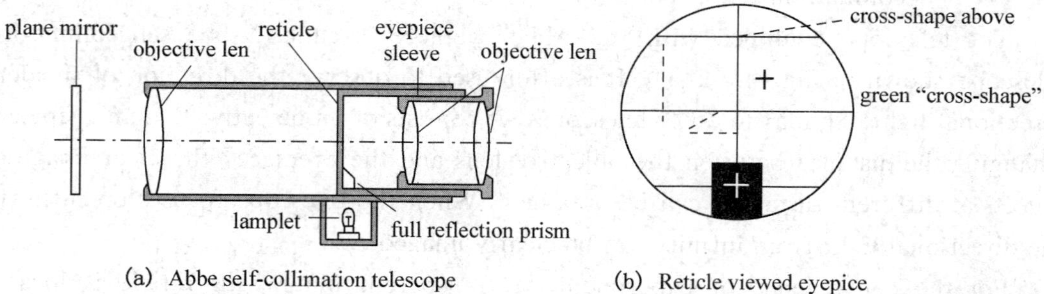

(a) Abbe self-collimation telescope (b) Reticle viewed eyepice

Fig. 4-5-2 Structure of abbe self-collimation telescope

light emitted by a small lamp is reflected by the prism and then makes the cross bright. Looking through the eyepiece, some parts of the reticle are covered by the small prism, leaving the uncovered parts to be identified.

The "cross" on the reticle of the Abbe-type self-collimation telescope and "cross-shape" window on the small prism are symmetrical with respect to the middle horizontal line. First place a plane mirror in front of the telescope. Light emitted by the "cross-shape" window is refracted by the objective lens, reflected by the plane mirror, and then refracted by the objective lens before passing through the objective lens into the telescope. Finally, a green "cross-shape" image will form on the reticle, shown as Fig. 4-5-3. If the reticle is right in the focal plane of the objective lens, there is little parallax between the green "cross-shape" image and cross-line. If the green cross-shape image is right on the "cross" on the reticle, the telescope is perpendicular to the plane mirror.

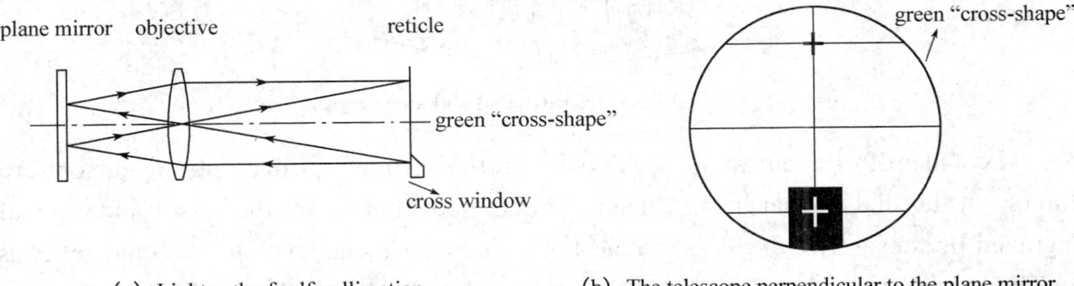

(a) Light path of self-collimation (b) The telescope perpendicular to the plane mirror

Fig. 4-5-3 Schematic of self-collimation

The telescope is adjusted in a similar way, namely making the green cross-shape image formed on the reticle plane, adjusting the image to coincide with cross-shape line and leaving no parallax with the the cross-line. Such a method is called self-collimation.

(2) Collimator

The structure of the collimator is shown in Fig. 4-5-4. One end of the collimator is equipped with an adjustable slit. Move the slit back and forth so that the slit is on the focal plane of the parallel convex lens. Under such conditions, the light passing through the slit will become a directional light after passing through the convex lens. The angle of the collimator can also be adjusted by screws on the mount.

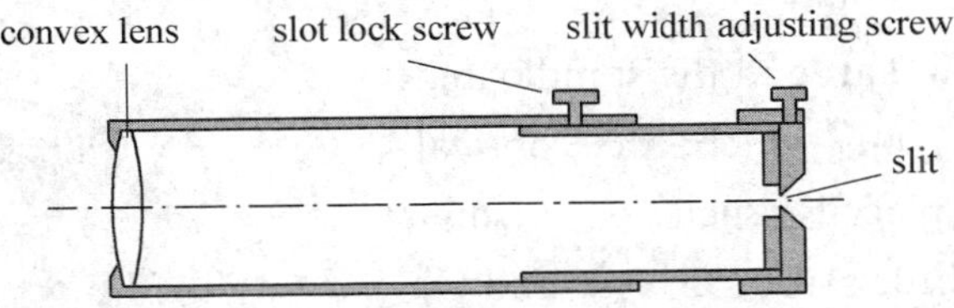

Fig. 4-5-4 Structure schematic of collimator

(3) Reading dials

A central shaft is fixed at the center of the base, and the dial and the cursor are sleeved on the central shaft so as to rotate around the central shaft. A thrust bearing support is arranged at the lower end of the dial to make the rotation light and flexible. A dial is engraved with 720 evenly divided lines, and each grid division indicates an angle of 30′. In order to eliminate the eccentric error of the dial, two cursors A and B with difference of 180 degrees are used for the reading. As shown in Fig. 4-5-5, the 30 grids on the cursors are equal to the angle of the 29 grids on the dial, so the minimum scale value of the cursor is 1′. The intial step of reading is to locate the position indicated by the cursor zero line.

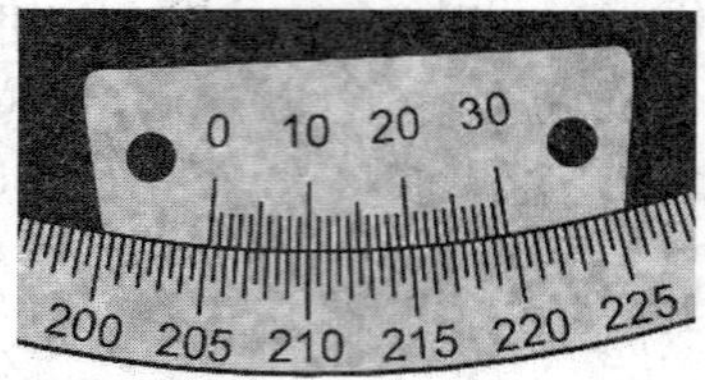

Fig. 4-5-5 Spectrometer dial and angle cursor

Next, identify the entire 30′ value between the zero line of dial and the cursor zero line of vernier dial and set it as α. Then, find the position where the cursor and the dial engraved lines exactly coincide and read the angular part smaller than 30′, and set it as β. The sum of two values ($\alpha+\beta$) is the measured angle value. Thus, the value in Fig. 4-5-5 is 205°11′.

In the experiment, the rotation angle of telescope is equal to that of the two cursors. The rotation angle of telescope can be obtained by measuring the angular coordinates of the zero mark of the two cursors before and after the telescope rotation. With the four collected values, the result can be calculated with the following formula:

$$\theta=\frac{1}{2}(\theta^{A}+\theta^{B})=\frac{1}{2}[(\varphi_2^{A}-\varphi_1^{A})+(\varphi_2^{B}-\varphi_1^{B})]$$

where θ is the rotation angle of the telescope, φ_1^{A} and φ_1^{B} are the initial angular coordinates for cursors A and B, while φ_2^{A} and φ_2^{B} are the corresponding post-rotation angular coordinates.

(4) Stage

The stage can rotate around the spindle and lift along the axis, on which the optical components can be placed, such as prisms, gratings and so on. Three screws G_1, G_2 and G_3 can be adjusted to change the tilt of the stage, as shown in Fig. 4-5-6. The stage can be fixed with a locking screw.

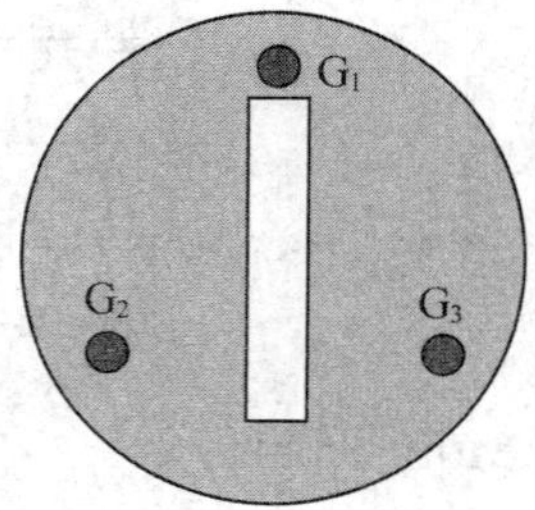

Fig. 4-5-6 The plane mirror is perpendicular to the line between G_2 and G_3

2. Spectrometer Adjustment

The requirements for the adjustment of the spectrometer are as follows: the collimator focuses on infinity so that it emits directional light; the telescope focuses on infinity to accept directional light; the optical axes of the collimator and the telescope are perpendicular to the major axis of the spectrometer. Tighten the dial lock screw before the adjustment, followed by a rough adjustment with a visual method. Then the fine-tuning will be carried out with four steps in detail.

(1) Adjust the telescope with self-collimation to make it focus on infinity.

① Turn on the power and observe the reticle black cross-shape line from the eyepiece. Then adjust the distance between the eyepiece and the reticle until the black cross-shape line is visible.

② Adjust the stage to make it close to its chassis. Place a plane mirror on the stage and make its mirror surface perpendicular to the line connecting the two screws G_2 and G_3 on the stage. The plane mirror should also be fixed through screw G_1, as shown in Fig. 4-5-6. In this way, it only needs to adjust G_2 or G_3 instead of adjusting G_1 to change the tilt of the plane mirror.

③ Find the green cross image. Adjust the objective lens until there is no parallax between the green cross image and black cross (self-collimation). Then move the green cross image to the upper cross of the reticle plate.

④ Place the reticle's cross line at the right position. Make sure that the stage together with the mirror rotates relative to the telescope so that the moving direction of the green cross image is parallel to the horizontal line of the cross lines. After the adjustment, lock the eyepiece tube.

⑤ Adjust the optical axis of the telescope to be perpendicular to the spectrometer spindle and ensure the plane mirror to face the telescope. Observe through the eyepiece to check the existence of the green cross-shape image. If not, adjust the screws "14" (not G_1 or G_2 at this moment) until the green cross image emerges in the field of view. Continue to make the adjustment until the green cross images reflected from both sides of the plane mirror appear in the field of view.

⑥ Use the "half-adjustment method" to make the telescope optical axis perpendicular to the spindle. Check whether the reflected green cross image is on the horizontal line or not. If not, adjust the platform screws G_2 or G_3 so that the green cross closer to the top line. Adjust the telescope again to make the green cross be on the horizontal line (at any position on the horizontal line). At this moment, the optical axis of the telescope is perpendicular to the plane mirror (but it doesn't mean that the telescope optical axis is perpendicular to the spectrometer spindle). Then, turn the plane mirror 180° to observe whether the green cross image from the other side is still on the horizontal line. If not, repeat the adjustment several times (usually 3 times are enough) with the "half-adjustment method" until the green cross-shape images reflected from the two sides are on the horizontal line. When the optical axis of the telescope is perpendicular to the spectrometer spindle, fix horizontally the position of telescope.

Note: The telescope adjustment screws must not be moved anymore during the following experiment.

(2) Adjust the collimator until it focuses on infinity

Use the light to project on the slit of the collimator and tune the telescope to align with the collimator. Observe the slit image from the eyepiece and adjust the distance between the slit and the lens until the slit image is clearly visible without parallax. At this moment, the slit is in the focal plane of the collimator objective lens and the light emitted by the objective lens is directional light.

① Rotate the slit and make it parallel to the vertical line of the reticle and and then lock it.

② Adjust the width of slit image until it is of roughly the same width as the black line in the reticle.

③ Adjust the collimator perpendicular to the spectrometer spindle.

Adjust the directional light tube to make the slit image equally divided by the the reticle horizontal line (when adjusting, set the slit image offset from the vertical line of the cross-shape line so that all the slit images can be seen in the field of view), as shown in Fig. 4-5-7. After the adjustment, get the horizontal position of the directional light tube fixed. Turn the stage so that the plane mirror is rotated 180 degrees about the axis.

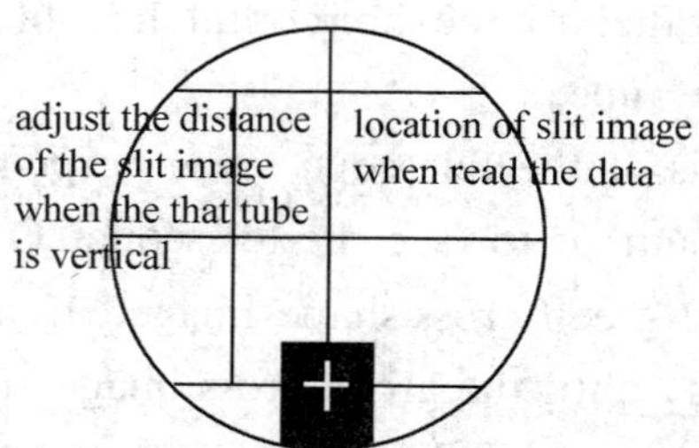

Fig. 4-5-7 The length of the slit image is bisectable

3. The Adjustment of Prism's Main Cross-Section

After the adjustment of the spectrometer, the prism to be measured is placed on the stage. Before the measurement, the main cross section of the prism must be perpendicular to the principal axis of the spectrometer so that the telescope is perpendicular to the side of the prism. When the telescope is perpendicular to the spectrometer spindle, the telescope cannot be adjusted. The specific steps of adjustment are as follows:

(1) Place the prism on the stage as shown in Fig. 4-5-8, where AB is perpendicular

to G_1G_2, BC is perpendicular to G_2G_3, and AC is perpendicular to G_1G_3. This position of prism makes it easy to set the AB plane perpendicular to the telescope only by tuning G_2 and to set the AC plane perpendicular to the telescope by only adjusting G_3. Given that the adjustment of G_1 will impact both the AB plane and the AC plane, it is unnecessary to tune G_1.

(2) Move the telescope to the positions perpendicular to the AB plane and the AC plane respectively, and adjust the corresponding green cross images to align with the upper horizontal line of the fork. Adjust the green cross-shape image of the AB plane and the AC plane by merely using G_2 and G_3 respectively.

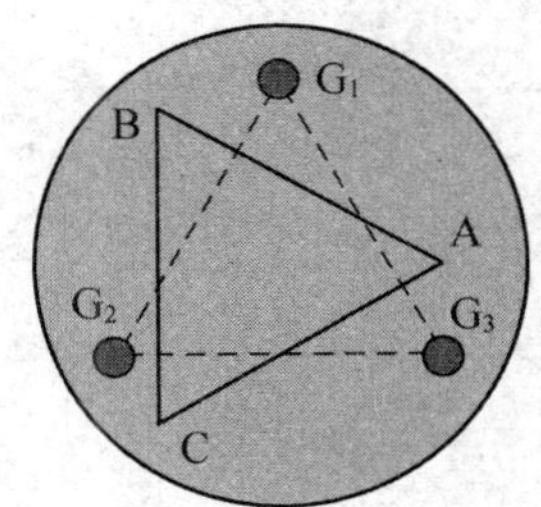

Fig. 4-5-8 Prism placement on the stage

Experimental Instruments

KF JJY1 type spectrometer, Plane mirror and Prism.

Experimental Contents and Steps

Measure the vertex angle of the prism.

Rotate the stage to make the apex of the prism face the parallelized light tube, as shown in Fig. 4-5-9. *A* directional beam of light (parallel to the main section) emitted by the collimator is divided by the apex of the prism. Light "1" is reflected by the AB side, and light "2" is reflected by the AC side (BC is the ground glass side). It can be shown that the angle between the two reflected beams is determined by the vertex angle *A* of the prism as follows:

$$A=\theta/2$$

Rotate the telescope to find the slit image from the AB side. Tune the telescope slightly to make the slit image coincide with the vertical line of the cross-shape line, and then record the readings φ_1^A and φ_1^B of the two callipers. Obtain the angular coordinates of another reflected light φ_2^A, φ_2^B with the same operation. Then the vertex angle can be measured as the following formula:

$$A=\frac{\theta}{2}=\frac{1}{4}(\theta_A+\theta_B)=\frac{1}{4}[(\varphi_2^A-\varphi_1^A)+(\varphi_2^B-\varphi_1^B)]$$

Rotate the stage slightly and repeat the measurement three times to get the average vertex angle.

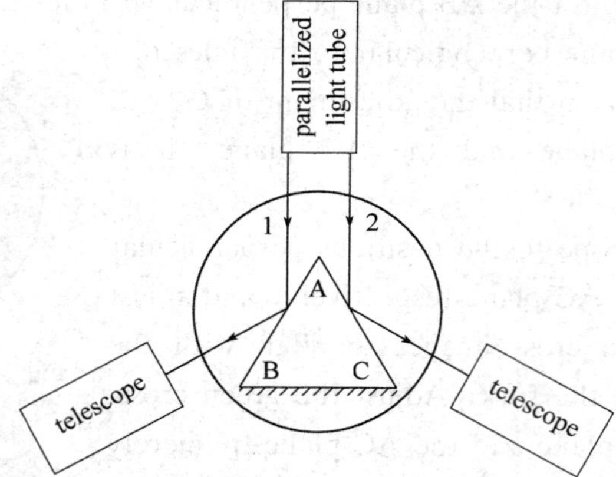

Fig. 4-5-9 Schematic diagram of measuring the vertex angle of prism

Precautions

To record the readings, the slit image should be aligned with the vertical cross-line.

In calculating the rotation angle of the telescope, the operator should pay attention to whether the telescope rotates beyond the zero point of the dial. For example, when the telescope rotates from position Ⅰ to position Ⅱ, the corresponding angular positions from the dial are recorded as follows:

Telescope location	Ⅰ	Ⅱ
Cursor A	175°45′(φ_1^A)	295°43′(φ_2^A)
Cursor B	355°46′(φ_1^B)	115°43′(φ_2^B)

Cursor A has not passed the zero point and the angle through which the telescope turned can be calculated as:

$$\theta^A=\varphi_2^A-\varphi_1^A=295°43'-175°45'=119°58'$$

Given that cursor B has passed the zero point, the angle the telescope turned should be calculated as:

$$\theta^B=\varphi_2^B-\varphi_1^B=(115°43'+360°)-355°46'=119°57'$$

Questions

1. Besides the understanding of the function of the main components of the spectrometer, the knowledge about the spectrometer's functions and its adjustment method, together with the requirements for the correct operations, are all necessary.

2. Considering the following questions: how to adjust the telescope focusing on infinity? how to judge whether the adjustment is done or not? and how to determine whether the directional light tube focuses on infinity?

3. The green cross image is visible in the telescope just above the upper cross of the reticle line while it moves below the upper cross of the reticle when the mirror turns 180°. Under such circumstance, what kind of measurements should be taken? To adjust the tilt of the telescope or to adjust the tilt of the platform? On the contrary, if the platform is turned 180° and the green cross-shape image is still above the the cross line, what should be done? To adjust the telescope or to adjust the platform?

4. How to adjust the collimator?

5. What are the purposes of setting up two reading callipers in the spectrometer?

4.6 电表的改装与校准

实验目的

1. 掌握将微安表头改装成电流表或电压表的原理及校准方法。
2. 熟悉滑线电阻器的两种用途。
3. 了解电表的标称误差及电表等级的意义。

实验原理

在测量电流或电压时,可能要各种量程的电流表或电压表,这些电流表或电压表一般都是由小量程的电表(俗称表头),经并联或串联适当阻值的电阻改装而成。

一、把微安表改装为电流表

将小量程的表头改装成大量程的电流表,实际上就是扩大原来电表的量程,通常是在表头上并联一个低值电阻 R_S,使得流过表头的电流只是总电流的一部分,大部分电流流过分流电阻 R_S,如图 4-6-1 所示。设表头内阻为 R_g,满偏电流为 I_g。并联电阻 R_S 后,当表头满偏时,电路中的总电流为 I,由欧姆定律可得

$$(I-I_g)R_S=I_gR_g \tag{4-6-1}$$

$$R_S=\frac{I_gR_g}{I-I_g}=\frac{R_g}{n-1} \tag{4-6-2}$$

式中,$n=\frac{I}{I_g}$为电流表量程扩大的倍数。

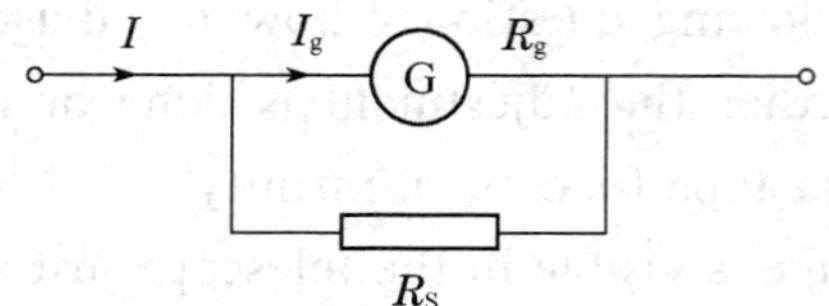

图 4-6-1　表头改装为电流表电路

本实验要求将量程为 $I_g=50\ \mu A$ 的微安表(内阻 R_g 见表头,约几千欧),改装为量程为 $I=15$ mA 的电流表。

二、把微安表改装为电压表

将表头串联一个高值电阻 R_p,就将表头改成了电压表,如图 4-6-2 所示。表头两端的电压只是总电压 U 的一部分。当表头满偏时,由欧姆定律可得:

$$U=I_g(R_g+R_p) \tag{4-6-3}$$

$$R_p=\frac{U-I_gR_g}{I_g}=(n'-1)R_g \tag{4-6-4}$$

式中,$n'=\dfrac{U}{I_gR_g}$为电压表量程扩大的倍数。

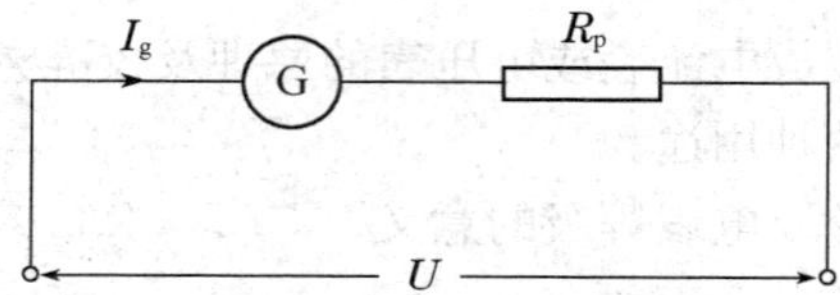

图 4-6-2　表头改装为电压表电路

本实验要求将量程为 $I_g=50\ \mu A$ 的微安表(内阻 R_g 见表头,约几千欧),改装为 $U=5$ V的电压表。

三、电表的标称误差和校准

标称误差是一种简化而实用的相对误差。为了确定标称误差,先将需要标准的电表和一个标准电表同时测量一定的电流(或电压),称为校准。校准结果得到电表各个刻度的绝对误差,选取其中最大的绝对误差除以量程,即为该电表的标称误差。即

$$标称误差=\frac{最大的绝对误差}{量程}\times 100\% \tag{4-6-5}$$

根据标称误差的大小,电表分为不同的等级。一般分为 0.05、0.1、0.2、0.5、1.0、1.5、2.5、4.0、5.0 共九级。电表等级常标在电表的板面上,如 0.5,表示该电表为 0.5 级,其标称误差不大于 0.5%。

在校准电流表时,设改装表的指示值为 I',标准表对应的指示值为 I,则绝对误差为 $\Delta I'=I-I'$。测量改装表各个整数刻度的 I'和相应的标准表指示值 I,以 I'为横坐标,$\Delta I'$

为纵坐标作出改装表的校准曲线，两个校准点之间用直线连接，如图4-6-3。在以后使用这个改装表时，根据校准曲线可以修正电表的读数，得到较为准确的结果。

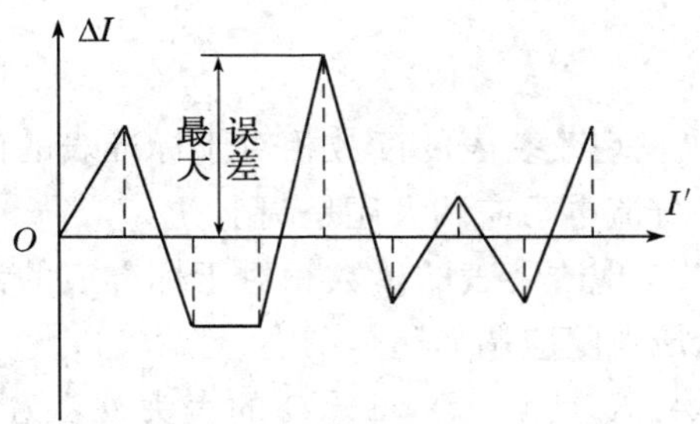

图4-6-3　改装电流表的校准曲线

校准电压表的方法与校准电流表的方法相同，校准曲线以改装电压表的读数 U' 为横轴，以绝对误差为 $\Delta U=U-U'$ 为纵轴，如图4-6-4所示。

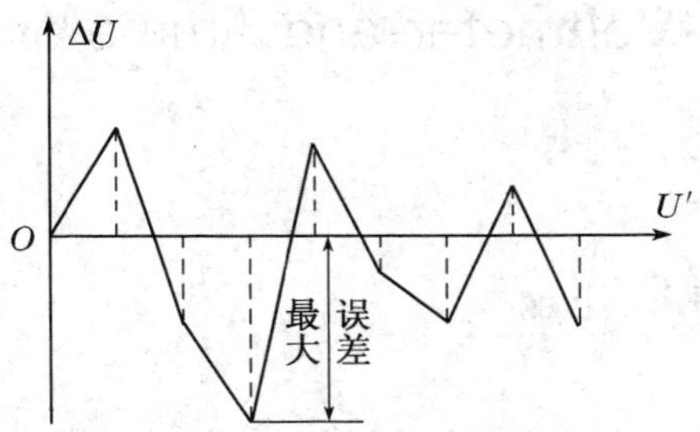

图4-6-4　改装电压表的校准曲线

实验器材

微安表（$I_g=50.0\ \mu A$，R_g 约为4000 Ω，具体数值贴在微安表背面，1.5级）；

电流表（多量程0.5级）、UJ31型电位差计；

电阻箱（0.02级，0～99 999.9 Ω）、滑线电阻（200 Ω）；

导线、开关、电源。

注意事项

1. 将表头改装为量程为15 mA的电流表：计算 R_S 值，设计校准电路，对改装表进行校准，绘制出校准曲线，确定改装电流表等级。

2. 把表头改装为量程为5 V的电压表：算得 R_P 值，设计校准电路，用电位差计对改装表进行校准，绘制出校准曲线，确定改装电压表等级。

思考题

1. 校准电流表时,如果发现改装表的读数相对于标准表的读数都偏高,试问要达到标准表的数值,此时改装表的分流电阻应调大还是调小?

2. 校准电压表时,如果发现改装表的读数相对于标准表的读数都偏低,试问要达到标准表的数值,此时改装表分压电阻应调大还是调小?

3. 如何将 $I_g=50.0\ \mu A$, $R_g=4.000\times10^3\ \Omega$ 的表头改装为量程为 500 mA 和 150 mA 的双量程电流表?画出设计线路图,并计算其分流电阻。

Experiment 4.6 Conversion of Galvanometer to Voltmeter and Ammeter

Experimental Objectives

1. To master the calibration method and principle of converting micro-meter meter into ammeter or voltmeter.

2. To be familiar with the two usages of the sliding line rheostat.

3. To understand the nominal error of meter and the meaning of meter level.

Experimental Principle

When measuring the current or voltage, the ammeters or voltmeters with various ranges are needed. These ammeters or voltmeters are usually refitted from small range meters (commonly known as meters) in parallel or in series with other resistors.

1. Converting a Microampere Meter into an Ammeter

To convert a small-range microampere meter into a big-range ammeter actually means expanding the range of the original meter.

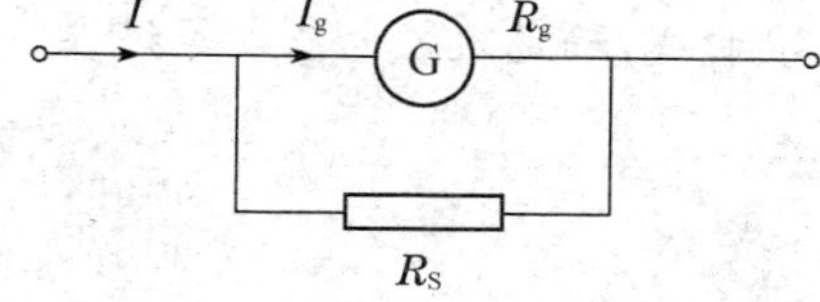

Fig. 4-6-1 Conversion of a galvanometer to an ammeter

As Fig. 4-6-1 shows, a low value resistor R_S is usually connected in parallel to the meter head so as to ensure only part of current to flow through the meter head and most of the current to flow through the resistor R_S. Suppose the head resistance as R_g and the full bias current as I_g. Set the total current flowing through the circuit as I when the meter head is fully biased, and the following formula can be obtained according to Ohm's Law:

$$(I-I_g)R_s=I_gR_g \tag{4-6-1}$$

$$R_s=\frac{I_gR_g}{I-I_g}=\frac{R_g}{n-1} \tag{4-6-2}$$

Where $n=I/I_g$ is the expanded multiple of the range of the ammeter.

This experiment aims to refit the micrometer with a range of $I_g=50\ \mu A$ (internal resistance can be seen in the meter head, around several thousand Ohms) into the ammeter with the range of $I=15$ mA.

2. Converting Microampere Meter into Voltmeter

As Fig. 4-6-2 shows, a voltmeter can be made by connecting a high value resistance in series with a meter head. The voltage of the head is only a part of the total voltage U. When the head is fully biased, the two values can be expressed by the Ohm's Law as:

$$U=I_g(R_g+R_p) \tag{4-6-3}$$

$$R_p=\frac{U-I_gR_g}{I_g}=(n'-1)R_g \tag{4-6-4}$$

where $n'=U/(I_gR_g)$ is the expanded multiple of the range of the voltmeter.

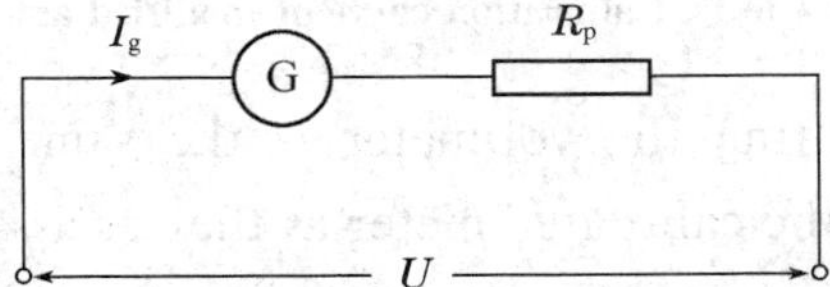

Fig. 4-6-2 Convertion of microampere meter to voltmeter

This experiment aims to modify the micrometer with a range of $I_g=50\ \mu A$ (internal resistance can be seen in the meter head, about several thousand Ohms) to a voltmeter with the range of $U=5$ V.

3. Nominal Error of Meter and Calibration

Nominal error is a simplified and practical relative error. Calibration is the first necessary step to determine nominal error, namely the operator uses a meter to be normalized and a standard meter to measure a particular current (or voltage) simultaneously. As a result of calibration, the absolute error of the meter at each scale can be obtained. Next choose the maximum absolute error to be divided by the range, the nominal error can be calculated as follows:

$$\text{standard error}=\frac{\text{maximum absolute error}}{\text{range}}\times 100\% \qquad (4\text{-}6\text{-}5)$$

The meter can be classified into different levels according to the value of standard error. There are nine levels of meter, namely, 0.05, 0.1, 0.2, 0.5, 1.0, 1.5, 2.5, 4.0, and 5.0. The level of meter is usually marked on the board of the meter, for example, 0.5 represents the meter's level of 0.5, which means the standard error of the meter is no more than 0.5%.

When calibrating the ammeter, set the value of the refitted meter as I' and the corresponding value of the standard meter as I, the absolute error is $\Delta I=I-I'$. Next, each integer scale I' and the corresponding value of the standard meter I need to be measured. Set I' as the abscissa and ΔI as the ordinate to make the standard curve of the modified meter. The standard points are linked by straight lines, as shown in Fig. 4-6-3. For the future application of the modified meter, this standard curve can be used to correct the meter's reading for more accurate results.

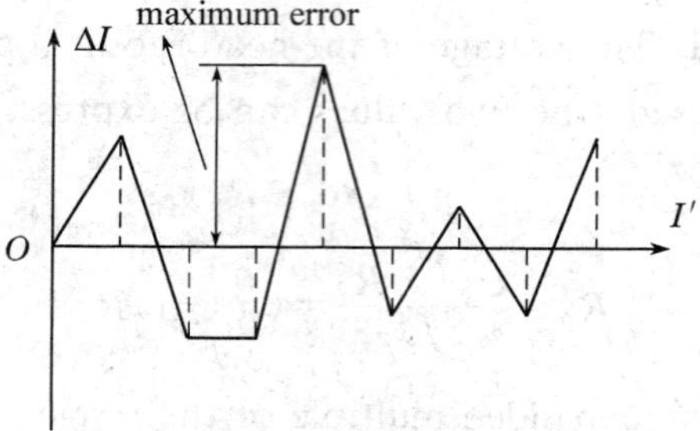

Fig. 4-6-3 Calibration curve of modified ammeter

The method of calibrating the voltmeter is the same as that of calibrating the ammeter. Set the value of the calibrated meter as the abscissa and the absolute error $\Delta U=U-U'$ for the ordinate to make the standard curve, as shown in Fig. 4-6-4.

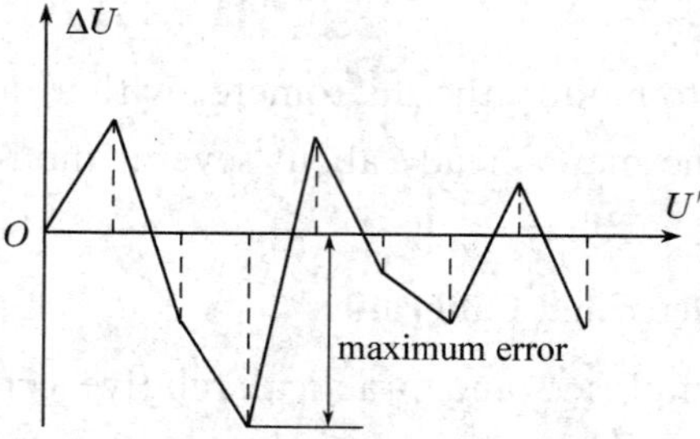

Fig. 4-6-4 Calibration curve of modified voltmeter

Experimental Apparatus

Micro-ampere meter (I_g=50 μA, R_g is about 4000 Ω, the specific value is posted on the back of the micro-ampere meter, level 1.5), ammeter (with several ranges, level 0.5), voltmeter (with several ranges, level 0.5), resistance box (level 0.02, 0～99,999.9 Ω), rheostat (200 Ω), wire, switch and power supply.

Fig. 4-6-5 Micro-amphere meter

Experimental Contents and Steps

1. The procedures of converting the meter head to an ammeter with the range of 15 mA include: calculating of the value of R_s, designing the standard circuit, calibrating the modified meter, making the standard curve and finally determining the level of modified ammeter.

2. The procedures of converting the meter head to a voltmeter with the range of 5 V include: calculating the value of R_p, designing the standard circuit, calibrating the modified meter with a voltmeter, making the standard curve, and finally determining the level of modified voltmeter.

Questions

1. When calibrating the ammeter, it is found that the reading value of modified ammeter is relatively higher than that of the standard meter. What should be done to achieve a standard value? To turn up the value of the shunt resistor of the modified ammeter or to turn it down?

2. When calibrating the voltmeter, it is found that the reading value of the modified ammeter is relatively lower than that of the standard meter. What should be done to achieve a standard value? To turn up the value of the voltage divider of the modified voltmeter or to turn it down?

3. How to convert a meter head ($I_g=50\ \mu A$, $R_g=4.000\times10^3\ \Omega$) to a double-range ammeter with the range of 500 mA and 150 mA? Design the circuit and calculate its shunt resistor?

4.7 测量超声波在空气中的传播速度

声波是在弹性媒质中传播的一种机械波，由于其振动方向与传播方向一致，故声波是纵波。振动频率在 20 Hz～20 kHz 的声波可以被人们听到，称为可闻声波；频率超过 20 kHz 的声波称为超声波。

对于声波特性的测量(如频率、波速、波长、声压衰减、相位等)是声学应用技术中的一个重要内容，特别是声波波速(简称声速)的测量，在声波定位、探伤、测距等应用中具有重要的意义。

本实验利用压电晶体换能器来测量超声波在空气中的速度。声速的测量方法可分为两类。第一类方法是根据 $v=L/t$，测量距离 L 和时间隔 t，即可算出声速 v；第二类方法是利用 $v=\lambda f$，测出频率 f 和波长 λ，就可计算出速度 v。本实验采用的共振干涉法和相位比交法均属于第二类方法。

实验目的

1. 学会用共振干涉法和相位比较法测量空气中的声速。
2. 学会用逐差法进行数据处理。
3. 了解声速与气体参数的关系。

实验原理

由于超声波具有波长短、易于定向发射等优点，所以在超声波段进行声速测量是比较方便的。超声波的发射和接收一般通过电磁振动与机械振动的相互转换来实现，最常见的是利用压电效应和磁致伸缩效应。

声波的传播速度 v 与其频率 f 和波长 λ 的关系为

$$v=\lambda f \tag{4-7-1}$$

由上式可知，测得声波的频率 f 和波长 λ，就可算出声速 v，其中声波频率 f 可通过频率计测得，本实验的主要任务是测出声波波长 λ。

一、超声波的获得和压电换能器

压电陶瓷声波换能器是由压电陶瓷片和两种金属组成。压电陶瓷片(如钛酸钡、锆钛酸铅等)由一种多晶结构的压电材料构成，在一定的温度下经极化处理后，具有压电效应。当压电陶瓷片受到与极化方向一致的应力时，在极化方向上产生一定的电场强度，且应力与电场强度呈线性关系；反之，当与极化方向一致的外加电压加在压电材料上时，材料的伸缩形变与外加电压也具有线性关系。因此，可以将正弦交流电信号使压电材料纵向长度伸缩而成为超声波波源，同样也可以使声压变化转变成电压信号来接收。

二、共振干涉法

实验装置如图 4-7-1 所示，图中 S_1 和 S_2 为压电陶瓷换能器，S_1 作为声波源，它被振荡频率可以调节的低频信号发生器输出的电信号激励后，由于逆压电效应发生受迫振动，并向周围空气定向发出一近似平面声波；S_2 为超声波接收器，声波传至它的接收面上时由压电效应变成电压信号，用电压表或示波器可以测量其大小。同时 S_2 也反射一部分声波。当 S_1 与 S_2 的表面互相平行时，S_1 发出声波与 S_2 反射的声波就在两个平面间相互干涉，产生驻波共振现象。

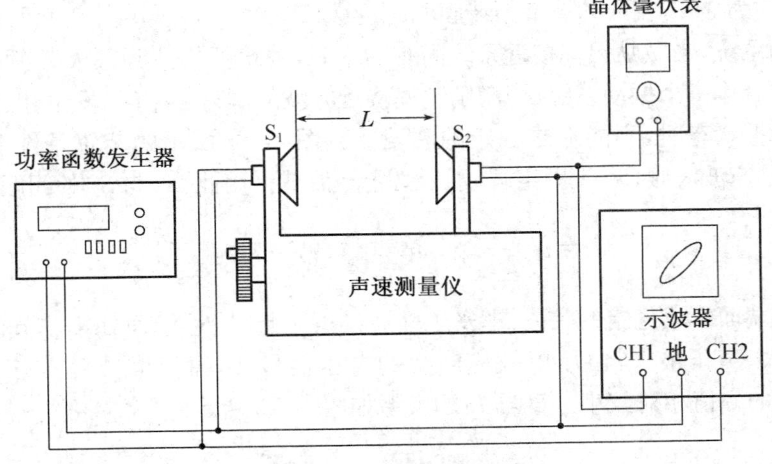

图 4-7-1　实验装置示意图

当两列振幅相同、频率相同、振动方向相同而传播方向相反的波叠加在一起,就可以形成驻波。设有振幅相同、频率相同沿 x 轴相向传播的两列声波,从两列波在各点位移相同的某时刻开始计时,且 x 轴原点 O 选在某最大位移处,此时两列的波动方程可写为

$$\begin{cases} y_1=A\cos 2\pi\left(ft-\dfrac{x}{\lambda}\right) \\ y_2=A\cos 2\pi\left(ft+\dfrac{x}{\lambda}\right) \end{cases} \tag{4-7-2}$$

根据波的叠加原理,两列波叠加形成的波的波动方程为

$$y=y_1+y_2=2A\cos 2\pi\left(\frac{x}{\lambda}\right)\cos 2\pi ft \tag{4-7-3}$$

这是一个驻波方程,图 4-7-2 描绘了不同时刻该驻波的波形图。

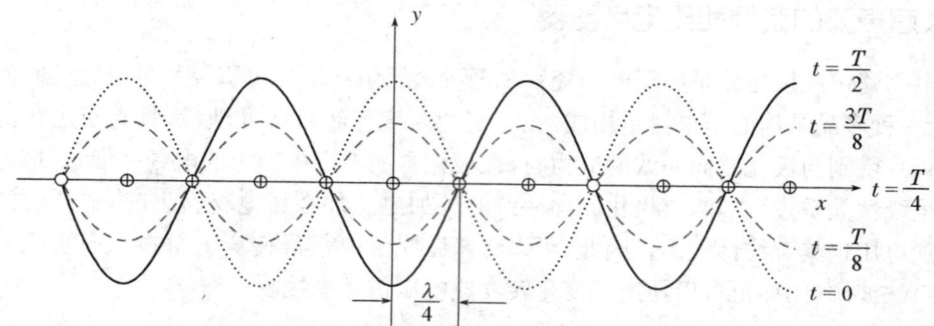

图 4-7-2 驻波在不同时刻的波形图

由方程可以看出,在 $x=\pm k\lambda/2(k=0,1,2,\cdots)$ 处振幅为 $\left|2A\cos 2\pi\left(\dfrac{x}{\lambda}\right)\right|=2A$,即振幅最大,这些位置称为“波腹”,上图中以⊕标记这些位置;在 $x=\pm(2k+1)\lambda/4(k=0,1,2,\cdots)$ 处,振幅 $\left|2A\cos 2\pi\left(\dfrac{x}{\lambda}\right)\right|=0$,这些位置称为“波节”,图中用○标记这些位置。在相邻的两波节间,各点的振动是同相的,同一波节两侧的各点的振动是反相的。相邻的波节间距、波腹间距均为 $\lambda/2$,相邻的波节与波腹间距为 $\lambda/4$。

一个振动系统,当激励频率接近系统固有频率时,系统振幅达到最大,这种状态称作共振。驻波场可看作一个振动系统,当激励频率接近驻波的固有频率时,发生驻波共振。当驻波系统偏离共振状态时,不能形成稳定的驻波。在图 4-7-1 所示的装置条件下,S_1 为自由端,端面必定是声波波腹;S_2 为固定端,端面必定是波节。因此,S_1 与 S_2 的间距 L 必须满足

$$L=n\frac{\lambda}{2}+\Delta,\ n=0,1,2,\cdots,\Delta<\lambda \tag{4-7-4}$$

才能在两端面形成稳定的驻波,即发生驻波共振。此时,S_2 端面处于波节位置,振幅最小,而形变最大、声压最大,由压电效应产生的电压信号 U 幅值也最大。当 L 不满足式(4-7-4)时,U 的幅值相对较小。U 与 L 的关系如图 4-7-3 所示。

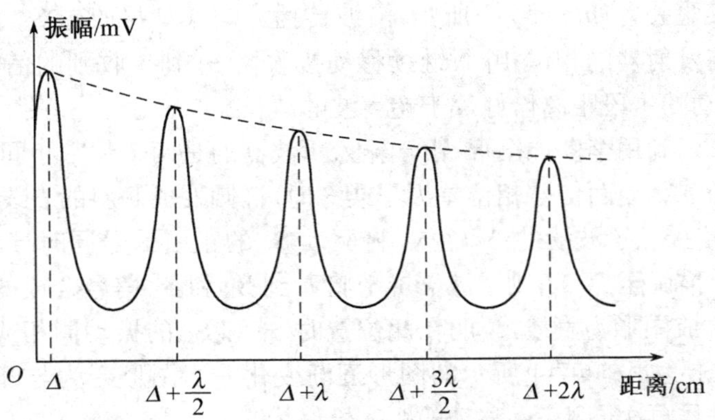

图 4-7-3　S_2 端面产生电压与 S_1S_2 间距的关系示意图

图中各极大值之间的距离均为 $\lambda/2$，由于衍射和其他损耗，各极大值幅值随距离增大而逐渐减小。只要测出与各极大值对应的接收器 S_2 的位置，就可以测出波长 λ。

若测出 20 个极大值的位置，并依次算出每经过 10 个 $\lambda/2$ 的距离

$$\Delta L_{11-1}=L_{11}-L_1=10\,\frac{\lambda}{2}$$

$$\Delta L_{12-2}=L_{12}-L_2=10\,\frac{\lambda}{2}$$

……

$$\Delta L_{20-10}=L_{20}-L_{10}=10\,\frac{\lambda}{2}$$

把等式两边各自相加，得

$$\sum_{i=1}^{10}\Delta L_{(10+i)-i}=100\,\frac{\lambda}{2}$$

$$\lambda=\frac{1}{50}\left(\sum_{i=1}^{10}\Delta L_{(10+i)-i}\right)$$

由低频信号发生器或频率计读得超声波的频率值 f 后，即可由下式求得声速

$$v=\frac{1}{50}\left(\sum_{i=1}^{10}\Delta L_{(10+i)-i}\right)\times f \tag{4-7-5}$$

若测不到 20 个极大值，则可少测几个。例如，测到 12 个极大值，可依次算出他们经 6 个 $\lambda/2$ 的距离，最后得

$$v=\frac{1}{18}\left(\sum_{i=1}^{6}\Delta L_{(6+i)-i}\right)\times f \tag{4-7-6}$$

三、相位比较法

波是振动状态的传播，也可以说是相位的传播。沿波传播方向的任何两点，当其相位与波源的相位间的相位差相同时，这两点间的距离就是波长的整数倍。利用这个原理，可以精

确地测量波长。实验装置如图 4-7-1 所示,沿波传播方向移动接收器 S_2,总可以找到一点,使接收到的信号与发射器的相位相同;继续移动接收器 S_2 到接收到的信号,再次与发射器的相位相同时,移动的这段距离恰好等于超声波的波长。

判断相位差可以利用李萨如图形,由于输入示波器的是两个频率相同的信号,因此李萨如图形是稳定的椭圆。当两信号相位差为零或 π 时,椭圆变成倾斜的直线。

将 S_2 信号直接接入示波器的 Y 输入,将驱动 S_1 的正弦信号同时接入示波器的 X 输入。将示波器的扫描调至 X-Y,则示波器屏上将看到 S_1 和 S_2 信号合成的李萨如图形。设 S_2 移动在某位置合成图形为直线,说明在此位置时 S_1 和 S_2 的振动同相或反相。逐渐移动 S_2,将看到李萨如图形如图 4-7-4 所示的图形逐渐变化。当图形变为与开始时完全相同的直线时,S_1 和 S_2 信号再次同相,S_2 移动的距离恰好为 λ。

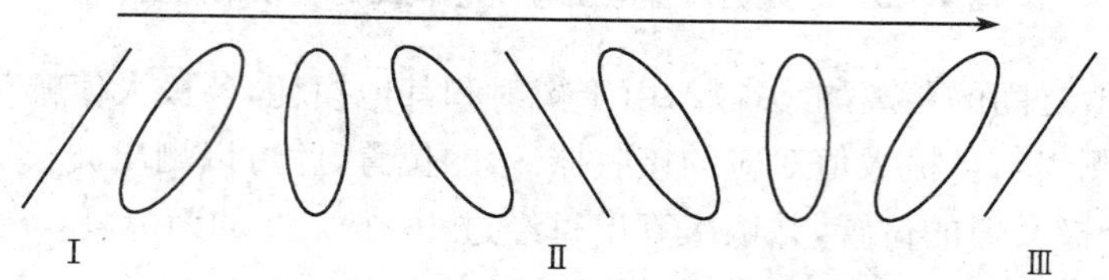

图 4-7-4 相位比较法中的李萨如图形

李萨如图形从Ⅰ变为Ⅱ,移动半个波长;从Ⅱ变为Ⅲ,移动半个波长

与共振干涉法相类似,可测得 20 个或 12 个相应的数值,以便进行数据处理。

实验器材

声速测量仪,功率函数信号发生器,晶体管毫伏表,示波器。

实验内容及步骤

一、共振干涉法测量声速

1. 按照图 4-7-1 所示连接线路(接晶体管毫伏表),将 YB1631 型功率函数发生器接功率输出端,置30 V_{P-P}挡,频率选择置 10~100 kHz 挡。

2. 开启电源,将幅度调到最高,在 40 kHz 附近缓慢调节频率(本实验换能器的固有频率在 40 kHz 左右),使毫伏表的示值最大(中间可能需要调节毫伏表的量程),此状态即为信号发生器的输出与发射端 S_1 处于共振(与驻波共振意义不同)状态。记下此时信号发生器的输出频率 f。

3. 驻波调节及测量,由远而近移动 S_2,以改变 L,初步观察 L 改变时,DA - 16 晶体管毫伏表上的电压变化情况(根据电压变化的情况适当选择晶体管毫伏表量程),会发现毫伏表的指针随着 L 的移动而左右偏转,其中每一次向右偏转到极大值都对应与各个声压波腹位置。

4. 观察到规律后，将 S_2 移动到 100 mm 附近，由远而近移动 S_2，依次记下各个毫伏表极大值时对应的 L（S_2 的位置坐标），测 12 组数据。

5. 记录室温 t。

二、相位比较法测量声速

1. 按图 4-7-1 连接线路（接示波器），将 YB1631 型功率函数发生器接功率输出端，置 $30V_{P\text{-}P}$ 档，频率选择置 10～100 kHz 挡。

2. 开启电源，将幅度调到最高，在 40 kHz 附近缓慢调节频率（本实验换能器的固有频率在 40 kHz 左右），使示波器的图形在 y 方向达到最大值（中间可能需要调节示波器输入通道的“Volts/Div”值），此状态即为信号发生器的输出与发射端 S_1 处于共振状态。记下此时信号发生器的输出频率 f。

3. 驻波调节及测量，移动 S_2，以改变 L，初步观察 L 改变时，示波器中图形的变化情况，图形的变化应该是直线—椭圆—直线—椭圆—直线的过程，如图 4-7-4 所示。

4. 观察到规律后，将 S_2 移动到 100 cm 附近，由远而近移动 S_2，依次记下各个直线时对应的 L（S_2 的位置坐标），测 12 组数据。

5. 记录室温 t。

自拟数据表格，根据实验测得的数据，用逐差法计算出声速 v，同时用下列校正公式算出 $v_{校}$：

$$v_{校}=331.25\sqrt{\left(1+\frac{t_{干}}{t_0}\right)\left(1+\frac{0.3192p_{w}}{p}\right)}\approx 331.45\sqrt{1+\frac{t}{t_0}} \tag{4-7-7}$$

式中，$t_0=273.15$ ℃；p_w 为水蒸气气压，p 为大气压，mmHg（1 mmHg＝133.322 Pa）。最后算出百分误差$=\frac{|\Delta v|}{v_{校}}\times 100\%$。

注意事项

1. 实验时应先找到换能器的谐振频率。

2. 在实验过程中要保持激振电压值不变。

3. 在转动手轮使接收端 S_2 移动的过程中，应朝一个方向转动，中途不可以后退，以免螺旋空程带来实验误差。

思考题

1. 什么是逐差法？它的优点是什么？

2. 为什么换能器要在谐振频率条件下进行声速测定？

3. 为什么在实验过程中改变 L 时，压电晶体换能器 S_1 和 S_2 的表面应保持互相平行？

不平行会产生什么问题?

4. 是否可以用此方法测定超声波在其他媒质(如液体和固体)中的传播速度?

Experiment 4.7 Ultrasonic Sound Velocity Measurement

A sound wave is a mechanical wave propagating in an elastic medium. Because the vibration direction is the same as the propagation direction, the sound wave is a longitudinal wave. The sound wave with the 20 Hz~20 kHz vibration frequency can be detected by human's ears, and it is referred to as an audible sound wave. The sound wave with frequency over 20 kHz is called an ultrasonic wave.

The measurement of the characteristics of an acoustic wave (such as frequency, wavelength, wave velocity, attenuation of sound pressure, phase, etc.) plays an important role in the application of acoustic technology. The measurement of sound velocity, in particular, is of great significance for the acoustic positioning, flaw detection and ranging.

In this experiment, the piezoelectric transducer is used to measure the velocity of ultrasonic waves in air. The method of measuring sound velocity can be divided into two categories. The first method is based on $v=L/t$. With the measured distance L and time interval t, the sound velocity v can be calculated. For the second method, the frequency f and wavelength λ need to be measured and then the sound velocity v can be calculated according to $v=\lambda f$. The resonance interferometry and phase comparison used in this experiment belong to the second category of methods.

Experimental Objectives

1. To measure the velocity of sound in air by means of resonance interferometry and phase comparison.
2. To learn to deal with the data with the successive difference method.
3. To understand the relationship between sound velocity and gas parameters.

Experimental Principle

Because the ultrasonic wave has the advantage of long wave length and easy directional emission, it is convenient to measure the sound velocity in the ultrasonic wave band. The transmitting and receiving of ultrasonic waves are realized by the mutual conversion between electromagnetic vibration and mechanical vibration.

The relationship between the propagation velocity v, frequency f and wavelength λ can be summarized as

$$v=\lambda f \tag{4-7-1}$$

Accordingly, the sound velocity v can be calculated with the measurement of the acoustic frequency f and the wavelength λ. Given that the acoustic frequency f can be measured directly by using the frequency meter, the main task of this experiment is to measure the acoustic wavelength λ.

1. Ultrasonic Acquisition and Piezoelectric Transducer

Piezoelectric ceramic transducer is composed of piezoelectric ceramic sheets and two kinds of metal. Piezoelectric ceramic sheets (such as bariumtitanate, lead zirconate titanate, etc.) are made of a piezoelectric material with a polycrystalline structure.

The sinusoidal AC signal can be used to extend the longitudinal length of the piezoelectric material into an ultrasonic wave source, and the sound pressure can also be converted into a voltage signal.

2. Resonance Interference Method

The experimental device is shown in Fig. 4-7-1, where S_1 and S_2 are the piezoelectric ceramic transducers. As the source of the sound wave, S_1 is excited by the electric signal output of low frequency signal generator with adjustable oscillation frequency. S_2 is an ultrasonic receiver. When the sound wave is transmitted to the receiving surface of S_2, the piezoelectric effect can be changed into a voltage signal. At the same time, S_2 also reflects parts of the sound wave. When the surfaces of S_1 and S_2 are parallel to each other, the sound waves emitted by S_1 and reflected by S_2 will interfere with each other in the two planes, resulting in the phenomenon of standing wave resonance.

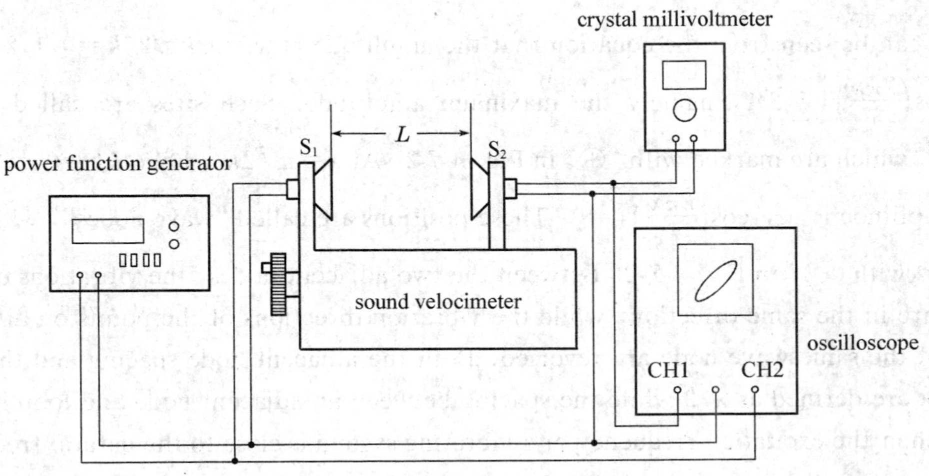

Fig. 4-7-1 Schematic diagram of the experiment device

When the two sequences of waves with the same amplitude, frequency and direction of vibration yet opposite propagation direction are superimposed together, the standing wave can be generated. Suppose two sequences of acoustic waves with the same amplitude and the same frequency propagate in the opposite directions along the x axis. The timing starts when the displacement of the two waves is the same at each point, and the origin o of the x-axis is selected at a certain maximum displacement, then the wave equations of the two columns can be written as

$$\begin{cases} y_1 = A\cos 2\pi\left(ft - \dfrac{x}{\lambda}\right) \\ y_2 = A\cos 2\pi\left(ft + \dfrac{x}{\lambda}\right) \end{cases} \tag{4-7-2}$$

The superposition of the two waves is

$$y = y_1 + y_2 = 2A\cos\left(\frac{2\pi x}{\lambda}\right)\cos 2\pi ft \tag{4-7-3}$$

The equation represents a standing wave. Fig. 4-7-2 depicts the waveform of the standing wave at different times.

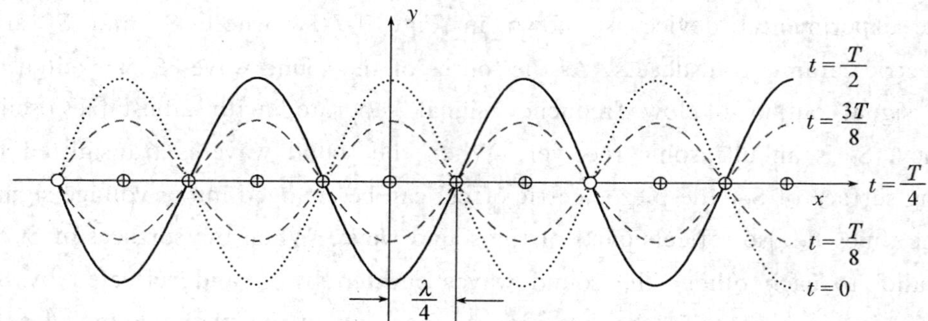

Fig. 4-7-2 The waveform of a standing wave at different time

It can be seen from the equation that the amplitude at $x = \pm k\lambda/2 (k = 0, 1, 2, \cdots)$ is $\left|2A\cos\left(\frac{2\pi x}{\lambda}\right)\right| = 2A$, namely the maximum amplitude. Such sites are called "wave loops", which are marked with "⊕" in Fig. 4-7-2. At $x = \pm(2k+1)\lambda/4 (k = 0, 1, 2, \cdots)$, the amplitude is $\left|2A\cos\left(\frac{2\pi x}{\lambda}\right)\right| = 0$. These positions are called "wave nodes", which are marked with "○" in Fig. 4-7-2. Between the two adjacent nodes, the vibrations of each point are in the same direction, while the vibration directions of the points on different sides of the same wave node are reversed. Both the adjacent node spacing and the loop spacing are defined as $\lambda/2$ while the spacing between an adjacent node and loop is $\lambda/4$.

When the excitation frequency of a vibrating system is close to the natural frequency of the system, the amplitude of the system can reach its maximum. Such a state is called resonance. Standing wave field can be regarded as a vibration system. When the excitation frequency is close to the natural frequency of the anchor, the standing wave resonance occurs. When the standing wave system deviates from the resonance state, stable standing waves cannot form. As shown in Fig. 4-7-1, S_1 is the free end and its end must be a wave loop; S_2 is a fixed end and its end face must be wave node. Therefore, the spacing between S_1 and L, S_2 must satisfy the following relation

$$L=n\frac{\lambda}{2}+\Delta,\ n=0,1,2,\cdots,\ \Delta<\lambda \tag{4-7-4}$$

Stable standing wave can be formed between S_1 and S_2, namely the occurrence of standing wave resonance. At this moment, S_2 must be in the position of the wave node with the minimum amplitude yet the maximum deformation and sound pressure. The voltage signal U produced by the piezoelectric effect also has the largest amplitude. If L does not satisfy Equation(4-7-4), the amplitude of U will become relatively small. The relationship between U and L is shown in Fig. 4-7-3.

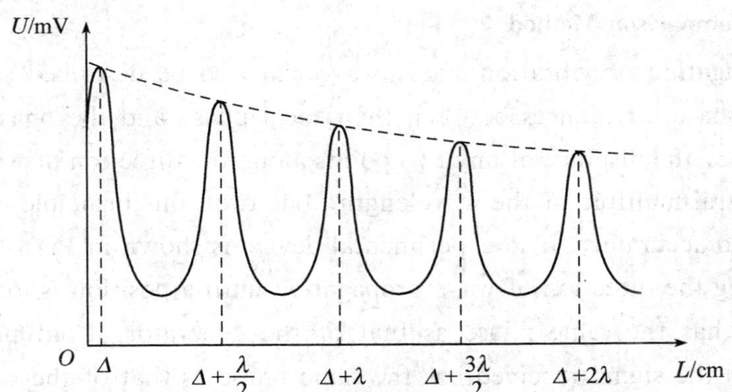

Fig. 4-7-3 Schematic diagram of the relationship between the voltage generated at the end of S_2 and the spacing between S_1 and S_2

According to the above graph, the distance between the adjacent maxima is $\lambda/2$, and the maximum amplitude decreases with the increase of the distance. Thus, once the positions of the corresponding maxima of the receiver S_2 are measured, the wavelength λ can be calculated.

For example, first measure the position of the 20 maxim, then calculate the distance between each of the $10\frac{\lambda}{2}$,

$$\Delta L_{11-1}=L_{11}-L_1=10\frac{\lambda}{2}$$

$$\Delta L_{12-2}=L_{12}-L_2=10\frac{\lambda}{2}$$

$$\cdots$$

$$\Delta L_{20-10}=L_{20}-L_{10}=10\frac{\lambda}{2}$$

Summing up both sides of above equations to get the following:

$$\sum_{i=1}^{10}\Delta L_{(10+i)-i}=100\frac{\lambda}{2},\qquad \lambda=\frac{1}{50}\left(\sum_{i=1}^{10}\Delta L_{(10+i)-i}\right)$$

After obtaining the frequency value f of the ultrasonic signal with the low frequency signal generator or the frequency meter, the velocity can be calculated with the following formula

$$v=\frac{1}{50}\left[\sum_{i=1}^{10}\Delta L_{(10+i)-i}\right]f \tag{4-7-5}$$

If 12 successive maxima are measured, Equation (4-7-5) can be rewritten as

$$v=\frac{1}{18}\left[\sum_{i=1}^{6}\Delta L_{(6+i)-i}\right]f \tag{4-7-6}$$

3. Phase Comparison Method

As a propagation of vibration state, wave can also be described as the spread of phase. If the phase differences between their local phase and the phase of the source remain the same, the distance of any two points along the direction of wave propagation will be an integral multiple of the wavelength. Based on this principle, the wavelength can be measured accurately. The experimental device is shown in Fig. 4-7-1. Move the receiver S_2 along the direction of wave propagation until a position is located where the received signal has the same phase as that of the generator. Continue to move the receiver S_2 until the signal received has the same phase as that of the transmitter for a second time. Consequently, the distance the receiver S_2 has moved is just equal to the wavelength of the ultrasonic wave.

The determination of phase difference can be achieved by using Lissajous figures. Given that the inputs of the two oscilloscopes are two signals with the same frequency signal, the Lissajous figure for the two waves is a stable elliptic. When the phase difference between two signals is 0 or π, the ellipse becomes a straight line.

The S_2 signal is directly connected to the Y input of the oscilloscope, and the sine signal driving S_1 is simultaneously connected to the X input of the oscilloscope. Adjust the oscilloscope's scan mode to "X-Y", and the Lissajous figure of S_1 and S_2 signal synthesis will be displayed on the oscilloscope screen. If the composite figure of S_2 moving at a certain position is a straight line, it means that the S_1 and S_2 vibrations in this position are either in phase or inverted. Move S_2 gradually and a gradual change in the pattern will be shown as in Fig. 4-7-4. When the graph becomes a straight line exactly the same as the initial line, the S_1 and S_2 signals are in the same phase again and the distance of S_2 movement is just λ.

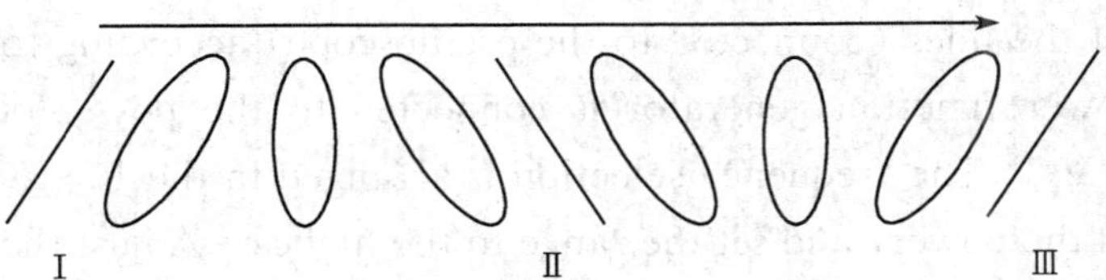

Fig. 4-7-4 Lissajous figures in phase comparison method

Experimental Instruments

Sound velocity measuring instrument, Power function signal generator, Crystal voltmeter and Oscilloscope.

Experimental Contents and Steps

1. Resonance Interferometry

(1) Connect the wire to the crystal millivoltmeter as shown in Fig. 4-7-1. Connect the YB1631 type power function generator to S_1 to power output set in the 30 V_{P-P}. The frequency is selected at the 10~100 kHz block.

(2) Turn on the power, and set the range to the highest level. Adjust the frequency in the vicinity of 40 kHz (the natural frequency of the transducer is about 40 kHz) gradually for the maximum value in the millivolt meter (the range of the millivolt meter may require some intermediate adjustment). Such a state means that the output of the signal generator and the transmitter S_1 are in a resonance state (different from the standing wave resonance). Record the output frequency of the signal generator as f.

(3) Adjust and measure the standing wave. Move the mobile S_2 from far to near to change L and observe how the voltage of the DA-16 transistor changes with L (select the range of the transistor based on the change of voltage). It is found that the pointer of the millivolt meter will deflect between left and right as L moves and the maximum value of each right deflection corresponds to the position of each sound pressure antinode.

(4) Move S_2 to the vicinity of 100 mm. Move S_2 from far to near and write down the L values (coordinate position of S_2) corresponding to the maximum values in the millivolt meter in turn. Measure and write down 12 sets of data.

(5) Measure and take down the room temperature t.

2. Phase Comparison Method for Measuring Sound Velocity

(1) Connect the lines (connected to the oscilloscope) according to Fig. 4-7-1. The YB1631 type power function generator is connected to the power output end and is placed in the 30 $V_{p\text{-}p}$. The frequency selection is arranged in the 10~100 kHz block.

(2) Turn on the power, and set the range to the highest. Adjust the frequency in the vicinity of 40 kHz (the natural frequency of the transducer is about 40 kHz) gradually for the maximum value of the oscilloscope graphics in the Y direction (It may be necessary to adjust the "Volts/Div" value of the oscilloscope input channel during this process). Such a state means that the output of the signal generator and the transmitter S_1 are in a resonance state (different from the standing wave resonance). Record the output frequency of the signal generator as f.

(3) Adjust and measure the standing wave. Move the mobile S_2 to change L. Observe how the figure in the oscilloscope changes when L changes. The graph should evolve with a cycle of line—ellipse—line—ellipse—line, as shown in Fig. 4-7-4.

(4) Move S_2 to the vicinity of 100 cm. Move S_2 from far to near and write down the L values (coordinate position of S_2) corresponding to each line graph in turn. Measure and write down 12 sets of data.

(5) Measure and Record the room temperature t.

Make a data table. Based on the measured data, calculate the sound velocity v by applying successive difference methods. Meanwhile, v_c can be calculated by using the following correction formula:

$$v_c = 331.25\sqrt{1+\frac{t}{t_0}} \tag{4-7-6}$$

Where $t_0 = 273.15$ ℃ and the percentage error of v can be calculated as:

$$E = \frac{|\Delta v|}{v_c} \times 100\%$$

Precautions

1. The resonance frequency of the transducer should be detected in advance.

2. It is required that the excitation voltage value should remain unchanged in the whole course of the experiment.

3. Turn the handwheel in one direction when turning it to move the receiving end of the S_2. Don't turn the handwheel backwards to avoid the experimental errors caused by spiraling empty space.

Questions

1. What is the successive difference method? What are the possible advantages of this method?

2. Why should the transducers be measured under the condition of resonance frequency?

3. Why should the S_1 and S_2 surfaces of the piezoelectric transducer be kept parallel to each other when the L is changed during the experiment? What kind of problems will be aroused if they are not parallel?

4. Can this method be applied to measure the propagation velocity of ultrasound in other media (such as liquids and solids)?

4.8　弦线上驻波实验

波动的研究几乎出现在物理学的每一领域中。如果在空间某处发生的扰动，以一定的速度由近及远向四处传播，则称这种传播着的扰动为波。机械扰动在介质内的传播形成机械波，电磁扰动在真空中或介质内的传播形成电磁波。不同性质的扰动的传播机制不同，但由此形成的波却有共同的规律性。本实验介绍一种利用驻波原理测量弦线上横波波长的方法，并验证弦线上横波波长与弦线中张力的平方根成正比，而与波源的振动频率成反比的关系。

实验目的

1. 观察在弦线上形成的驻波。
2. 验证频率不变时，横波的波长与弦线中张力的关系。
3. 验证张力不变时，横波的波长与波源振动频率的关系。

实验原理

在一根拉紧的弦线上，其中张力为 T、线密度为 μ，则沿弦线传播的横波应满足下述运动方程

$$\frac{\partial^2 y}{\partial t^2}=\frac{T}{\mu}\cdot\frac{\partial^2 y}{\partial x^2} \tag{4-8-1}$$

式中,x 为波在传播方向(与弦线平行)的位置坐标;y 为振动位移。

将式(4-8-1)与波动方程$\frac{\partial^2 y}{\partial t^2}=v^2\frac{\partial^2 y}{\partial x^2}$相比较,即可得到波的传播速度

$$v=\sqrt{\frac{T}{\mu}} \tag{4-8-2}$$

若波源的振动频率为 f,横波波长为 λ,由于 $v=f\lambda$,代入式(4-8-2)可得波长与张力及线密度之间的关系

$$\lambda=\frac{1}{f}\sqrt{\frac{T}{\mu}} \tag{4-8-3}$$

为了用实验证明式(4-8-3)成立,将该式两边取对数,得

$$\lg\lambda=\frac{1}{2}\lg T-\frac{1}{2}\lg\mu-\lg f$$

若固定频率 f 及线密度 μ,而改变张力 T,并测出各相应波长 λ,作 $\lg\lambda\sim\lg T$ 图,若得一直线,计算其斜率值(如为 1/2),则证明了 $\lambda\propto T^{1/2}$ 的关系成立。同理,固定线密度 μ 及张力 T,改变振动频率 f,测出各相应波长 λ,作 $\lg\lambda\sim\lg f$ 图,如得一斜率为 -1 的直线,就可验证 $\lambda\propto f^{-1}$ 的关系。

弦线上的波长可利用驻波测量。两列振幅和频率相同、振动方向一致的波在同一直线上相向传播,当它们相遇时,会发生干涉而形成驻波。弦线上的波形成驻波时,弦线上出现一些静止点,即驻波的波节。相邻的两波节间距为半个波长。

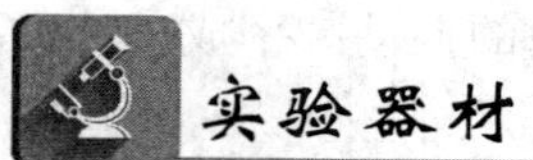

实验器材

FD-SWE-Ⅱ弦线上驻波实验仪、平台、固定滑轮、可调滑轮、砝码盘、米尺、弦线、砝码、电子天平。

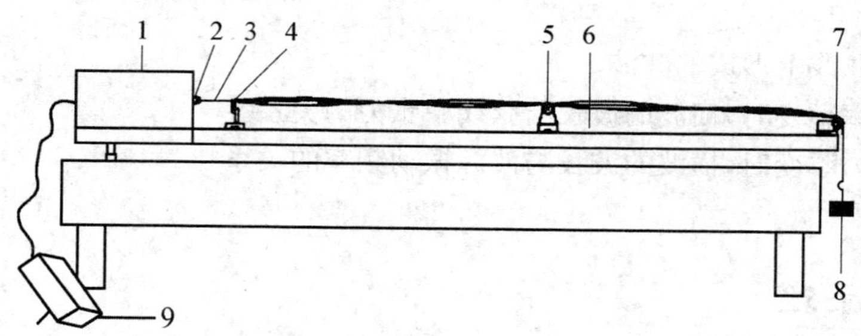

图 4-8-1　实验装置图

1—机械振动源;2—振动簧片;3—弦线;4—可动刀口支架;5—可动滑轮支架;6—标尺;7—固定滑轮;8—砝码与砝码盘;9—变压器。

实验装置如图 4-8-1 所示，金属弦线的一端系在能水平方向振动的可调频率数显机械振动源的振簧片上，频率变化范围从 0～200 Hz 连续可调，频率最小变化量为 0.01 Hz，弦线一端通过滑轮悬挂一砝码盘；在振动装置（振动簧片）的附近有可动刀口，在实验装置上还有一个可沿弦线方向左右移动并撑住弦线的动滑轮。这两个滑轮固定在实验平台上，其产生的摩擦力很小，可以忽略不计。若弦线下端所悬挂的砝码（包含砝码盘）的质量为 m，张力 $T=mg$。当波源振动时，即在弦线上形成向右传播的横波；当此波传播到可动滑轮与弦线相切点时，由于弦线在该点受滑轮两臂阻挡而不能振动，故波在切点被反射形成向左传播的反射波。这种传播方向相反的两列波叠加即形成驻波。当振动端簧片与弦线固定点至可动滑轮与弦线切点的长度 L 等于半波长的整数倍时，即

$$L=n\frac{\lambda}{2} \tag{4-8-4}$$

可得到振幅较大而稳定的驻波，振动簧片与弦线固定点为近似波节，弦线与动滑轮相切点为波节。其中 n 为任意正整数。利用式(4-8-4)即可测量弦线上横波波长。

实验时，将变压器（黑色壳）输入插头与 220 V 交流电源接通，输出端（5 芯航空线）与主机上的航空座相连接。打开数显振动源面板上的电源开关①（振动源面板如图 4-8-2 所示）。面板上数码管显示振动源振动频率×××.××Hz。根据需要按频率调节②中▲（增加频率）或▼（减小频率）键，改变振动源的振动频率，调节面板上幅度调节旋钮 4，使振动源有振动输出；当不需要振动源振动时，可按面板上复位键③复位，数码管显示全部清零。

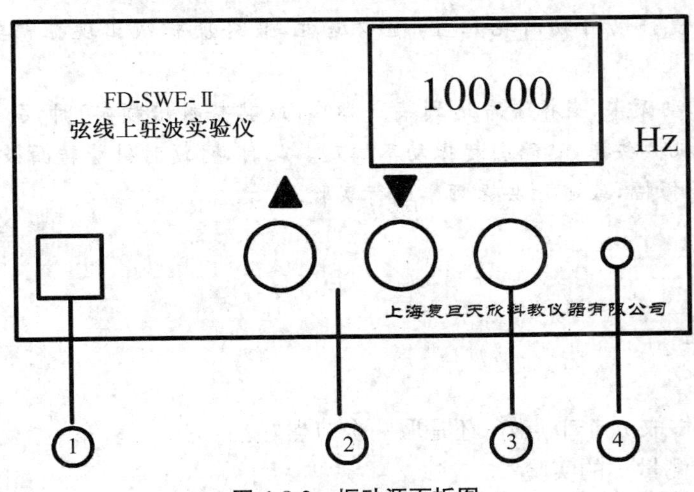

图 4-8-2　振动源面板图

1—电源开关；2—频率调节；3—复位键；4—幅度调节。

实验内容及步骤

一、验证横波的波长与弦线中的张力的关系

固定一个波源振动的频率,在砝码盘上添加不同质量的砝码,以改变同一弦上的张力。每改变一次张力(即增加一次砝码),均要左右移动可动滑轮的位置,使弦线出现振幅较大而稳定的驻波。用实验平台上的标尺测量 L 值,即可根据式(4-8-4)计算出波长 λ。作 $\lg \lambda \sim \lg T$ 图,求其斜率。

二、验证横波的波长与波源振动频率的关系

在砝码盘上放上一定质量的砝码,以固定弦线上所受的张力,改变波源振动的频率,用驻波法测量各相应的波长,作 $\lg \lambda \sim \lg f$ 图,求其斜率。最后,得出弦线上波传播的规律结论。

注意事项

1. 实验中,若要准确求得驻波的波长,必须在弦线上调出振幅较大且稳定的驻波。在固定频率和张力的条件下,可沿弦线方向左、右移动可动滑轮的位置,找出"近似驻波状态",然后缓慢移动可动滑轮位置,逐步逼近,最终使弦线出现振幅较大且稳定的驻波;

2. 调节振动频率,当振簧片达到某一频率(或其整数倍频率)时,引起整个振动源(包括弦线)的机械共振,从而引起振动不稳定。此时,可逆时针旋转面板上的输出信号幅度旋钮,减小振幅,或避开共振频率进行实验。

思考题

1. 测量 L 时,取 n 个节点好,还是取一个节点好?
2. 如何设计测量 f 的实验?
3. 弦线的质量及伸长对实验有何影响?
4. 弦线的粗细和弹性对实验各有什么影响,应如何选择?

Experiment 4.8 Investigating the Standing Wave on a String

Volatility is a most frequently studied topic in each field of physics. If a perturbation occurs somewhere in space and propagate from far and near at a certain speed, such a perturbation is known as a wave. The mechanical disturbance propagates in the medium to form a mechanical wave while the electromagnetic disturbance propagates in a vacuum or in the medium to form an electromagnetic wave. The propagation mechanisms of various disturbances may differ, but common regularities can be found from their waves formed. This section introduces a method to measure the wavelength of the transverse wave on the chord by using the standing wave principle. With this method, this experiment aims to verify if the wavelength of the transverse wave on the string is proportional to the square root of the tension in the chord and inversely proportional to the vibration frequency of the wave source.

Experimental Objectives

1. To observe the standing wave formed on the string.
2. To verify the relationship between the wavelength and the tension on the string with the constant frequency.
3. To find and verify the relationship between the wavelength of the shear wave and the vibration frequency of the wave source when the tension is constant.

Experimental Principle

For a tightened string, set the tension of the string and its line density as T and μ respectively, and the transverse wave propagating along the string should satisfy the following formula:

$$\frac{\partial^2 y}{\partial t^2}=\frac{T}{\mu}\cdot\frac{\partial^2 y}{\partial x^2} \tag{4-8-1}$$

where x is the position of the wave in the propagation direction (parallel to the chord), and y is the vibration displacement. The propagation velocity of the wave can be written as

$$v=\sqrt{\frac{T}{\mu}} \tag{4-8-2}$$

If the vibration frequency of the wave source is f, the wavelength of the wave on the string is λ, the relationship between the wavelength and the tension, the linear density can be obtained by substituting $v = f\lambda$ into Equation (4-8-2), resulting the following:

$$\lambda=\frac{1}{f}\sqrt{\frac{T}{\mu}} \tag{4-8-3}$$

In order to verify Equation (4-8-3), take the logarithm of both sides of the formula and then get the following result:

$$\lg \lambda=\frac{1}{2}\lg T-\frac{1}{2}\lg \mu-\lg f \tag{4-8-4}$$

Suppose that the frequency f and the linear density μ are kept unchanged, yet the tension T changes. Then measure the corresponding wavelength λ so as to construct the $\lg \lambda$-$\lg T$ curve. If a straight line is reached, calculate the slope value (such as 1/2), and thus relationship of $\lambda \propto \sqrt{T}$ can be verified. Similarly, the relationship of $\lambda \propto 1/f$ can be established with the fixed frequency, the fixed tension T and the changing vibration frequency f. After measuring the corresponding wavelength λ, the $\lg \lambda$-$\lg f$ curve can be plotted. If the slope of the straight line is -1, the relationship of $\lambda \propto f^{-1}$ can be verified.

The wavelengths on the string can be measured by using standing waves. Two trains of waves with the same amplitude and frequency and the same vibration direction are counter-propagating in the same straight line. When the two waves meet and interfere with each other, the standing waves form. Some static points will emerge on the string, and these points are called nodes. The adjacent two-node spacing is half a wavelength.

Experimental Apparatus

The apparatus includes: a FD-SWE-string standing wave current tester, a platform, a fixed pulley, an adjustable pulley, a weight plate, a meter stick, string, a weight and an electronic balance.

As illustrated in Fig. 4-8-1, one end of the metal string is attached to the vibrating reed with mechanical vibration source, which can be modulated in the horizontal direction with the adjustable frequency digital display. With the frequency range of 0～200 Hz, this mechanical vibrating reed is continuously adjustable and the minimum frequency is 0.01 Hz. One end of the string is hung with a weight plate by a fixed pulley and a movable knife edge is near the vibrating device (vibrating reed). Besides, there is a live pulley which can move along the string and support the chord. The two pulleys are fixed on the experimental platform and thus the friction generated is very small and negligible. Suppose the mass of the weight (including the weight plate) hung at the lower end of the string as m, the tension can be calculated as $T=mg$. When the wave source vibrates, the transverse wave propagating to the right is formed on the string. When the wave propagates to the tangent point of the movable pulley and the string, where the string cannot vibrate due to the pulley block, it is reflected at the tangent point and forms a reflection wave propagating to the left. The superposition of two trains of waves with opposite propagation direction forms a standing wave. Set the distance from the fixed point of vibrating reed and the string to the tangent point of the live pulley and the string as L. When the length of L is an integer multiple of the half wavelength, their relation can be summarized as follows:

$$L=n\frac{\lambda}{2} \tag{4-8-5}$$

The stability of the standing wave can be obtained according to the above equation. Additionally, the fixed point of vibration reed and chord is identified as the approximate node, and the tangent point of chord and movable pulley is set as the node. Here n is any positive integer, and the shear wavelength on the string can be measured by using the Equation (4-8-5).

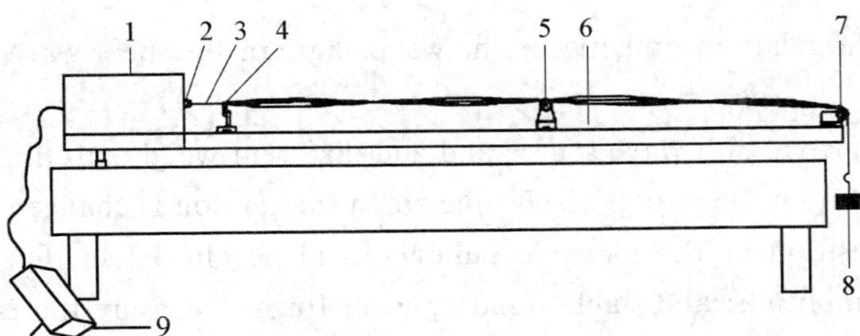

Fig. 4-8-1 Experimental device diagram

1—mechanical vibration source; 2—vibration reed; 3—string; 4—movable blade holder; 5—movable pulley holder; 6—scale plate; 7—fixed pulley; 8—weight and weight plate; 9—transformer.

In the experiment, the transformer (black shell) input plug is connected to the 220 V AC power and the output (five-core line) is connected to the host on the air seat. The power switch on the digital vibration source panel (shown in Fig. 4-8-2) should be turned on. The panel digital tube should display a vibration source frequency of ×××.×× Hz. Notably, the vibrating frequency of the source needs to be adjusted by selecting the ▲ (increasing frequency) or the ▼ (decreasing frequency) key according to the frequency adjustment. Additionally, the vibration knob on the panel should be adjusted to enable the vibration source to have a vibration output. When the vibration source is not required, the reset button on the panel should be pressed and the digital tube display should be cleared.

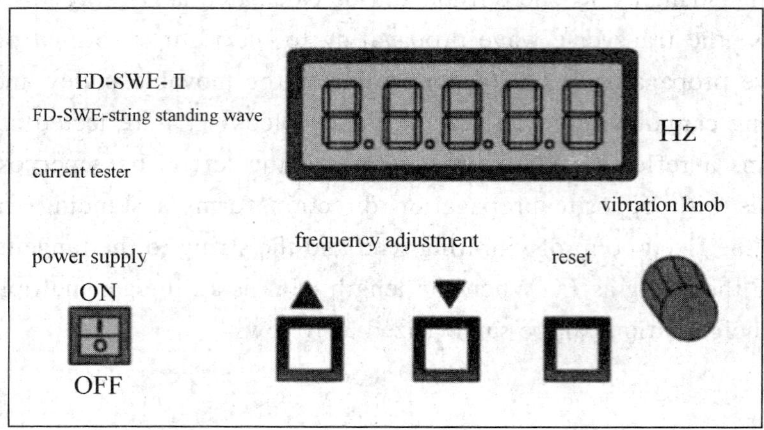

Fig. 4-8-2 Vibration source panel

Experimental Procedure

1. Verify the relationship between the wavelength of the shear wave and the tension in the string.

Fix the frequency of a wave source and add different weights to the weight plate to change the tension on the string. Each time when the tension is changed (increasing the weight), the position of the movable pulley should be altered so that the string can generate large amplitudes and stable standing wave forms. Measure the L value with the experimental platform scale and apply Equation (4-8-5) to calculate the wavelength λ. After drawing the lg λ-lg T graph, its slope can be identified.

2. Verify the relationship between the wavelength of the shear wave and the vibration frequency of the wave source.

Place a weight with a certain mass on the plate so as to fix the tension on the string. Change the frequency of the wave source vibration and measure the corresponding wavelength by using the standing wave method. After creating the lg λ-lg f graph, its slope can be identified. Finally, summarize the law of the wave propagation on the string based on all the collected data.

Precautions

1. The stable standing wave with large amplitudes is a must to accurately determine the wavelength of the standing wave. Under the fixed frequency and tension, change the position of movable pulley towards left or right along the chord line so as to identify the "approximate standing wave state". Then move the movable pulley very carefully until the stable standing wave with large amplitudes is be formed.

2. Adjust the vibration frequency. The vibrating reed with certain frequency (or integer multiple frequency) will cause the mechanical resonance of the whole vibration source (including the string), resulting in vibration instability. To solve the problem, turn the output signal amplitude knob on the panel counterclockwise so as to reduce the amplitude or avoid the resonance frequency.

1. How many nodes are necessary for a better measurement of L? And why?
2. How to design a measurement for f?
3. How will the mass and elongation of the string affect the experiment result?
4. What are the possible effects of string thickness and string elasticity on the experiment and how to make a choice?

4.9　薄透镜焦距的测量

光学仪器种类繁多，而透镜是光学仪器中最基本的元件。反映透镜特性的一个重要特点是焦距。在不同的使用场合，由于使用目的的不同，需要选择不同焦距的透镜或透镜组。要测定透镜的焦距，常用的方法有平面镜法和物距像距法。对于凸透镜还可用移动透镜二次成像法（又称共轭法），应用这种方法，只需要测定透镜本身的位移，测法简便，测量的准确度较高。为了能正确地使用光学仪器，必须掌握透镜成像的规律，学会光路的调节技术和焦距的测量方法。

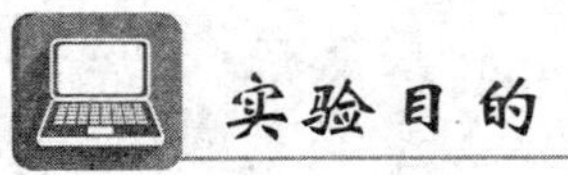

1. 学习测量薄透镜焦距的几种方法。
2. 掌握简单光路的分析和调整方法。
3. 了解透镜成像原理，观察透镜成像的像差。

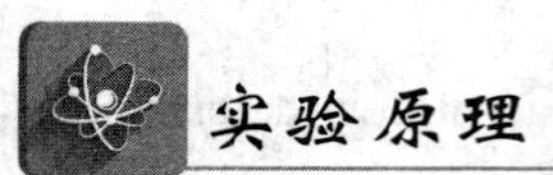

实验原理

一、薄透镜成像公式

透镜分为凸透镜和凹透镜两大类。凸透镜(也称为正透镜或会聚透镜)对光线起会聚作用,当一束平行于透镜主光轴的光线通过透镜后,将会聚于主光轴上。会聚点 F 称为该透镜的焦点,透镜光心 O 到焦点 F 的距离称为焦距 f,如图 4-9-1(a)所示。焦距越短,会聚本领越大。凹透镜(也称负透镜或发散透镜)对光线起发散作用,即一束平行于透镜主光轴的光线通过透镜后将散开。把发散光的延长线与主光轴的交点 F 称为该透镜的焦点,透镜光心 O 到焦点 F 的距离称为它的焦距 f,如图 4-9-1(b)所示。焦距越短,发散本领越大。

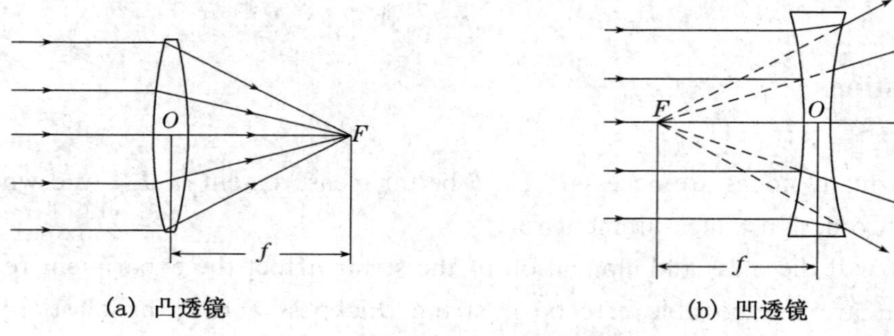

(a) 凸透镜　　(b) 凹透镜

图 4-9-1　薄透镜的焦点和焦距

当透镜中心厚度 d 与其焦距 f 相比为甚小时,这种透镜称为薄透镜。如图 4-9-2 所示,在近轴光线的条件下,薄透镜(包括凸透镜和凹透镜)成像的规律可表示为

$$\frac{1}{u}+\frac{1}{v}=\frac{1}{f} \tag{4-9-1}$$

式中,u 表示物距;v 为像距;f 为透镜的焦距。u、v 和 f 均从透镜的光心 O 点算起。物距 u 恒取正值,像距 v 的正负由像的实虚来确定。实像时,v 为正;虚像时,v 为负。凸透镜的 f 取正值;凹透镜的 f 取负值。

为了便于计算透镜的焦距 f,式(4-9-1)可改写为

$$f=\frac{uv}{u+v} \tag{4-9-2}$$

只要测得物距 u 和像距 v,便可算出透镜的焦距 f。

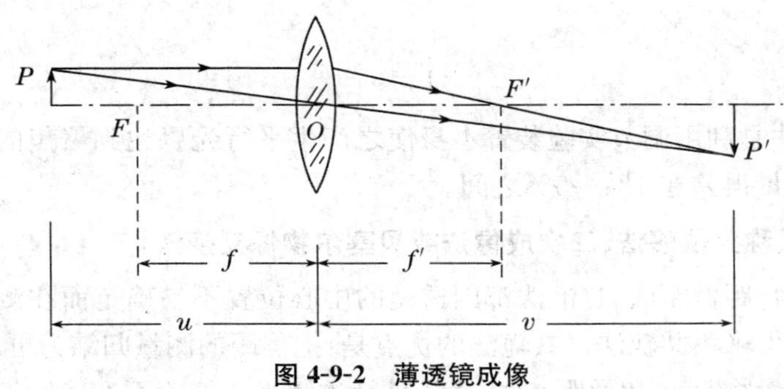

图 4-9-2　薄透镜成像

二、凸透镜焦距的测定

1. 粗略估测法

以太阳光或较远的灯光为光源，用凸透镜将其发出的光线聚成一光点（或像），此时，$u\to\infty$，$v\approx f$，即该点（或像）可认为是焦点，而光点到透镜中心（光心）的距离，即为凸透镜的焦距。此法的测量误差约在 10%。由于这种方法误差较大，大都用在实验前作粗略估计，如挑选透镜等。

2. 物距像距法

物体发出的光线，经过凸透镜折射后将成像在另一侧。在实验中分别测出物距 u 和像距 v，将其代入式(4-9-2)中即可求出该透镜的焦距 f。

3. 自准法

当光点（物）处在凸透镜的焦点平面上时，它发出的光线通过透镜后将为一束平行光。若用与主光轴垂直的平面镜将此平行光反射回去，反射光再次通过透镜后仍会聚于透镜的焦平面上，其会聚点将在光点相对于光轴的对称位置上。

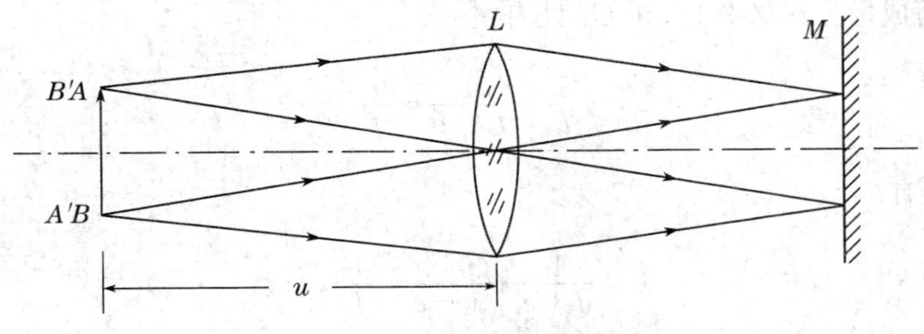

图 4-9-3　凸透镜自准法成像

如图 4-9-3 所示，在待测透镜 L 的一侧放置被光源照明的“1”字形物屏 AB，在另一侧放一与主光轴垂直的平面反射镜 M，移动透镜（或物屏），当物屏 AB 正好位于凸透镜之前的焦平面时，物屏 AB 上任一点发出的光线经透镜折射后，将变为平行光线，然后被平面反射镜反射回来。再经透镜折射后，仍会聚在它的焦平面上，即原物屏平面上，形成一个与原物大小相等方向相反的倒立实像 $A'B'$。此时物屏到透镜之间的距离，就是待测透镜的焦

距，即

$$f=u \tag{4-9-3}$$

由于这个方法是利用调节实验装置本身使之产生平行光以达到聚焦的目的，所以称之为自准法，该法测量误差在 1%～5%之间。

4. 共轭法(又称为位移法、二次成像法或贝塞尔物像交换法)

粗略估测法、物距像距法、自准法都因透镜的中心位置不易确定而在测量中引进误差，为避免这一缺点，可采用共轭法。共轭法的优点是：把焦距的测量归结为可以精确测定的量的测量，避免在测量物距 u 和像距 v 时，由于估计透镜光心位置不准确所带来的误差(因为在一般情况下，透镜的光心并不跟它的对称中心重合)。

如图 4-9-4 所示，取物屏和像屏之间的距离 L 大于等于 4 倍焦距($L \geqslant 4f$)，且保持不变，沿光轴方向移动透镜，当它在 O_1 处时，屏上将出现一个放大的清晰的像(设此时物距为 u，像距为 v)；当它在 O_2 处(设 O_1、O_2 之间的距离为 e)时，在屏上又得到一个缩小的清晰的像。

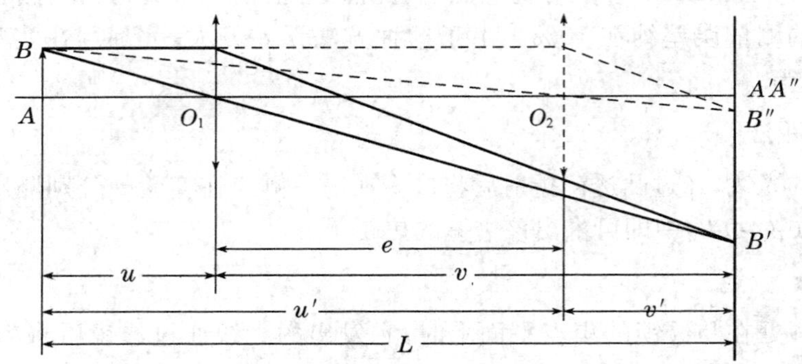

图 4-9-4　共轭法测凸透镜的焦距

按照透镜成像公式(4-9-1)，在 O_1 处有

$$\frac{1}{u}+\frac{1}{L-u}=\frac{1}{f} \tag{4-9-4}$$

在 O_2 处有

$$\frac{1}{u+e}+\frac{1}{v-e}=\frac{1}{f} \tag{4-9-5}$$

因式(4-9-4)和式(4-9-5)等号右边相等，而 $v=L-u$，故可解得

$$u=\frac{L-e}{2} \tag{4-9-6}$$

将式(4-9-6)代入式(4-9-4)，得

$$\frac{2}{L-e}+\frac{2}{L+e}=\frac{1}{f} \tag{4-9-7}$$

即

$$f=\frac{L^2-e^2}{4L} \tag{4-9-8}$$

可见，只要在光具座上确定物屏、像屏以及透镜二次成像时其滑座边缘所在位置，就可较准确地求出焦距 f。这种方法无须考虑透镜本身的厚度，测量精确度可达到1%。

三、凹透镜焦距的测定

凹透镜是发散透镜，不能直接成像。所以要测量凹透镜的焦距，必须借助于一凸透镜。具体的方法有以下两种。

1. 成像法(又称为物距像距法)

如图4-9-5所示，先使物 AB 经凸透镜 L_1 后形成一倒立缩小的实像 $A'B'$，然后在 L_1 和 $A'B'$ 之间放入待测凹透镜 L_2，如果 $u_2<f$，就能使虚物 $A'B'$ 产生一实像 $A''B''$。分别测出 L_2 到 $A'B'$ 和 $A''B''$ 之间距离 u_2、v_2，根据式(4-9-8)即可求出 L_2 的像焦距 f。

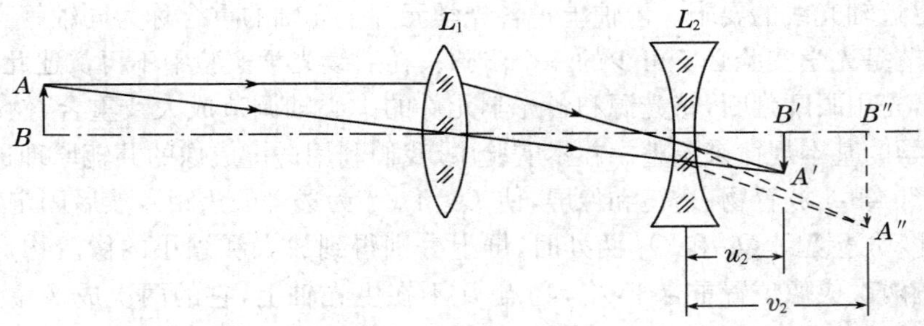

图4-9-5 凹透镜成像法

2. 自准法

如图4-9-6所示，在光路共轴的条件下，先去掉凹透镜 L_2，移动凸透镜 L_1，使物屏上物 AB 发出的光经凸透镜 L_1 成缩小的实像 $A'B'$，然后放置并移动凹透镜 L_2，当 $O_2B'=f$ 时，虚物 $A'B'$ 就在物屏上得到一个与其大小相等的倒立实像。由光的可逆性原理可知，由 L_2 射向平面镜 M 的光线是平行光线，点 B' 是凹透镜 L_2 的焦点。记录凹透镜 L_2 和实像 $A'B'$ 的位置，可直接测出 f。

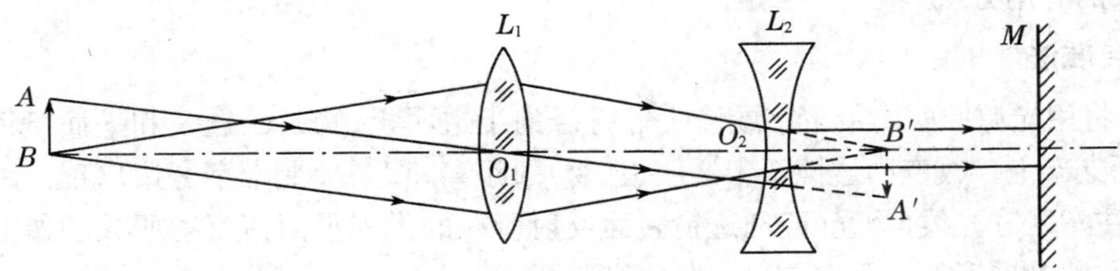

图4-9-6 凹透镜自准法

实验器材

光具座,凸透镜,凹透镜,光源,物屏,平面反射镜,水平尺等

实验内容及步骤

一、光具座上各光学元件同轴等高的调节

薄透镜成像公式(4-9-1)仅在近轴光线的条件下才能成立。对于一个透镜的装置,应使发光点处于该透镜的主光轴上,并在透镜前适当位置上加一光栏,挡住边缘光线,使入射光线与主光轴的夹角很小,对于由 n 个透镜等元件组成的光路,应使各光学元件的主光轴重合,才能满足近轴光线的要求。习惯上把各光学元件主光轴的重合称为同轴等高。显然,同轴等高的调节是光学实验必不可少的一个步骤。在后续光学实验中不再赘述此要求。

调节时,先用眼睛判断,将光源和各光学元件的中心轴调节成大致重合,然后借助仪器或者应用光学的基本规律来调整。在本实验中,我们利用透镜成像的共轭原理进行调整。

(1) 按图 4-9-4 放置物、透镜和像屏,使 $L>4f$(f 为透镜的焦距),然后固定物和像屏。

(2) 当移动透镜到 O_1 和 O_2 两处时,屏上分别得到放大和缩小的像。物点 A 处在主光轴上,它的两次成像位置重合于 A',物点 B 不在主光轴上,它的两次成像位置 B',B''分开。当 B 点在主光轴上方时,放大的像点 B'在缩小的像点B''的下方。反之,则表示 B 点在主光轴的下方。调节物点的高低,使经过透镜两次成像的位置重合,即达到了同轴等高。

(3) 若固定物点 A,调节透镜的高度,也可出现(2)中所述的现象。根据观察到的透镜两次成像的位置关系,判断透镜中心是偏高还是偏低。最后将系统调成同轴等高。

二、凸透镜焦距的测量

测量之前,将待测透镜安装好,以实验室中远处的窗子、室内的物品等作为物,经过透镜折射后成像在屏上。测出透镜至屏的距离,即为透镜焦距的近似值。这是一种能迅速提供大致结果的有用方法。

1. 自准法

(1) 将用光源照明的带十字或箭头等符号的半透明板(物)、凸透镜和平面镜依次装在光具座的支架上。改变凸透镜至半透明板(物)的距离,直至板上十字旁边出现清晰的十字像为止[注意区分光线(物光)经凸透镜表面反射所在的像和平面镜反射所成的像],测出此时的物距,即为透镜的焦距。画出此时的光路图。

(2) 在实际测量时,由于对成像清晰程度的判断总不免有一定的误差,故常采用左右逼近法读数,先使透镜由左向右移动,当像刚清晰时停止移动,记下透镜位置的读数,再使透镜自右向左移动,在像刚清晰时又可读得一数,取这两次读数的平均值作为成像清晰时凸透镜的位置。重复以上测量步骤5次，求其平均值和平均误差。

(3) 固定凸透镜，然后改变平面镜和凸透镜之间的距离，观察成像有无变化，并加以解释。

(4) 稍微改变平面镜的法线和光轴的相对位置，例如使平面镜上下倾斜或左右偏转，观察像与物相对位置的偏移和平面镜转角变化之间有何关系。画出光路图并加以分析。

2. 物距像距法

(1) 在物距 $u>2f$ 和 $2f>u>f$ 的范围内，各取两个 u 值，又取 $u=2f$，用左右逼近读数法分别测出相应的像距。按式(4-9-2)算出焦距 f。过程中应同时观察像的特点(如大小，取向等)分别画出光路图，并作出说明。

(2) 取 $u<f$，观察能否用屏得到实像？应当怎样观察才能看到物像？试画出光路图并加以说明。

(3) 将以上所得数据和观察到的现象进行比较，列表说明物距 $u=\infty$、$u=2f$、$2f>u>f$，$u=f$ 和 $u<f$ 时所对应的像距 v 和成像特征。

3. 共轭法

(1) 按图4-9-4将被光源照明的刻有十字的板、透镜和像屏装在光具座支架上。取板和像屏的间距 $L>4f$(f 为透镜的焦距)。

(2) 移动透镜，当像屏上出现清晰的放大像和缩小像时，记录透镜所在位置 O_1、O_2 的读数(用左右逼近法读数)。测出 O_1、O_2 的距离 e。由式(4-9-8)算出透镜的焦距。

(3) 多次改变板和像屏的距离 L，测出相应的 e。对于每一组 L、e，分别算出焦距 f，然后求其平均值和不确定度。

注意：间距 L 不要取得太大，否则，将使一个像缩得很小，以致难以确定凸透镜在哪一个位置上时成像最清晰。

三、凹透镜焦距的测量(选做)

在凹透镜焦距测量中，需要两个透镜共轴，首先将物点A调到凸透镜的主光轴上。然后增加凹透镜(凹透镜支座需采用二维可调节支座，以便于左右调节)，同样根据轴上物点的像总在轴上的道理，调节直至凹透镜中心在凸透镜主光轴上。

1. 成像法测凹透镜焦距

(1) 如图4-9-5所示，调节各元件共轴后，暂不放入凹透镜，并使物屏和像屏距离略大于 $4f$。移动凸透镜 L_1，使像屏上出现清晰的、倒立的、大小适中的实像 $A'B'$，记下 $A'B'$ 所在位置的读数。

(2) 保持凸透镜 L_1 的位置不变，将凹透镜 L_2 放入 L_1 与像屏之间，移动像屏，使屏上重新得到清晰、放大、倒立实像 $A''B''$，记录 $A''B''$ 所在位置的读数。

(3) 采用左右逼近法记录凹透镜 L_2 所在位置的读数，算出物距 u_2 和像距 v_2，代入式(4-9-2)求出焦距 f。

(4) 改变凹透镜位置，重复测3次，求焦距 f 的平均值及其不确定度。

2. 自准法测凹透镜焦距

(1) 如图 4-9-6 所示,调节各元件共轴后,暂不放入凹透镜,取物屏与凸透镜的距离约等于 $2f$。

(2) 移动像屏,使像屏上出现清晰的、倒立的、缩小的实像 $A'B'$,采用左右逼近测读法测定像屏的位置,记下像屏位置的读数。

(3) 保持凸透镜 L_1 的位置不变,将凹透镜 L_2 取代像屏,平面镜紧贴近凹透镜,向凸透镜方向移动凹透镜 L_2 和平面镜,在物屏上得到一个与物大小相等的倒立实像,采用左右逼近测读法测定凹透镜的位置,记录凹透镜 L_2 的位置读数。

(4) 改变凸透镜位置,要求重复 3 次,求出焦距 f 的平均值及其不确定度。

注意事项

1. 使用光学元器件要注意问题。例如,光学器件的镜面不要用手触及,光学器件易碎,要轻拿轻放,用完后光学器件要规整、整齐,放回原处等。

2. 以"1"字屏中叉丝为物体中心,以其清晰成像确定光学元件所处位置。建议将"1"字屏倒立,只观察叉丝到"1"字顶部成的像。

3. 多次测量时,可以采取左右逼近的读数方法。

4. 凹透镜需采用二维可调节支座,以便于左右调节,保证其透镜中心在物与凸透镜确定的光轴上。

1. 透镜成像公式$\frac{1}{u}+\frac{1}{v}=\frac{1}{f}$成立的条件是什么?为什么要调节光学系统共轴?
2. 自准法测凸透镜焦距需要满足什么条件?成像特点是什么?
3. 在实际测量时,为什么常常采取左右逼近法读数?如何进行操作?

Experiment 4.9 Measurement of Focal Lengths of Thin Lenses

There are various kinds of optical instruments. Lens is the most basic element in optical instruments and focal length is a most important characteristic of the lens. Lenses or lens groups with different focal lengths are needed for different occasions or different purposes. The common methods to measure the focal length of lens are the plane lens method and the object distance image distance method. For a convex lens, the moving lens secondary imaging method (also known as the conjugate method) can also be used. This method has the advantage of convenience and accuracy, and its application only needs to measure the displacement of the lens. For the correct operation of the optical instruments, the operator must have a good understanding of the law of lens imaging, and learn how to adjust the optical instruments and how to measure the focal length.

Experimental Objectives

1. To learn the methods of measuring the focal length of a thin lens.
2. To master the analysis and adjustment method for a simple optical path.
3. To understand the principle of lens imaging and to observe the aberration of lens imaging.

Experimental Principle—Measurement of Focal Length of Convex Lens

1. Rough Estimation Method

Take the sunlight or the distant light as the source and gather the emitting light into a light point (or image) by using the convex lens. Under this condition, the object distance is $u \to \infty$, and $v=f$, that is, the point (or image) can be regarded as the focus, and the distance from the light point to the lens center (optical center) is the focal length of the convex lens. Given that the measurement error of this method is around 10%, rough estimations, like how to select the lens, are necessary before the experiment.

2. Object Distance Image Distance Method

The light emitting from an object can be refracted by a convex lens and be imaged on the other side. Measure the object distance u and image distance v respectively, and substitute them into the corresponding formula to calculate the focal length f of the lens.

3. Auto Collimation Method

When the light point (object) is on the focal plane of the convex lens, its emitting light will be a beam of parallel light after passing through the lens. If the plane mirror perpendicular to the main optical axis is used to reflect the parallel light back, the reflected light will still converge on the focal plane of the lens after passing through the lens again. Its convergence point will be at the symmetrical position of the light point relative to the optical axis.

As shown in Fig. 4-9-1, place a "1" shaped object screen AB illuminated by the light source on one side of the lens L to be tested, put a plane mirror M perpendicular to the main optical axis on the other side, and then move the lens (or object screen). When the object screen AB is just located on the focal plane in front of the convex lens, the emitting light at any point of the object screen AB will become a parallel light after being refracted by the lens. Then the parallel light will be reflected back by the plane mirror.

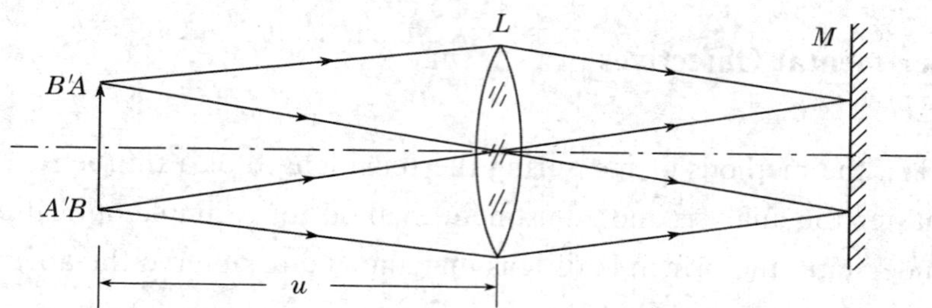

Fig. 4-9-1　Convex lens imaging by autocollimation

After being refracted by the lens, the parallel light will still converge on its focal plane, namely the original screen plane, and form an inverted real image $A'B'$ with the same size and opposite direction as the original. In this way, the distance between the object screen and the lens is the focal length of the lens to be measured, i.e.

$$f=u \tag{4-9-1}$$

This method is called the auto collimation method.

The application of this method can produce parallel light to realize light focus through adjusting the experimental device, thus this method is also called auto collimation method. The measurement error of this method is around 1%~5%.

4. Conjugate Method (also known as displacement method, secondary imaging method, or Bessel object image exchange method)

Due to the difficulty of determining the center position of the lens, the methods of rough estimation, object distance, image distance and auto collimation may bring errors in the measurement. The conjugate method can be adopted to overcome this weakness. By using the conjugate method, the focal length is converted to a length unit that can be precisely measured. It can avoid the errors caused by the inaccurate estimation of the optical center position of lens in the measurement of object distance u and image distance v, because the optical center of the lens does not coincide with its symmetrical center.

As shown in Fig. 4-9-2, the distance L between the object screen and the image screen is more than or equal to four times the focal length ($4f$) and remains unchanged. Then, move the lens along the optical axis direction. When the lens arrives at O_1, an enlarged clear image will be shown on the screen (set the object distance as u and the image distance as v); when the lens arrives at O_2 (set the distance between O_1 and O_2 as e), a reduced clear image will be displayed on the screen.

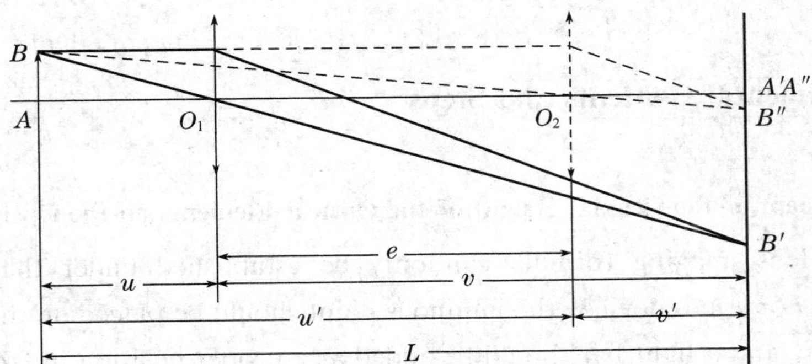

Fig. 4-9-2 Measurement of focal length of convex lens by conjugate method

According to the lens imaging formula, the equation at O_1 can be summarized as

$$\frac{1}{u}+\frac{1}{L-u}=\frac{1}{f} \tag{4-9-2}$$

The Equation at O_2 can be summarized as

$$\frac{1}{u+e}+\frac{1}{v-e}=\frac{1}{f} \tag{4-9-3}$$

The right sides of Equations (4-9-2) and (4-9-3) are equal, and $v=L-u$, so we can get

$$u=\frac{L-e}{2} \tag{4-9-4}$$

Substituting Equation(4-9-4) into Equation(4-9-2), we can get

$$\frac{2}{L-e}+\frac{2}{L+e}=\frac{1}{f} \tag{4-9-5}$$

That is

$$f=\frac{L^2-e^2}{4L} \tag{4-9-6}$$

It can be seen that as long as the positions of the object screen, image screen and the slide edges for lens in secondary imaging are determined on the optical bench, the focal length f can be accurately calculated. The thickness of the lens itself does not need to be considered when using this method and its accuracy can reach 1%.

Experimental Instruments

Optical bench, convex lens, concave lens, light source, object screen, plane mirror and a level ruler.

Experimental Contents and Steps

1. Adjustment of the Coaxial Height of the Optical Elements on the Optical Base

The thin lens imaging formula can only be established under the condition of paraxial light. For a lens device, the luminous point should be placed on the main optical axis of the lens, and a light bar should be added at a proper position in front of the lens to block the edge light. Given the small angle between the incidental light and the mian optical axis, the main optical axis of the optical elements, including N lenses and other elements, should be coincidental so that the paraxial light can be provided.

Traditionally, the coincidence of main optical axes of optical elements is called coaxial equal height. As an essential step in optical experiments, the adjustment of a coaxial equal height will be explained in detail only in this section and will not repeated in the subsequent optical experiments.

At the initial stage, adjust the central axis of light source and optical elements to roughly coincide with eye judgment, and then continue the adjustment by using optics instruments according to the basic laws of applied optics. In this experiment, the conjugate principle of the lens imaging will be applied to make the adjustment.

(1) Place the object, lens and image screen according to Fig. 4-9-2 with $L>4f$ (f is the focal length of the lens), and then fix the object and image screen.

(2) Move the lens to O_1 and O_2 so that the enlarged and reduced images are formed on the screen respectively. Object point A is on the main optical axis and its two imaging positions coincide with A'. Object point B is not on the main optical axis, and its two imaging positions B', B'' are separated. When point B is above the main optical axis, the enlarged image point B' is below the reduced image point B''. Otherwise, point B is below the main optical axis. Adjust the height of the object point so that the positions of the two images through the lens coincide, which means the coaxial equal height has been achieved.

(3) Besides the above operation procedures, similar results could be achieved by simply adjusting the height of the lens with object point A fixed. Based on the observed position relationship between the two images of the lens, judge whether the center of the lens is on the high side or on the low side. Thereafter, adjust the device to the coaxial equal height accordingly.

2. Measurement of Focal Lengths of Convex Lenses

Before the measurement, install the lens to be measured and take the distant windows and indoor objects in the laboratory as test objects, which are refracted by the lens and imaged on the screen. Observe the images on the screen and measure the distance between the lens and the screen, namely the approximate value of the lens focal length. With such a method, approximate results can be quickly obtained.

(1) Autocollimation method

① Install the translucent plate (object) with the light-illuminated symbol of "+" or arrow, convex lens and plane mirror on the bracket of the light fixture base in turn. Adjust the distance between the convex lens and the translucent plate (object) until the clear image of "+" appears next to the "+" on the plate [note the difference between the image reflected by the convex lens surface and the image reflected by the plane mirror]. Measure the object distance, namely the focal length of the lens, and draw the light path diagram.

② Given the possible errors of judging the clarity of imaging in the actual measurement, the left-right approaching method is often used for reading.

First, move the lens from left to right, pause when the image is just clear and record the reading of the lens position. Then move the lens from right to left, and record the position value when the image is just clear. Take the average value of the two readings as the position of the convex lens with the clear image. Repeat the above measurement steps five times to obtain the average value and average error.

③ Fix the convex lens, then change the distance between the plane lens and the convex lens. Observe whether the image changes and then explain the possible reasons.

④ Slightly change the normal direction of the plane mirror and the relative position of the optical axis. For example, tilt the plane mirror up and down or deflect it left and right. Observe the relative positions offset between the image and the object and the change of the plane mirror angle, and then probe their possible relationships. Draw the light path chart and analyze it.

(2) Object distance image distance method

(1) In the range of object distance $u>2f$ and $2f>u>f$, take two independent values for u, and set $u=2f$, then measure the corresponding image distance respectively by using the left and right approaching reading method. Calculate the focal length f according to the formula. Observe the characteristics of the image (such as size and orientation), draw the light path chart and make the explanation.

② Set $u<f$ and check out whether the real image can be obtained with the screen. How to observe the object image? Draw the light path chart and explain it.

③ Compare the above data with the observed phenomena and make a data list to illustrate the corresponding image distance v and image features with the object distance of $u=\infty$, $u=2f$, $2f>u>f$, $u=f$ and $u<f$.

(3) Conjugate method

① According to Fig. 4-9-2, install the lens, image screen and board engraved with the light-illuminated word “X” on the light fixture bracket, and record the distance between the board and image screen as $L>4f$ (f is the focal length of the lens).

② Move the lens and record the readings of the lens position at O_1 and O_2 when there are clear enlarged and reduced images on the image screen (read with the left and right approaching method). Measure the distance of e between O_1 and O_2 and calculate the focal length of the lens according to the Equation (4-9-6).

③ Make several changes of the distance L between the board and the image screen and record the corresponding e. For each group of L and e, calculate the focal length f respectively and then calculate its average value and uncertainty.

Note: the space L should not be too large, or it will make an image shrink so small that it is difficult to determine the position where the clearest image through the convex lens can be identified.

Questions

1. What are the conditions for the establishment of the lens imaging formula

$$\frac{1}{u}+\frac{1}{v}=\frac{1}{f}$$

and why should the coaxial optical system be adjusted?

2. What are the requirements to measure the focal length of convex lens with the autocollimation method? What are the image features?

3. Why should the left and right approaching method be adopted for the measurement? And how to operate the measurement?

4.10 利用霍尔效应测磁场

置于磁场中的载流体,如果电流方向与磁场垂直,则在电流和磁场的方向会产生一附加的横向电场,1879 年美国霍普金斯大学研究生霍尔在研究金属导电机构时发现了这种电磁现象,故称霍尔效应。后来曾有人利用霍尔效应制成测量磁场的磁传感器,但因金属的霍尔效应太弱而未能得到实际应用。随着半导体材料和制造工艺的发展,人们又利用半导体材料制成霍尔元件,由于它的霍尔效应显著而得到实用和发展,现在广泛用于非电量电测,电动控制,电磁测量和计算装置方面。在电流体中的霍尔效应也是目前在研究中的“磁流体发电”的理论基础。近年来,霍尔效应实验不断有新发现。1980 年原西德物理学家冯·克利青(K · Von Klitzing)研究二维电子气系统的输运特性,在低温和强磁场下发现了量子霍尔效应,这是凝聚态物理领域最重要的发现之一。目前对量子霍尔效应正在进行深入研究,并取得了重要应用,例如用于确定电阻的自然基准,可以极为精确地测量光谱精细结构常数等。

在磁场，磁路等磁现象的研究和应用中，霍尔效应及其元件是不可缺少的，利用它观测磁场直观、干扰小、灵敏度高、效果明显。

实验目的

1．测量霍尔器件在螺线管不同位置处的霍尔电压。

2．描绘螺线管中的磁感应强度分布曲线。

实验原理

一、霍尔效应

霍尔效应是由于运动电荷在磁场中受到洛伦兹力作用而产生的。半导体中的电流是由载流子（电子或空穴）的定向运动形成的。如图 4-10-1 所示，将一块厚度为 d、宽度为 b、长度为 l 的半导体材料制成的霍尔器件放在磁场 B 中，设控制电流 I_c 沿 x 轴正向流过半导体，若半导体内的载流子为电子，其电荷量为 e，平均迁移速度为 v，则载流子在磁场中受到洛伦兹力为

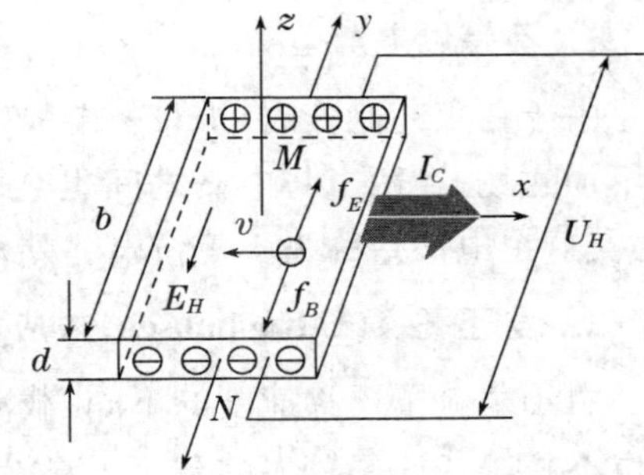

图 4-10-1　霍尔效应示意图

$$f_B = evB \tag{4-10-1}$$

在 f_B 作用下，电子流发生偏转，部分电子聚集到薄片的横向端面 N 上，而使横向端面 M 上出现了剩余正电荷。由于电荷的聚集而形成了一个横向电场 E_H，方向由 M 指向 N，电场对载流子产生一个方向和 f_B 相反的静电场力 f_E，其大小为

$$f_E = eE_H \tag{4-10-2}$$

由于受到静电场力 f_E 的作用，电荷的进一步聚集受到阻碍。开始时，f_E 较小，电子能继续聚集，电场 E_H 逐渐增强。直到作用在载流子上洛仑兹力与电场力达到动态平衡状态 $f_B = f_E$，即

$$evB = eE_H = eU_H/b \tag{4-10-3}$$

这时，MN 间的电势差即霍尔电压为

$$U_H = vbB \tag{4-10-4}$$

控制电流 I_c 与载流子电荷 e、载流子浓度 n、迁移速度 v 及霍尔片的截面积 bd 之间的

关系为

$$I = nevbd \tag{4-10-5}$$

于是,

$$U_H = \frac{IB}{ned} = R_H \frac{IB}{d} = K_H IB \tag{4-10-6}$$

式中,n 为载流子浓度;$R_H = 1/ne$ 为霍尔系数;而 $K_H = 1/ned$ 为霍尔元件的灵敏度。对于一定的霍尔元件,K_H 是常数,通常用实验方法测定。本实验中各组霍尔元件的灵敏度的值由实验室给出,单位为 mV/(mAT)。

从式(4-10-6)可以看出,如果知道霍尔元件的灵敏度,用仪器分别测出控制电流 I_c 和霍尔电压 U_H,就可以算出磁感应强度 B 的大小。这就是用霍尔效应测磁场的原理。

二、消除霍尔元件副效应对测量结果的影响

在测量霍尔电压 U_H 时,不可避免地会产生一些副效应,由于这些副效应产生的电势差叠加在霍尔电压上,形成了测量中的系统误差,这些副效应分别为:

1. 不等位电势差 U_0

由于在工艺制作时,很难将电势电极焊在同一等势面上,因此当电流流过霍尔片时,即使不加磁场,在两电势电极之间也会产生一电势差 U_0。$U_0 = IR_x$(R_x 为沿 x 轴方向的电阻)。这个电势差称为不等位电势差,显然它只与电流 I_c 方向有关,而与磁场 B 方向无关。

2. 爱廷豪森(Ettinghausen)效应 U_E

由于载流子迁移速度的不同,载流子所受的洛伦兹力的不想等。做圆周运动的轨道半径也不相等。速率较大的载流子将沿较小半径的圆轨道运动,从而导致霍尔片一面出现快载流子多,温度高;而另一面则出现慢载流子多,温度低。两端面之间由于温度差而出现温差电动势 U_E,该效应称为爱廷豪森效应。U_E 的方向与 IB 乘积成正比,方向随 I_c、B 同时换向而改变。

3. 能斯特(Nernst)效应 U_N

由于霍尔元件的电流引出线焊点的接触电阻不同,通以电流以后,发热程度不同而产生不同的焦耳热,引起两电极间的温差电动势,此电动势又产生温差电流(称为热扩散电流)Q,热电流在磁场的作用下将发生偏转,结果在 x 方向上产生附加的电势差 U_N,U_N 的方向与磁场 B 方向有关,与电流 I_c 方向无关。

4. 里吉-勒迪克(Righi-leduc)效应 U_R

上述的热扩散电流的载流子迁移速度不尽相同,在磁场作用下,除了在 x 方向产生电势差外,还将在 x 方向上引起霍尔片两侧的温差,此温差又在 x 方向上产生附加温差电动势 U_R,它只和磁场 B 方向有关,和电流 I_c 方向无关。

以上的几种副效应所产生的电势差总和,有时甚至远大于霍尔电压,形成测量中的系统误差,致使霍尔电势差难以测准。实际测量值为综合效应的结果,为了减少和消除这些效应

引起的附加电压，我们巧妙地利用这些附加电压与霍尔元件的控制电流 I_c、磁场 B 的关系，通过改变 I_c 和 B 的方向，使 U_0、U_N、U_R 从计算中消失。而 U_E 的方向始终与 U_H 的方向保持一致，在实验中无法消去，但一般 U_E 的大小比 U_H 小得多，由它带来的误差忽略不计。

于是，实际测到的电压为

$$U = U_H + U_0 + U_N + U_R$$

当控制电流 I_c 和磁场 B 的方向同时改变时，测得的电压变为

$$U' = U_H - U_0 - U_N - U_R$$

则可以消去几个副效应电势差而得到霍尔电压

$$U_H = (U + U')/2 \tag{4-10-7}$$

实验器材

霍尔效应实验仪，电位差计，稳压电源，平衡指示仪(检流计)，毫安表 2 只，电阻箱，标准电池，电池盒(1.5 V×2)导线若干。其中螺线管长度 265 mm，2400 匝。

实验内容及步骤

1. 按实验装置图(图 4-10-2)连接导线。

2. 接通励磁电流、霍尔电流的双刀双掷开关，调节螺线管的励磁电流 I_m = 550 mA，调节电阻箱使霍尔片的控制电流 I_c = 8.0 mA。

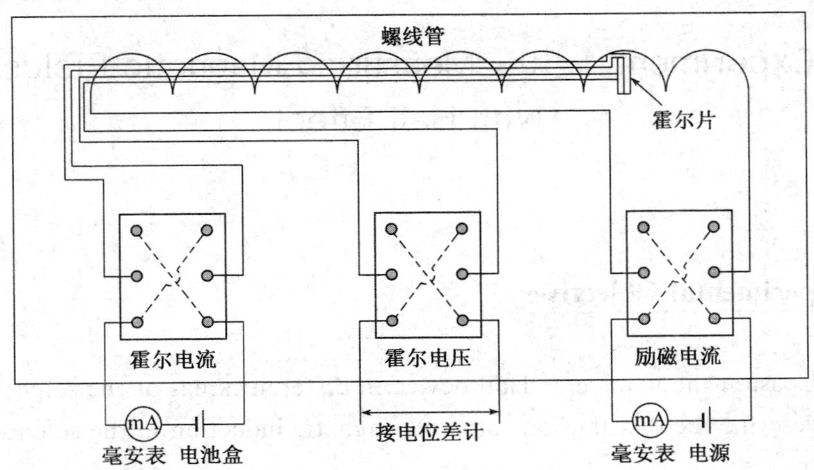

图 4-10-2　霍尔效应实验装置图

3. 使霍尔片处于螺线管中的位置 $x=0$,合上霍尔电压的双刀双掷开关,用电位差计测量该位置的电压 U(电位差计的使用方法参照物理实验的《电位差计的调节与使用》),再分别测量 $x=2,4,6,8,10,12$ 处的电压值。

4. 将励磁电流、霍尔电流的双刀双掷开关分别换向,再分别测量 $x=12,10,8,6,4,2,0$ 位置处的电压值 U'。

5. 按式(4-10-7)计算不同位置处的霍尔电压 U_H,并且按照 $B=\frac{U_H}{K_H I_c}$ 计算处在对应的磁感应强度。

6. 作出螺线管中磁感应强度的分布曲线。

7. 找出曲线中与纵坐标的交点作为螺线管中心位置磁感应强度的实验值 $B_{0实}$,并计算该位置的理论值 $B_{0理}=\mu_0 nI_m$($\mu_0=4\pi\times10^{-3}$ G·A^{-1},n 为单位长度螺线管的匝数),把实验值与理论值相比较,计算百分误差。

注意事项

1. 两块电流表千万不要接错,因为励磁电流 I_m 和控制电流 I_c 的大小相差悬殊;

2. 因实验中无法判断待测电压的正负,若遇电位差计无法测量,可把霍尔电压双刀双掷开关换向后再测。

思考题

1. 若霍尔片平面与磁场不垂直,对测量有什么影响,为什么?

2. 测量中如何消除各种副效应引起的测量误差?

Experiment 4.10 Measuring Magnetic Fields with Hall Effect

Experimental Objectives

1. To measure the voltage of Hall device in different areas of the solenoid.

2. To describe the distribution curve of magnetic induction in the solenoid.

Experimental Principle

1. Hall Effect

The Hall effect is caused by the movement of the electric charge acted by the Lorentz force in the magnetic field. The current in the semiconductor is formed by the directional motion of the charge carriers (electron or hole). Put a Hall device (thickness d, width b, length l) made of semiconductor material in the magnetic field B. The direction of magnetic field is along the z axis. Assume that the control current I_c flows forward through the semiconductor along the x axis. If the carrier in the semiconductor is an electron with the charge of e and the average migration velocity is v, the Lorentz force for the carrier in the magnetic field can be expressed as

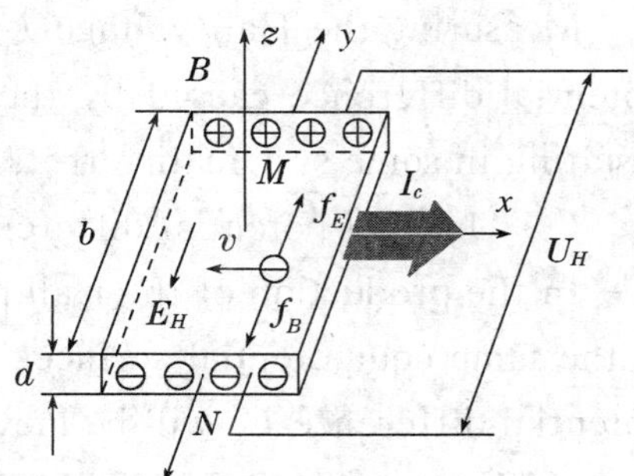

Fig. 4-10-1　Schematic diagram of Hall effect

$$f_B = evB$$

The electron flow is deflected under the effect of f_B. Some electrons gather on the transverse face N of the slice, resulting in the residual positive charge forming on the transverse face M. The agglomeration of electric charges generates a transverse electric field E_H with direction from M to N. The electric field produces an electrostatic field force f_E opposite to f_B on the carrier, and its size can be expressed as

$$f_E = eE_H$$

The further charge accumulation is hindered due to the electrostatic field force f_E. Since the initial f_E is quite small, electrons can continue to accumulate and the electric field E_H can gradually increase until the dynamic equilibrium has been achieved between the Lorentz force on carrier and the electric field force, namely $f_B = f_E$, i.e.

Then the electric potential difference between M and N, namely the Hall voltage is

$$U_H = vbB$$

The relationship formula of I_c, e, n, v and bd can be summarized as

$$I_c = nevbd$$

Thus

$$U_H = \frac{I_c B}{ned} = R_H \frac{I_c B}{d} = K_H I_c B$$

According to the above formular, suppose the sensitivity value of Hall components is a given condition, and the control current I_c and Hall voltage U_H can be measured with relative instruments, then the magnitude of magnetic induction B can be calculated. Such is the principle of magnetic field measurement with the Hall effect.

2. Eliminate the Influence of Hall Component's Side Effect on the Measurement Results

Measuring the Hall voltage U_H will inevitably bring about some side effects. The potential difference caused by the side effects will be superimposed on Hall voltage, resulting in some systematic errors of measurement.

(1) Unequal potential difference U_0

In the production of the Hall piece, it is difficult to weld all the potential electrodes at the same equipotential surface. When the current flows through the Hall piece, the potential difference U_0 will be inevitably generated between the two potential electrodes even without the magnetic field. This potential difference is called the unequal potential difference and its equation is $U_0 = I_c R_x$. It can be seen that potential difference is independent of the magnetic field B and only related to the current I_c.

(2) Ettinghausen effect U_E

The Lorentz force on the carrier varies with the carrier migration velocity. The orbital radius of different carriers in circular motion may differ and the carrier with higher velocity will move along the circular orbit with smaller radius. As a result, one Hall side with more fast-carrier has higher temperature, while the other Hall side with more slow-carrier has lower temperature. The temperature difference between the two sides generates the electromotive force U_E, and such a process is called the Ettinghausen effect. The direction of U_E is proportional to the product of IB and changes with the commutation of I_c and B.

(3) Nernst effect U_N

Because of the different contact resistance of current lead solder joint, different Joule heats caused by temperature difference eventually result in the electromotive force between the two poles. The electromotive force can generate the electromotive current (known as the diffusion current) Q. This thermal current will deflect under the effect of the magnetic field, resulting in the additional potential difference U_N in the x direction. The direction of U_N is related to the direction of the magnetic field B, and independent of the direction of current I_c.

(4) Righi-leduc effect U_R

The carrier diffusion rates of the above thermal diffusion currents are different. Under the effect of the magnetic field, besides the potential difference in the x direction, the temperature difference between the two sides of the Hall plate in the x direction will be generated. This temperature difference may also generate an additional thermo-electromotive force U_R in the x direction. U_R is related to the direction of the magnetic field B and independent of direction of current I_c.

The sum of the potential differences caused by the above-mentioned side effects are sometimes much larger than the Hall voltage, resulting in the systematic errors of measurement. It is difficult to measure the Hall potential difference accurately and the actual value is the final result of various comprehensive effects. To reduce and eliminate the additional voltage caused by these various effects, all relative factors, including the additional voltages, the control current I_c of the Hall element, the magnetic field B and their possible relationship, could be taken into consideration. For example, U_0, U_N, and U_R can be deleted from the calculation by changing the direction of I_c and B. As for U_E, its direction is always consistent with that of U_H and it cannot be totally eliminated in the experiment. Given that its value is much smaller than U_H, the errors it brings out can be negligible.

Therefore, the actual measured voltage is

$$U = U_H + U_0 + U_N + U_R$$

When changing the I_c and B, the voltage to be measured is

$$U' = U_H - U_0 - U_N - U_R$$

The final Hall voltage is

$$U_H = (U + U')/2$$

Experimental Instruments

Hall effect tester, potentiometer, regulated power supply, balance indicator, milli-ammeter (2 sets), resistance box, standard battery, battery box (1.5V × 2), wire, solenoid (265 mm, 2400). Fig. 4-10-2 is the experimental device.

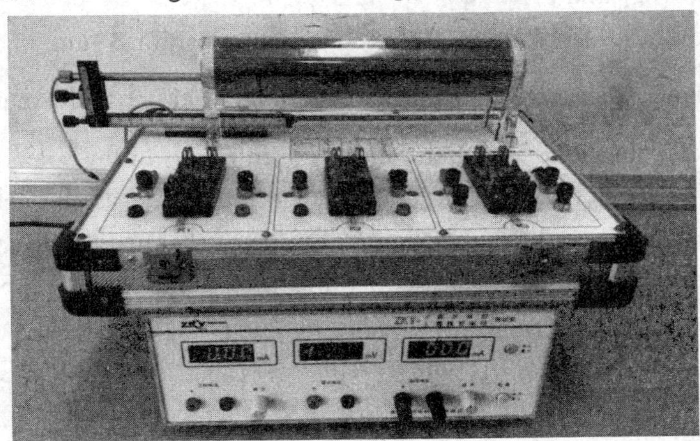

Fig. 4-10-2　The display of experimental device

Experimental Contents and Steps

1. Connect the wires according to the experimental device diagram (Fig. 4-10-3)

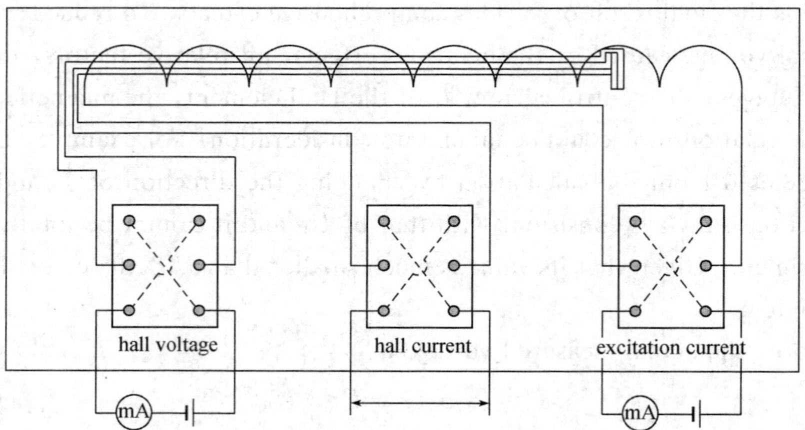

Fig. 4-10-3 Experimental device of Hall effect

2. Turn on the excitation current and the double-pole, double-throw switch of Hall current, adjust the solenoid's excitation current $I_m=550$ mA, and adjust the resistance box to set the Hall control current as $I_c=8.0$ mA.

3. Place the Hall plate in the position of the solenoid $x=0$, and turn on the double-pole and double-throw switch of Hall voltage. Measure the voltage U at this location by using the potentiometer. Then measure the voltage at the position of $x=2$, 4, 6, 8, 10 and 12 cm respectively.

4. Commutate the excitation current and the double-pole-and-double-throw switch of Hall current respectively. Then measure and record the voltage U' at the position of $x=12$, 10, 8, 6, 4,2, and 0.

5. Calculate the Hall voltage U_H at different positions, and make the further calculation for the corresponding magnetic induction based on the formula of $B=\frac{U_H}{K_H I_c}$.

6. Plot the distribution of magnetic induction intensity in the solenoid.

7. Locate the intersection of the curve and the ordinate as the center position of solenoid, measure and record the experimental value of the magnetic f induction as B_{0t}. Calculate the theoretical value $B_{0th}=\mu_0 n I_m$ and then compare the experimental value with the theoretical value. Finally, calculate the percentage error on the foundation of the comparison.

Precautions

1. It is forbidden to connect the two currents meter incorrectly, because there is a huge difference between the excitation current I_m and the control current I_c.

2. It is difficult to judge whether the voltage to be measured is positive or negative. If the potentiometer cannot be measured, the DPDT switch of Hall voltage could be commutated for a second try.

1. If the Hall plate is not perpendicular to the magnetic field, what kinds of impacts will be aroused on the measurement? And why?

2. How to eliminate the measurement errors caused by various side effects?

4.11 用复摆法测定金属环的转动惯量

实验任务

根据复摆原理测定一只匀质金属圆环(图 4-11-1)绕垂直于圆环面的对称轴的转动惯量 $J_{C实}$。

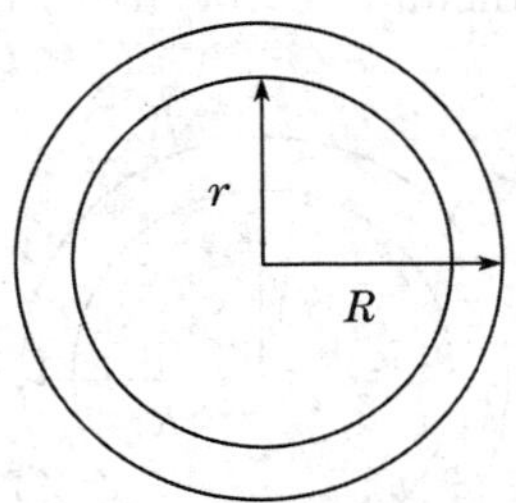

图 4-11-1　待测金属圆环

实验器材

秒表,电子天平,游标卡尺,螺旋测微器,支架。

实验步骤

1. 查阅有关参考资料,写出较详细的实验原理,测量主要公式。

2. 自行确定实验方案,说明直接测量量、测量方法、次数和步骤。

3. 根据测量数据计算出 $J_{C\text{实}}$ 和圆环绕其对称轴的转动惯量理论值 $J_{C\text{理}}$;计算百分误差,并分析产生误差的主要因素。

4. 根据实验目的及上述要求,自拟实验步骤,并设计数据记录表格,指出实验中应注意的事项。

5. 将上述 1,2,4 三项要求的内容,写成设计性实验的预备报告,在实验前五天左右交教师审阅。经认可后,才能进行实验。

6. 在规定的实验时间内,独立地完成实验内容,并进行数据处理,按实验报告要求,写出完整的报告。

Experiment 4.11 Determine the Moment of Inertia of the Metal Ring with Compound Pendulum Method

Experimental Objectives

To measure the moment of inertia for a homogeneous metal ring (Fig. 4-11-1) around the symmetry axis perpendicular to the torus based on the compound pendulum theory.

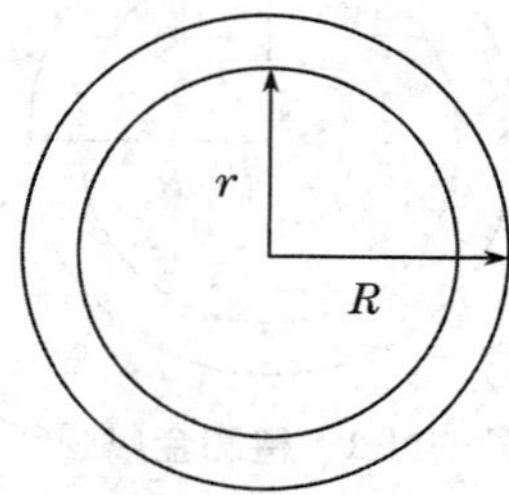

Fig. 4-11-1 Metal ring to be measured

Experimental Instruments

Stopwatch, electronic balance, vernier caliper, spiral micrometer and bracket.

Experimental Contents and Procedures

1. With reference to the relative materials, take down the experimental principles and the main formula in detail.

2. After designing the experimental program, make a clear explanation about the measurement quantities, the measuring methods, the measuring times and the measuring steps in detail.

3. Calculate the practical value of the moment of inertia of the circular ring around the symmetry axis and its relative theoretical value. Then calculate the percentage error and analyze the main factors for the errors.

4. Based on the purpose of the experiment and the above requirements, make the arrangement about the experiment steps, design the data-recording form and stress the precautions of the experiment.

5. Following the above requirements ①, ②, and ④, write a preliminary report for a experiment design and submit it for the teacher's review five days before the experiment. Conduct the experiment only after teachers' approval.

6. Conduct the experiment independently within the specified time. After the data processing, write a complete experiment report as required.

4.12 凯特摆测量重力加速度

1818年凯特提出的倒摆，经雷普索里德做了改进后，成为当时测量重力加速度 g 最精确的方法。波斯坦大地测量研究所曾用5个凯特摆用了8年时间(1896～1904年)，测得当地的重力加速度 $g=(981.274\pm0.003)\mathrm{cm/s^2}$，许多地区的 g 值都曾以此为根据。凯特摆测量重力加速度的方法不仅在科学史上有着重要的价值，而且在实验设计上亦有值得学习的技巧。

实验目的

1. 了解凯特摆的实验设计思想和技巧。

2. 掌握一种比较精确的测量重力加速度的方法。

实验原理

图 4-12-1 是复摆的示意图，设一质量为 m 的刚体，其重心 G 到转轴 O 的距离为 h，绕 O 轴的转动惯量为 I，当摆幅很小时，刚体绕 O 轴摆动的周期 T 为

$$T=2\pi\sqrt{\frac{I}{mgh}} \tag{4-12-1}$$

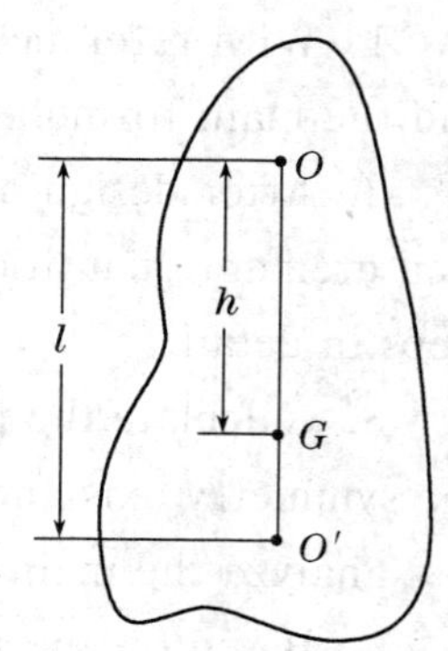

图 4-12-1 复摆示意图

式中，g 为当地的重力加速度。

设复摆绕通过重心 G 的轴的转动惯量为 I_G，当 G 轴与 O 轴平行时，有

$$I=I_G+mh^2 \tag{4-12-2}$$

代入式(4-12-1)得

$$T=2\pi\sqrt{\frac{I_G+mh^2}{mgh}} \tag{4-12-3}$$

对比单摆周期的公式 $T=2\pi\sqrt{\frac{l}{g}}$，可得

$$l=\frac{I_G+mh^2}{mh} \tag{4-12-4}$$

式中，l 为复摆的等效摆长。因此只要测出得摆周期和等效摆长便可求得重力加速度。

复摆的周期我们能测得非常精确，但利用式(4-12-4)来确定 l 是很困难的。因为重心 G 的位置不易测定，因而重心 G 到悬点 O 的距离 h 也难以精确测定。同时由于复摆不可能做成理想的、规则的形状，其密度也难绝对均匀，想精确计算 I_G 也是不可能的。利用复摆上两点的共轭性可以精确求得 l。在复摆重心 G 的两旁，总可找到两点 O 和 O'，使得该复摆以 O 悬点的摆动周期 T_1 与以 O' 为悬点的摆动周期 T_2 相同，那么可以证明 OO' 就是我们要求的等效摆长 l。

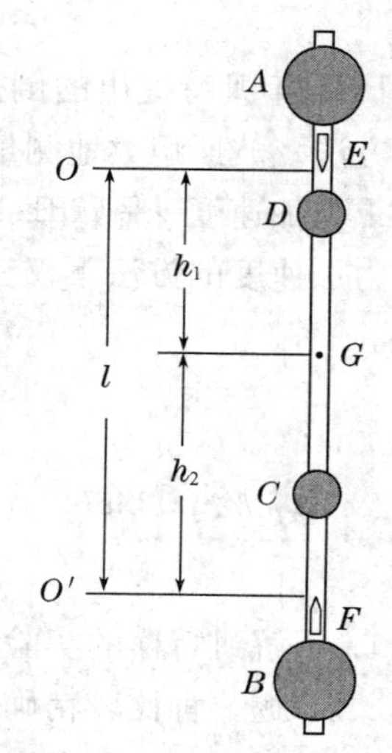

图 4-12-2 凯特摆摆杆示意图

图 4-12-2 是凯特摆摆杆的示意图。凯特摆由底座、压块、支架、V 形刀承和一根长 1 m 的金属摆杆组成。金属摆杆上嵌有两个对称的刀口 E 和 F，作悬挂之用，一对大小形状相同、但质量不同的大摆锤 A、B 分别位于摆杆的两端，另一对小摆锤 D、C 位于刀口 E 和 F 的内侧，摆锤 A、D 由金属制成，摆锤 C、B 由塑料制成。摆杆各部分处于对称状态，其目

的在于抵消实验时空气浮力的影响以及减小阻力的影响，调节刀口 E 和 F 可以改变等值单摆长 l。在实验中当两刀口位置确定后，调节摆锤 A、B、C、D 的位置，可以改变摆杆系统的质量分布。设 h_1 和 h_2 分别为悬点 O 和 O' 到摆杆体系重心的距离。当4个摆锤调节到某一合适的位置时，以 O 为悬点的摆动周期 T_1 和以 O' 为悬点的摆动周期 T_2 相等，即 $T_1 \approx T_2$。由式(4-12-3)可得：

$$T_1 = 2\pi\sqrt{\frac{I_G + mh_1^2}{mgh_1}} \tag{4-12-5}$$

$$T_2 = 2\pi\sqrt{\frac{I_G + mh_2^2}{mgh_2}} \tag{4-12-6}$$

由式(4-12-5)和式(4-12-6)消去 I_G 可得

$$\frac{4\pi^2}{g} = \frac{T_1^2 + T_2^2}{2l} + \frac{T_1^2 - T_2^2}{2(2h_1 - l)} = a + b \tag{4-12-7}$$

式中，l、T_1、T_2 都是可以精确测定的量，而 h_1 则不易测准。由此可知，a 项可以精确求得，而 b 项不易精确求得。但当 $T_1 = T_2 = T$ 且 $|2h_1 - l|$ 的值较大时，b 项的值相对 a 项是非常小的，这样 b 项的不精确对测量结果产生的影响就微乎其微。这样，重力加速度为

$$g = \frac{8\pi^2 l}{T_1^2 + T_2^2} = \frac{4\pi^2 l}{T^2} \tag{4-12-8}$$

实验器材

凯特摆、光电探头、米尺和 VAFN 多用数字测试仪。

实验内容及步骤

1. 将光电探头放在摆杆下方，调整它的位置和高度，让摆针在摆动时经过光电探测器。电信号由B插口输入到数字测试仪中，数字测试仪的功能选择旋钮放在“振动计数”挡，时标旋钮放在“0.1 ms”挡，计停开关置于“停止”，然后接通电源。

2. 让摆杆作小角度的摆动，待其摆动若干次稳定后，按下数字测试仪的“复位”按钮。此时测试仪开始自动记录一个周期的时间。显示屏左边显示摆动的次数(即周期数)，右边显示摆动数个周期的时间数值。

3. 调节4个摆锤的位置，使 T_1 与 T_2 逐渐靠近，一般粗调用大摆锤，微调用小摆锤。当 T_1 和 T_2 比较接近估算值 T 时，最好移动小塑锤，使 T_1 与 T_2 的差值小于0.001 s。

4. 当周期的调节达到要求后，将测试仪的计停开关拨到“计数”挡，测量凯特摆正、倒摆动10个周期的时间，$10T_1$ 和 $10T_2$ 各测量5次取平均值。

5. 将摆杆从刀承上取下，平放在刀口上，使其平衡，平衡点即重心 G 的所在，测出 $OO'(l)$，$GO(h_1)$或 $GO(l-h_1)$的值，代入式(4-12-7)中计算 g 值。

6. 推导不确定度公式，计算 u_g。

注意事项

刀口必须与摆垂直，两刀口必须平行和对称；刀与刀承应是线接触。

思考题

1. 凯特摆测重力加速度，在实验设计上有什么特点？避免什么量的测量？降低了哪个量的测量精度？实验上如何实现？

2. 结合误差计算，你认为影响凯特摆测量 g 精度的主要因素是什么？将所得的实验结果与当地的重力加速度的公认值相比较，能得到什么结论？若有偏差，试分析之。

3. 摆的角振幅(即摆杆的偏转角)的大小，对实验结果有无影响？能否进行理论修正？

Experiment 4.12 Measuring the Gravitational Acceleration with the Kate Pendulum

The inverted pendulum was initially invented by Kate in 1818. With Repsol's improvement, the inverted pendulum became the most accurate method to measure gravitational acceleration g. It took the Boston Institute for Geodesy 8 years to measure the local Gravitational acceleration as $g=(981.274\pm0.003)\mathrm{cm/s^2}$, which has become a widely used basis of g value in many areas. The method of measuring Gravitational acceleration with the Kate pendulum has an important scientific value, and its design can provide great insightful experimental techniques as well.

Experimental Objectives

1. To understand the principles and techniques of the experimental design.
2. To master a more accurate method of measuring Gravitational acceleration.

Experimental Principle

Fig. 4-12-1 illustrates the principle for a compound pendulum. Set the mass of a rigid body as m, the distance from the center of gravity G to the shaft O as h and the moment of inertia around the O axis as I. When the amplitude of the swing is very small, the cycle of the rigid body around the O axis T can be summarized as:

$$T=2\pi\sqrt{\frac{I}{mgh}} \tag{4-12-1}$$

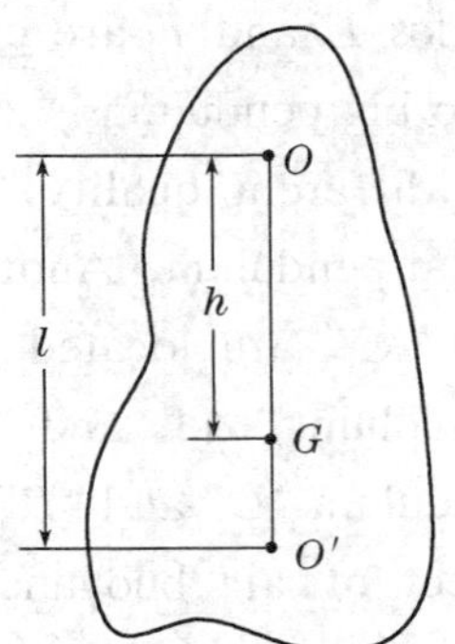

Fig. 4-12-1 Illustration of a compound pendulum

Where g is the local Gravitational acceleration.

Set the moment of inertia of pendulum through the gravity G as I_G. When the G axis and O axis are in parallel, the moment of inertia can be expressed as

$$I=I_G+mh^2 \tag{4-12-2}$$

Substituting Equation (4-12-2) into Equation (4-12-1), it is easy to get

$$T=2\pi\sqrt{\frac{I_G+mh^2}{mgh}} \tag{4-12-3}$$

Compared with the formula of the pendulum cycle $\left(T=2\pi\sqrt{\frac{l}{g}}\right)$, the following can be obtained,

$$l=\frac{I_G+mh^2}{mh} \tag{4-12-4}$$

Where l is called the equivalent length of the compound pendulum. With the equivalent pendulum length and period of the pendulum, the gravitational acceleration can be easily obtained.

We can measure the period of the pendulum precisely, but it is difficult to determine l by using Equation (4-12-4). It is partly due to the fact that the distance h from the center of gravity G to the shaft O is hard to be measured, given the difficulty of measuring the center of gravity G. Another reason might be attributed to the inaccurate measurement of I_G because of the pendulum's irregular shape and its uneven density. Yet the precise calculation of l can be obtained by taking advantage of the conjugate between two points of the pendulum. Choose the two points O and O' on both sides of the center of gravity G. If the periods of the pendulum at different hanging points O and O', namely T_1 and T_2, are the same, it can be proved that the length of line segment OO' is the equivalent pendulum length of the pendulum l.

Fig. 4-12-2 is a schematic diagram of the Kate pendulum. A Kate pendulum consists of a base, a briquette, a stent, a V-type knife bearing and a one-meter-long metal pendulum. Two symmetrical hanging blades E and F are embedded in the metal pendulum. Two big pendulums "A" and "B", with the same shape but different quality, are located at both ends of the metal pendulum. Another pair of small pendulums "D" and "C" are located medial to blades "E" and "F". Pendulum "A" and "D" are made of metal while pendulum "C" and "B" are made of plastic. To offset the effect of air buoyancy and reduce the effect of air resistance, every part of the pendulum is in a symmetrical state. Adjust blades "E" and "F" to change the equivalent pendulum length of the pendulum l. After the determination of the two blades' positions, change the mass distribution of the pendulum system by adjusting the positions of pendulum "A", "B", "C", and "D". Set the distance from hanging points O and O' to the center of gravity of the pendulum system as h_1 and h_2. Adjust the positions of the four pendulum blobs appropriately until the periods of the pendulum T_1 and T_2 based on hanging points O and O' respectively are the same, which means $T_1 \approx T_2$. T_1 and T_2 can be expressed as Equation (4-12-3).

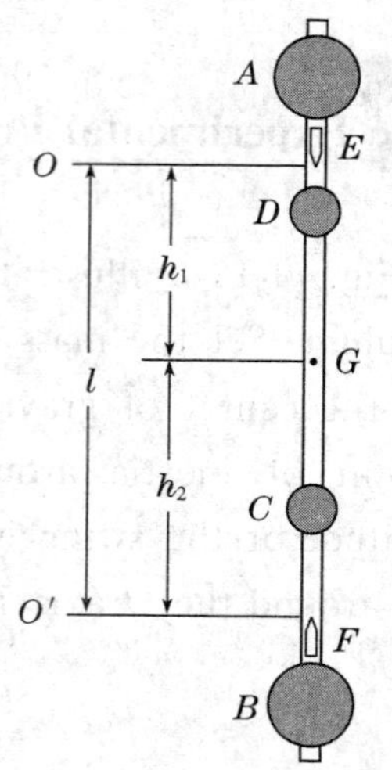

Fig. 4-12-2 Schematic diagram of the kate pendulum

$$T_1 = 2\pi\sqrt{\frac{I_G + mh_1^2}{mgh_1}} \tag{4-12-5}$$

$$T_2 = 2\pi\sqrt{\frac{I_G + mh_2^2}{mgh_2}} \tag{4-12-6}$$

Combining Equation(4-12-5) with Equation (4-12-6), we can obtain the following equation by eliminating I_G,

$$\frac{4\pi^2}{g} = \frac{T_1^2 + T_2^2}{2l} + \frac{T_1^2 - T_2^2}{2(2h_1 - l)} = a + b \tag{4-12-7}$$

Where l, T_1, and T_2 can be measured precisely while h_1 is hard to be measured. It can be judged that polynomial a can be calculated precisely while it is hard to calculate polynomial b. However, if $T_1 = T_2 = T$ and $|2h_1 - l|$ is relatively large, the value of polynomial b will be much smaller than that of polynomial a, which means that the inaccuracy value of b only has the minimal impact on the final results. In a word, the gravitational acceleration can be expressed as,

$$g = \frac{8\pi^2 l}{T_1^2 + T_2^2} = \frac{4\pi^2 l}{T^2} \tag{4-12-8}$$

Experimental Apparatus

Kate pendulum, Photoelectric probe, Meters and VAFN multi-purpose digital tester.

Experimental Contents and Steps

(1) Put the VAFN multi-purpose digital tester under the pendulum, and adjust its position and height until the pendulum needle passes through the photoelectric probe in the swing. Input an electric signal from the B port to the digital tester, and place the function selection button of the digital tester in the "Vibration Count" scale. Set the time scale knob at "0.1 ms" and the stop switch at "stop", then turn on the power.

(2) Let the pendulum swing at a small angle. After the swing is stable for several times, press the "reset" button of the digital tester. Thereafter, the multi-purpose digital tester has begun to automatically record the time of one cycle. On the left side of the display screen, the number of the pendulum swings (i. e. the number of cycles) is shown, while the time value of several swings is displayed on the right side.

(3) Adjust the positions of the four pendulums until T_1 and T_2 gradually become closer. Generally, big pendulum is used for coarse adjustment while small pendulum is adopted for fine adjustment. When T_1 and T_2 are close to the estimated value T, it is better to move the small plastic hammer so that the difference between T_1 and T_2 is less than 0.001 s.

(4) When the cycle adjustment meets the requirements, turn the stop switch of the tester to the "count" position, and record the time of Kate swing forward and backward for ten cycles. Make five measurements for 10 T_1 and 10 T_2 respectively, and then take the average value.

(5) Take down the pendulum from the knife notch and place it on the edge of the knife to make it balanced. The balance point is where the center of gravity g is. Measure the value of OO' (l), GO (h_1) or GO ($l-h_1$) and substitute them into Equation (4-12-7) to calculate the g value.

Precautions

The knife edge must be perpendicular to the pendulum, and the two knife edges must be parallel and symmetrical. The knife and the knife bearing should be in line contact.

Questions

1. What are the characteristics of the experimental design of Kate pendulum in measuring the acceleration of gravity? What quantity measurement is avoided? Which measurement accuracy is reduced? How to achieve these targets in the experiment?

2. Given the error calculation, what do you think are the main factors that impact the accuracy of measuring g with Kate pendulum? Comparing the experimental results with the accepted values of the local gravitational acceleration, what conclusions can you draw? If there is any deviation, please try to analyze it.

3. Does the angular amplitude value of the pendulum (i.e. the deflection angle of the pendulum rod) have an effect on the experimental results? Can you make a theoretical revision?

4.13 金属线膨胀系数的测定

一般物体都具有“热胀冷缩”的特性,因此,在工程结构设计,机械和仪表的制造,材料选择和加工(如焊接)中都必须考虑这一特性,否则,结构的稳定性和仪表的质量都会受到影响,甚至可能产生严重的后果。

固体受热后,在一定的温度变化范围内,一维方向上的膨胀为线膨胀。在相同条件下,不同材料的固体其线膨胀程度不同。这种差异可以用不同材料的线胀系数来反映。线胀系数是选用材料的一项重要指标,对于新材料的研制少不了对其线胀系数的测定。

测量固体的线膨胀系数,实验上归结为测量在某一温度范围内固体的相对伸长量。此相对伸长量的测量与杨氏弹性模量的测定相同,有光杠杆、测微螺旋和千分表等方法。而加热固体的办法,也有通入蒸汽法和电热法。一般认为,用电热丝通电加热,光杠杆法测量相对伸长量,是比较经济又准确可靠的方法。

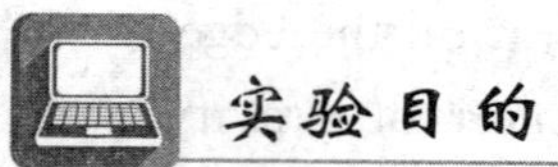

实验目的

1. 应用光杠杆测量微小伸长量。
2. 测量金属杆的线膨胀系数。

实验原理

一般固体的体积或长度，随温度的升高而膨胀，这就是固体的热膨胀。设物体的温度改变 Δt 时，其长度改变量为 ΔL，如果 Δt 足够小，则 Δt 与 ΔL 成正比，并且也与物体原长 L 成正比，因此有

$$\Delta L = \alpha L \Delta t \tag{4-13-1}$$

式中，比例系数 α 为固体的线膨胀系数，单位是℃$^{-1}$。

线膨胀系数的物理意义是温度每升高 1 ℃，固体的伸长量与它在 0 ℃时长度的比值。设在温度为 0 ℃时，固体的长度为 L_0，当温度升高为 t 时，其长度为 L，则有

$$(L-L_0)/L_0 = \alpha t$$

即

$$L = L_0(1+\alpha t) \tag{4-13-2}$$

设金属杆在温度为 t_1，t_2 时，其长度分别为 L，$L+\Delta L$，则可写出

$$L = L_0(1+\alpha t_1) \tag{4-13-3}$$

$$L+\Delta L = L_0(1+\alpha t_2) \tag{4-13-4}$$

由式(4-13-3)和式(4-13-4)消去 L_0 可得到

$$\alpha = \frac{\Delta L}{L(t_2-t_1)-\Delta L t_1} \tag{4-13-5}$$

由于伸长量 $\Delta L \ll L$，式(4-13-5)可近似为

$$\alpha = \frac{\Delta L}{L(t_2-t_1)} \tag{4-13-6}$$

由式(4-13-6)测得 L、ΔL、t_1 和 t_2，就可求得 α 的值。实际测量时，L、t_1、t_2 都比较容易测量，但 ΔL 很小，一般长度仪器不易测准。本实验中用光杠杆和望远镜标尺组来测量。

待测金属棒直立在仪器的大圆筒中，金属棒受热膨胀时，其下端位置不变，上端将升高。将光杠杆的前脚放在固定平台凹槽内，后脚放在金属棒的上顶端。由光杠杆的性质可得出

$$\Delta L = \frac{b}{2D}\Delta n$$

式中，b 为光杠杆后脚到前脚连线的距离；Δn 为金属棒膨胀前后尺读望远镜中看到的标尺移动的距离；D 为光杠杆平面镜到标尺的距离。因此金属的线膨胀系数可表示为

$$\alpha = \frac{b}{2DL(t_2-t_1)}\Delta n \tag{4-13-7}$$

实验器材

GXC 型固体线胀系数测定仪、数字温度计、卷尺、刻度尺、望远镜等。

一、线膨胀仪

线膨胀仪是采用电热法来测定金属棒的线膨胀系数的,它主要包括以下部分:给被测材料加热的加热器,安装加热器、散热罩的支架和放置光杠杆的平台。支架及平台与底座牢固地连接在一起,如图 4-13-1 所示。加热器中的加热管道上有电热丝,接通电源即可逐渐升温,并有温场均匀特点。加热管道内可放置待测材料棒和温度计。

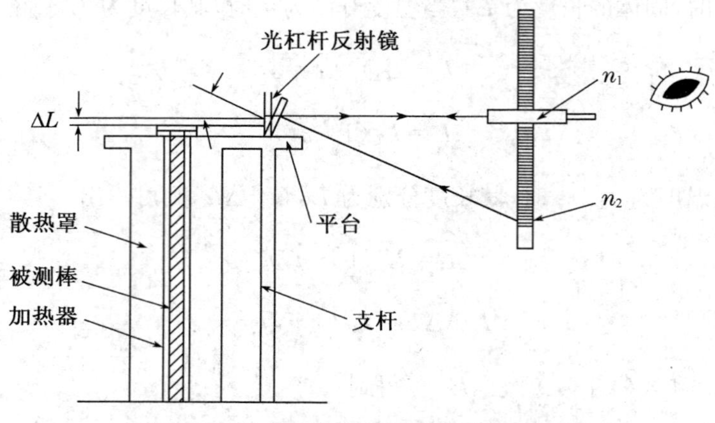

图 4-13-1　光杠杆法测线膨胀系数示意图

二、使用方法

测量铜管长度 L,然后把其慢慢放入加热管道内,直到被测棒的下端接触底面,把数字温度计的测温探头小心地放入铜管孔内;将光杠杆的前边刀口(或两前足)放在平台的凹形槽内,后脚尖立于被测杆顶端,并使光杠杆平面镜法线大致与望远镜同轴,且平行于水平底座,见装置图 4-13-2 所示。

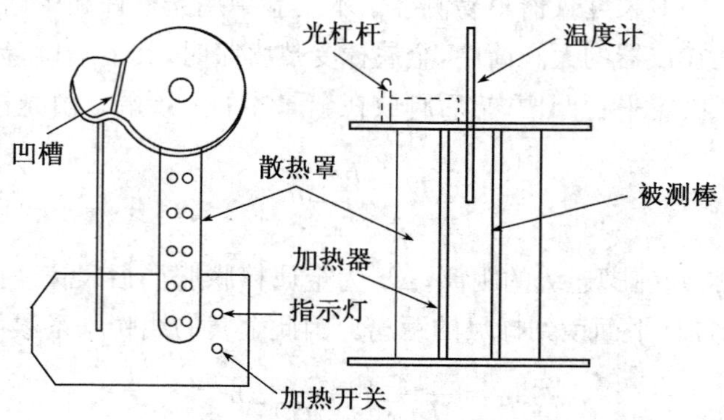

图 4-13-2　线膨胀仪结构示意图

实验内容及步骤

1. 用卷尺测量铜管长度 L，将被测棒放入线膨胀仪，数字温度计的测温探头小心地放入铜管孔内。

2. 将光杠杆的前边刀口（或两前足）放在平台的凹形槽内，后足尖立于被测杆顶端，并使光杠杆平面镜大致垂直，平面镜与望远镜等高。

3. 调整望远镜（参见“4.1 拉伸法测金属丝的杨氏弹性模量”）。

4. 打开数字温度计及线膨胀系数测定仪上的加热开关，并适当调节加热速度；记录测量数据，从 30 ℃开始，每隔 10 ℃记录标尺读数，一直到 100 ℃。关闭加热电源，等温度回到 100 ℃时开始记录降温过程数据，每隔 10 ℃记一次，直到 30 ℃。

根据测出数据，用逐差法求出 α，并和公认值比较求出百分误差（已知所测量的铜的线胀系数为 1.85×10^{-5} ℃$^{-1}$）。

注意事项

1. 在测量过程中，不能碰动线膨胀仪。光杠杆及望镜稍有移动实验得从头做起；

2. 实验中要注意观察和分析问题，尽可能减少引起测量误差的因素，想到防止的措施和改进的方法。

思考题

1. 本实验对各长度量分别用不同仪器测量，是根据什么原则考虑的？哪一个量的测量误差对结果的影响最大？

2. 两根材料相同，粗细长度不同的金属棒，在同样的温度变化范围内，它们的线膨胀系数是否相同？膨胀量是否相同？为什么？

Experiment 4.13 Measurement of Metal Linear Expansion Coefficients

"Thermal expansion and cold contraction" is a basic characteristic of general objects. In the design of the engineering structure and in the manufacture of machinery and instrumentation, this characteristic must be taken into account for the material selection and processing (such as welding). Otherwise, the stability of the structure and the quality of the instrument will be affected, resulting in some serious problems. When a solid is heated, it will expand in one dimension whin a certain temperature range, such an expansion is called linear expansion. The densities of different solid materials differ under the same conditions. These differences can be reflected by the linear expansion coefficients of different materials. Linear expansion coefficients is an important index for the materials selection, thus it is necessary to measure the linear expansion coefficient for the development of new materials. In the experimental design, to measure the linear expansion coefficient of solid is to measure the relative elongation of solid within a certain temperature range. Similar to Young's modulus of elasticity, the measurement of relative elongation can be classified into optical lever measurement, micrometer screw measurement and dial indicator measurement and etc.. The approach to heat the solid include the steam heating method and the electrothermal method. Generally, the combination of electric wire heating and measuring relative elongation with optical light lever is a more economical, accurate and reliable method to measure the relative elongation.

Experimental Objectives

1. To measure tiny elongation with the optical lever method.
2. To measure the linear expansion coefficient of the metal rod.

Experimental Principle

The volume or length of a solid generally grows as the temperature increases, which is called the thermal expansion of the solid. Set the temperature of the object changes by Δt and the length of the variable changes by ΔL, if Δt is small enough, Δt is proportional to ΔL. Additionally, it is also proportional to the original length L of the object. Then, the following equation can be summarized

$$\Delta L = \alpha L \Delta t \tag{4-13-1}$$

Here the proportion coefficient α is called the linear expansion coefficient, and its unit is $℃^{-1}$. The linear expansion coefficient physically stands for the ratio of the elongated length of the solid to its length at 0 ℃ for every 1 ℃ increase in temperature. Suppose the length of the solid is L_0 at the temperature of 0 ℃ and its length is L at the temperature of t, the following equation can be summarized

$$L = L_0(1 + \alpha t) \tag{4-13-2}$$

Based on the above equation, if the length of the rod at temperature of t_1 and t_2 is L and $L + \Delta L$ respectively. We obtain

$$L = L_0(1+\alpha t_1) \tag{4-13-3}$$

$$L+\Delta L = L_0(1+\alpha t_2) \tag{4-13-4}$$

Erasing L_0 from Equations (4-13-3) and (4-13-4), the coefficient of thermal expansion can be written as

$$\alpha = \frac{\Delta L}{L(t_2 - t_1) - \Delta L t_1} \tag{4-13-5}$$

Given that the elongation is $\Delta L \ll L$, Equation (4-13-5) can be approximated as

$$\alpha = \frac{\Delta L}{L(t_2 - t_1)} = \frac{\Delta L}{L\Delta t} \tag{4-13-6}$$

To determine the coefficient α, the value of L, ΔL, t_1 and t_2, needs to be measured. It is relatively easy to conduct the measurement for L, t_1 and t_2 in the experiment, yet ΔL is too small to be precisely measured. Consequently, the optical lever and telescope scale are to be used to measure ΔL in this experiment.

Place the metal rod to be tested in the large cylinder of the instrument.

When the metal rod is heated and expanded, the position of its lower end will remain unchanged yet its upper end will rise. Put the front foot of the optical lever in the fixed platform groove and the rear foot on the top of the metal bar. Based on the nature of the optical lever, the following equation can be summarized:

$$\Delta L = \frac{b}{2D}\Delta n$$

Where b is the distance from the rear foot of the optical lever to the front foot, Δn refers to the distance of the scale movement seen in the telescope before and after the expansion of the metal rod, and D stands for the distance from the light beam to the scale. Therefore, the linear expansion coefficient of the metal can be expressed as

$$\alpha = \frac{b}{2DL\Delta t}\Delta n \tag{4-13-7}$$

Experimental Instruments

A GXC solid linear expansion coefficient tester, a digital thermometer, a tape, a scale, a telescope and so on.

1. Linear Expansion Instrument

The linear expansion meter determines the linear expansion coefficient of the metal rod by utilizing the electrothermal method. It mainly consists of a heater, a bracket for installing heater and heat sink, and a platform to place the optical lever. The bracket and platform are securely connected to the base, as illustrated in Fig. 4-13-1. The heater in the heating pipe on the electric wire connected to the power can be gradually warmed, providing a uniform temperature field. The material rod and thermometer can be placed in the heating pipe (Fig. 4-13-2).

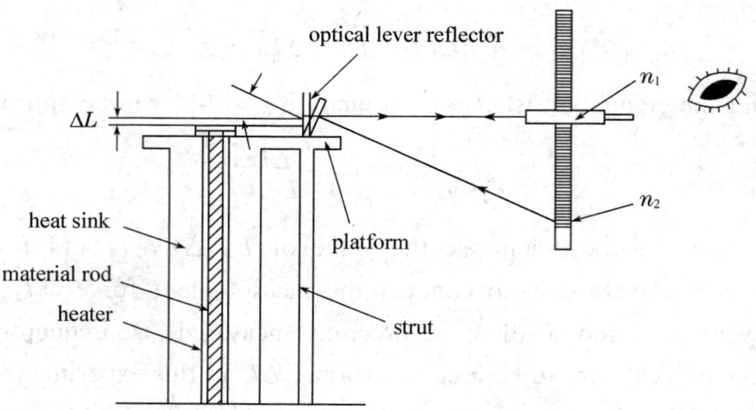

Fig. 4-13-1 Schematic diagram of measuring the linear expansion coefficient with optical lever method

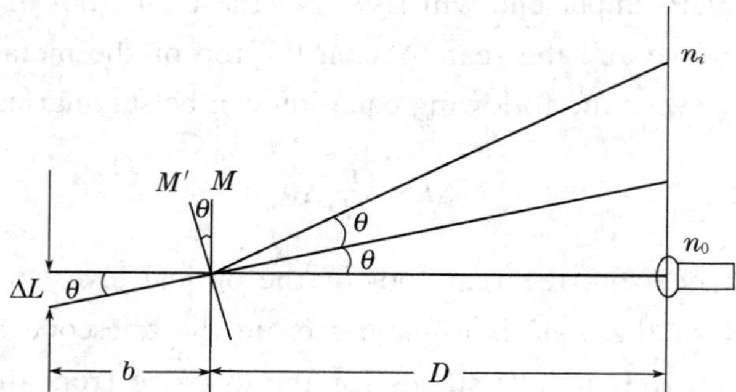

Fig. 4-13-2 Schematic diagram for measuring the slight change in the rod's length

2. Instructions

Measure the length of the copper pipe L, and then slowly place it into the heating pipe until the lower end of the rod to be measured is at the bottom. Place the digital thermometer temperature probe carefully into the copper tube hole. Place the front knife edge (or two fore feet) of the optical lever in the concave groove of the platform. Put the rear foot on the top of the pole to be measured, with the normal line of optical lever plane mirror roughly coaxial with the telescope and parallel to the horizontal base. The device is illustrated in Fig. 4-13-3.

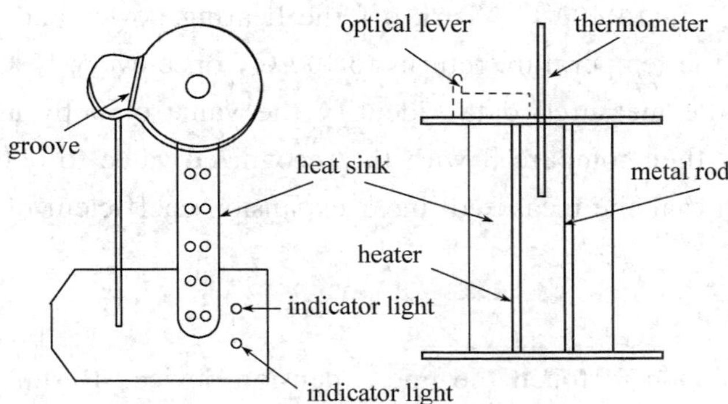

Fig. 4-13-3 Schematic diagram of linear expansion instrument

Experimental Contents and Steps

1. Measure the length of the copper tube with a tape measure, then place the test rod into the line expansion device. Next, place the temperature probe of digital thermometer carefully into the copper tube.

2. Place the front knife edge (or two forefeet) of the optical lever in the concave groove of the platform. Put the rear foot on the top of the pole to be measured, with the optical lever plane mirror roughly vertical and the plane mirror at the equal height as the telescope.

3. Adjust the telescope.

(1) Ensure the telescope's mirror and the eye of the telescope tube to be outside the upper side of the gap and within sight. This means that one can observe the image of the scale line in the mirror. If not, it should be adjusted until the fuzzy image of the scale appears in the telescope's central field of view.

(2) Adjust the telescope eyepiece until the "+" sub-fork on the reticle can be clearly seen. Adjust the handwheel of the objective lens for the clear scale until there is no relative displacement between the horizontal line of the fork and the scale (i. e., no parallax) when the eye moves up and down the eyepiece. Additionally, the parallex in the experimental operation can be relaxed to no more than half a grid.

(3) Make a fine tune of the plane mirror inclination (not to move the three supporting points of the plane mirror), so that the distance between the horizontal line of the cross lines in the telescope and the ruler zero scale line is within the range of ± 2 cm.

4. Turn on the digital thermometer and the heater switch of the linear expansion coefficient tester, and adjust the heating speed properly. Record the measurement data

every 10 ℃ from 30 ℃ to 100 ℃. Turn off the heating power and record the cooling process data when the temperature returns to 100 ℃, once every 10 ℃ until 30 ℃.

According to the measured data, identify the value of α by using the successive difference method, then compare it with the recognized value to obtain the percentage error. (It is known that the measured linear expansion coefficient of copper is 1.489×10^{-5} ℃$^{-1}$)

Precautions

1. It is forbidden to touch the line expansion device. Restart the experiment even with the slight movement of the optical lever and the telescope.

2. Enough attention should be paid to observation and problem analysis. Try to reduce the factors that may cause measurement errors. Always be ready for preventive measures and improvement methods.

Questions

1. What are the principles according to which different instruments are adopted to measure different length quantities? Which measurement error has the greatest impact on the final results?

2. As for the two metal rods of the same material with different thickness and lengths, are their coefficients of linear expansion the same within the same temperature range? Do the two metal rods have the same expansions? And why?

4.14 三棱镜折射率的测定

光从真空射入介质发生折射时，入射角与折射角的正弦值之比 n 叫作介质的“绝对折射率”，简称“折射率”。折射率是物质重要的光学特性常数，精确测定折射率的方法很多，对固体介质，常用最小偏向角法或自准直法；液体介质常用临界角法(阿贝折射仪)；气体介质则用精密度更高的干涉法(瑞利干涉仪)。这里介绍最小偏向角法，并以此作为分光计调整和使用的练习。

实验目的

1. 掌握最小偏向角法测量介质折射率的原理。

2. 测量三棱镜的折射率。

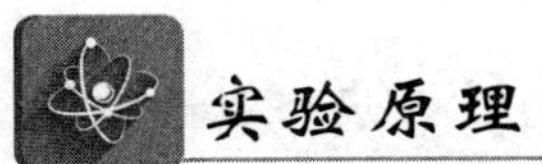

实验原理

如图 4-14-1 所示，设三棱镜顶角 A 的大小为 α，棱镜折射率为 n。光线 S 代表一束单色平行光，以入射角 i_1 投射到棱镜的 AB 面上，经棱镜两次折射后以角 i_4 从另一面 AC 射出来，成为光线 S'。经棱镜两次折射，光线传播方向总的变化可用入射光线 S 和出射光线 S' 延长线的夹角 δ 来表示，δ 称为偏向角。由几何关系可得

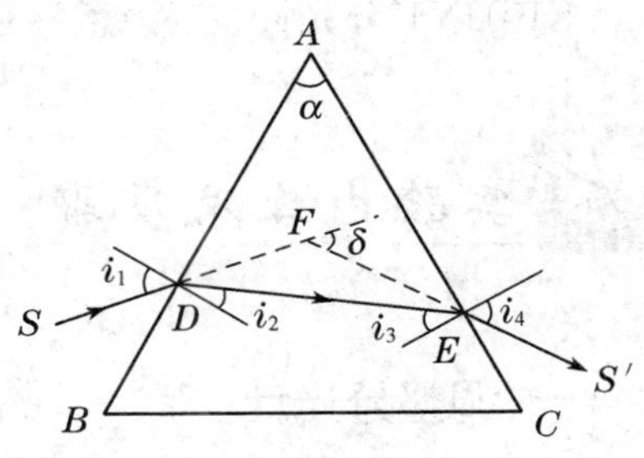

图 4-14-1　最小偏向角法原理

$$\alpha=i_2+i_3 \tag{4-14-1}$$

$$\delta=(i_1-i_2)+(i_4-i_3)=i_1+i_4-\alpha \tag{4-14-2}$$

由折射定律

$$\sin i_1=n\sin i_2 \tag{4-14-3}$$

$$n\sin i_3=\sin i_4 \tag{4-14-4}$$

以上公式表明，对于给定棱镜，若顶角 α 和折射率 n 已定，偏向角 δ 随入射角 i_1 变化，则 δ 是 i_1 的函数，可以证明(证明见本实验附录)。当 $i_1=i_4$ 时，即入射光线 S 出射光线 S' 对称地分布在棱镜两侧时，偏向角 δ 有最小值 $\delta_{\min}$，此值称为最小偏向角。此时，i_2 与 i_3 也相等，且有

$$i_2=i_3=\frac{\alpha}{2},\ i_1=\frac{\alpha+\delta_{\min}}{2} \tag{4-14-5}$$

可以得出三棱镜的折射率为

$$n=\frac{\sin\left(\frac{\alpha+\delta_{\min}}{2}\right)}{\sin\frac{\alpha}{2}} \tag{4-14-6}$$

用分光计测出棱镜的顶角 α 和最小偏向角 $\delta_{\min}$，由式(4-14-6)可求得棱镜的折射率 n。其中顶角的测量方法参见《分光计的调节和使用》一文。最小偏向角法是测折射率的基本方法，测量 n 的准确度与分光计的精度密切相关，多用于测固体折射率。

实验用的光源是汞灯，不是单色光，而是由几种波长的光组成的复色光。各种波长的光折射率不同，经三棱镜折射后，各种波长的光将被分开，再经分光计望远镜成像，在分划板上成一系列平行的单色狭缝像。

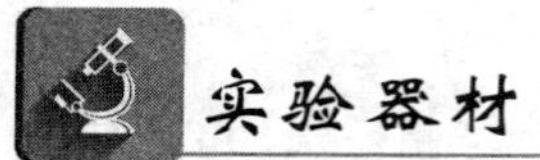

KF-JJY1′分光计、三棱镜、平面镜。

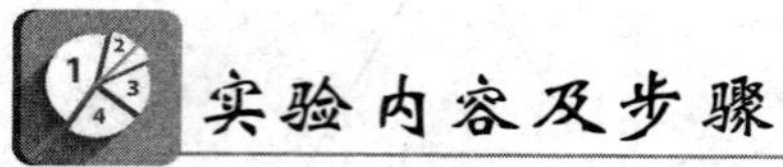

一、调整分光计

1. 用自准直法调节望远镜,使其聚焦于无穷远。
2. 调节望远镜光轴与分光计主轴垂直。
3. 调节平行光管,使其聚焦于无穷远。
4. 调节平行光管光轴与分光计主轴垂直。

二、调整三棱镜的主截面与分光计主轴垂直

三、用反射法测三棱镜顶角

四、测量三棱镜的最小偏向角

1. 松开望远镜止动螺钉和游标盘止动螺钉,把载物台及望远镜转至如图 4-14-2 所示角坐标 2 位置,再左右微微转动望远镜,使棱镜出射的汞灯各种颜色汞光谱线(每条线对应一种波长),出现在目镜可视范围内,如图 4-14-3 所示。

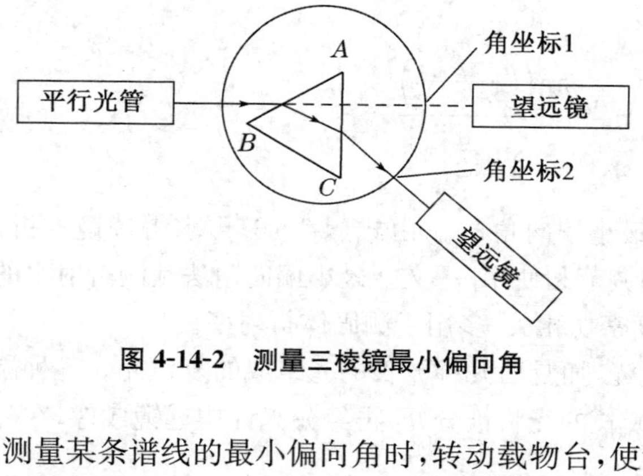

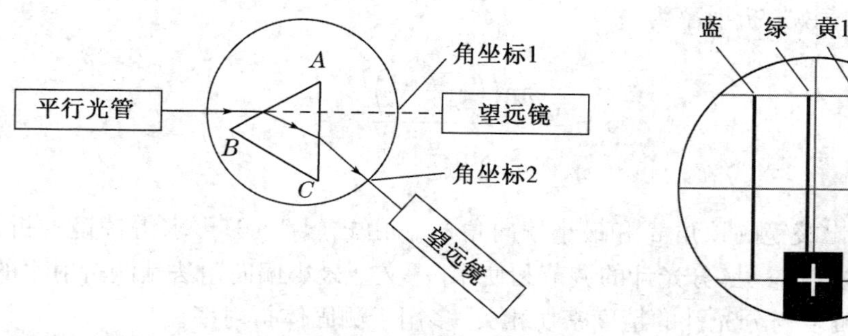

图 4-14-2 测量三棱镜最小偏向角

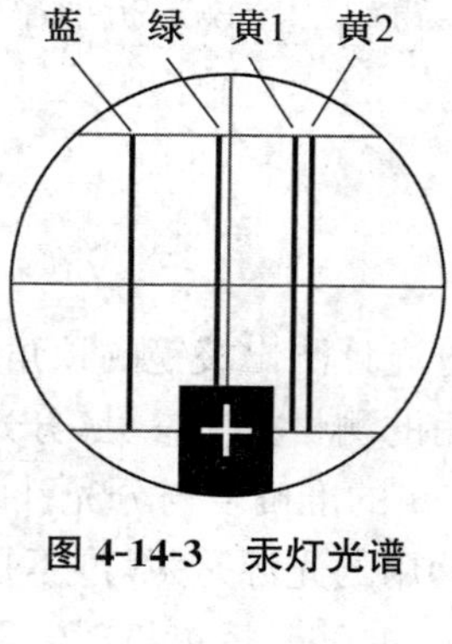

图 4-14-3 汞灯光谱

2. 测量某条谱线的最小偏向角时,转动载物台,使该条光谱线向右移动。当光谱线超出目镜可视范围时,转动望远镜使该条光谱线回到目镜可视范围;

3. 继续转动载物台,直至该条光谱线恰好开始向左移,将此谱线对准叉丝竖线,记录此时的角坐标 2;

4. 重复步骤1～3，将汞灯的蓝、绿、黄1、黄2各条谱线对应的角坐标2测出；

5. 取下三棱镜，将望远镜转动至图4-14-2所示的角坐标1的位置，将狭缝像对准叉丝竖线，记录此时的角坐标1；

6. 各条谱线的角坐标2减去角坐标1就是所测谱线的最小偏向角

$$\delta_{\min}=\frac{1}{2}\left[|\theta_2-\theta_1|+|\theta_2'-\theta_1'|\right] \tag{4-14-7}$$

式中，θ_2、θ_1分别为上述步骤中角坐标2和角坐标1在分光计一侧游标的读数；θ_2'、θ_1'分别为另一侧游标的读数。

测出蓝、绿、黄1、黄2的最小偏向角，计算出相应的折射率和不确定度。

注意事项

1. 测量时转座与度盘止动螺钉应锁紧，这样望远镜转动时刻度盘也随之同步转动。

2. 读数时，狭缝像必须与叉丝竖线对齐。

思考题

1. 实验中如何寻找最小偏向角？
2. 为什么汞灯光源发出的光经过三棱镜后会形成光谱？

Experiment 4.14 Measurement of Prism Refractive Index

Refraction occurs when light propagates into the medium from a vacuum. The ratio n of the sine of the incident angle and the refraction angle is called the "absolute refractive index" of the medium, shortened as the "refractive index". It is an optical property constant that describes matter.

There are many ways to accurately measure the refractive index. The least deflection angle method or self-collimation method are commonly used for the solid media; the critical angle method (Abbe refractometer) is usually adopted for the liquid medium; the more sophisticated method of interferometry (Rayleigh interferometer) is much suitable for the gaseous media. This section introduces the minimum deflection angle method and its application in the experiment of spectrometer adjustment.

Experimental Objectives

1. To master the principle of measuring the medium refractive index with the minimum deflection angle method.

2. To learn how to measure the refractive index of the prism.

Experimental Principle

Suppose that the vertex angle A of a prism is a and its refraction index is n, as shown in Fig. 4-14-1. The light ray S represents a bundle of monochromatic parallel light, which is projected on the AB plane of the prism at the incident angle i_1.

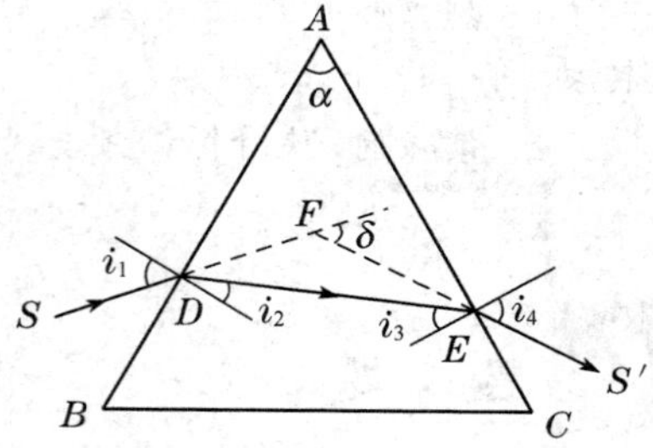

Fig. 4-14-1 The principle of the minimum deflection angle method

After being refracted twice by the prism, the light ray S shoots out from the other surface AC at the incident angle i_4 to become the light ray S'. The light has refracted twice through the prism. The whole change of the light propagation direction can be represented by the angle δ between the incident light S and the extended line of the outgoing light S'. δ is called the deflection angle and it can be calculated with the geometric equation as follows:

$$\alpha = i_2 + i_3 \tag{4-14-1}$$

$$\delta = (i_1 - i_2) + (i_4 - i_3) = i_1 + i_4 - \alpha \tag{4-14-2}$$

According to the law of refraction, we get

$$\sin i_1 = n \sin i_2 \tag{4-14-3}$$

$$n \sin i_3 = \sin i_4 \tag{4-14-4}$$

As can be seen from the above Equections, for a given prism, if the apex angle α and the refractive index n have been fixed, the deflection angle δ will vary with the incident angle of incidence i_1 and δ can be expressed as a function of i_1. It can be proved (see the appendix of this experiment) that if $i_1 = i_4$, that is, when the incident light S and the outgoing light S' are symmetrically distributed on both sides of the prism, the deflection angle δ will have the minimum value $\delta_{\min}$, which is referred to as the minimum deflection angle. Under this occasion, the values of i_2 and i_3 are equal and their relationship can be expressed as:

$$i_2 = i_3 = \frac{\alpha}{2}, \quad i_1 = \frac{\alpha + \delta_{min}}{2} \tag{4-14-5}$$

The refractive index of the prism can be calculated as follows:

$$n = \frac{\sin\left(\frac{\alpha + \delta_{min}}{2}\right)}{\sin\frac{\alpha}{2}} \tag{4-14-6}$$

Measure the apical angle α and the minimum deflection angle δ_{min} of the prism with the spectrometer. The refractive index n of the prism can be obtained from Equation (4-14-6). The measurement method for the vertex angle can be found in the experiment section 4.5 "Spectrometer Adjustment and Use". The minimum deflection angle method is the basic method to measure the refractive index. The accuracy of measuring n is closely related to the precision of the spectrometer. It is widely used to measure the refractive index of solids.

The experimental light source is a mercury lamp, which provides polychromatic light composed of several wavelengths of light, instead of the monochromatic light. Due to different refractive indices, light of various wavelengths will be separated after being refracted by the prism. Thereafter, light of various wavelengths is imaged by a spectrometer telescope and then a series of parallel monochromatic slit images will emerge on the reticle.

Experimental Instruments

KF JJY1′-type spectrometer, prism and mirror plane.

Experimental Contents and Steps

1. Spectrometer adjustment.

(1) Adjust the telescope with self-collimation to make it focus on infinity.

(2) Adjust the telescope optical axis to make it perpendicular to the spectrometer spindle.

(3) Adjust the parallel light pipe to make it focus on infinity.

(4) Adjust the optical axis of the collimator to make it perpendicular to the spectrometer spindle.

2. Adjust the prism's main cross-section to be perpendicular to the spectrometer

spindle.

3. Measure the vertex angle of prism with the reflection method.

4. Measure the minimum deflection angle of prism.

(1) Release the telescope stop screw and the vernier plate set screw, and move the stage and telescope to the angular coordinate 2, as shown in Fig. 4-14-2. Tune the telescope repeatedly until the colorful spectrum lines (each line corresponds to a wavelength) of the mercury light refracted by the prism appear within the visual range of the eyepiece, as shown in Fig. 4-14-3.

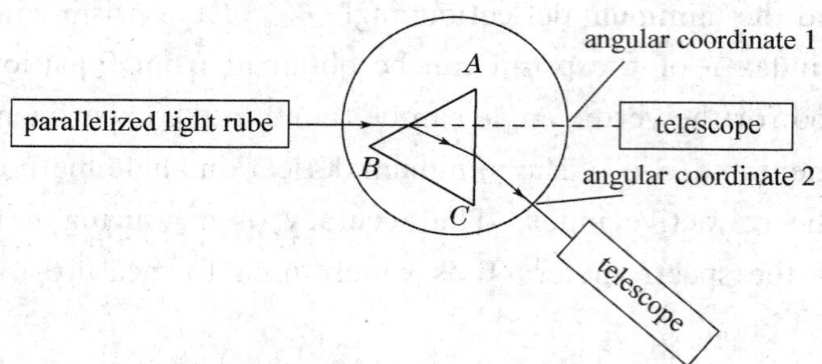

Fig. 4-14-2 Measuring the minimum deflection angle of prism

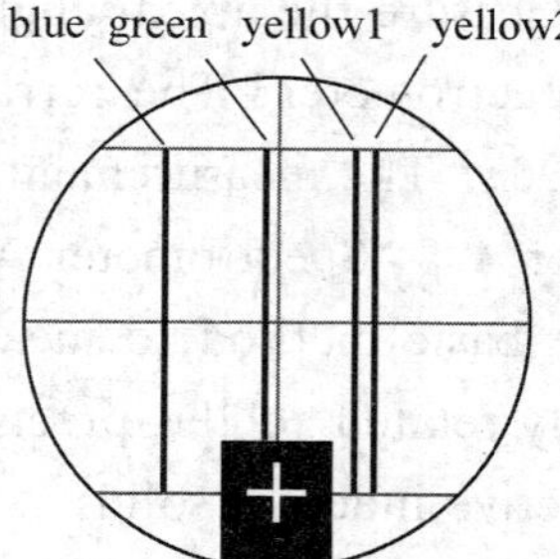

Fig. 4-14-3 Mercury lamp spectrum

(2) When measuring the minimum deflection of a spectral line, turn the stage to make the spectral line move to the right. When the spectral line goes beyond the eyepiece's visible range, move the telescope to make the spectral line seen within the eyepiece's visible range.

(3) Continue to move the stage until the line begins to move to the left. Align this line with the vertical line and record the angular coordinate 2.

(4) Repeat steps 1～3 to measure the spectral coordinate 2 corresponding to the mercury lamp's spectral lines of blue, green, yellow 1 and yellow 2.

(5) Remove the prism and move the telescope to the position of angular coordinate 1 in Fig. 4-14-2. Align the slit image with the vertical bar and record the angular coordinate 1.

(6) The minimum deviation of the measured line can be obtained by subtracting the angular coordinate 1 from the angular coordinate 2 of each line,

$$\delta_{\min}=\frac{1}{2}[|\theta_2-\theta_1|+|\theta_2'-\theta_1'|] \tag{4-14-7}$$

Where θ_2 and θ_1 are the angular coordinates 2 and 1 respectively mentioned in the above steps. θ_2 and θ_1 are pre-and-post-rotation readings from one side vernier of spectrometer, while θ_2' and θ_1' are the readings from the other.

With the minimum deflection angles of the spectral lines of blue, green, yellow 1 and yellow 2, calculate the corresponding refractive index and its uncertainty.

Precautions

1. During the measuring process, the transposition and the dial stop screw must be locked so as to make the dial rotate synchronously with the telescope.

2. During the reading process, the slit image must be aligned with the cross-shaped vertical line.

Questions

1. How to find the minimum deflection angle in the experiment?
2. Why does the light emitted by the mercury lamp form a spectrum after passing through the prism?

Appendix

Proof of the minimum deflection angle formula:

From Equations (4-14-1) to (4-14-4), we find a total of four equations and five variables: δ, i_1, i_2, i_3 and i_4. Therefore, only one of the five variables is independent. δ can be seen as a function of i_1, a and n are fixed values.

Applying i_1 derivation to Equations (4-14-1)~(4-14-4) respectively, we get

$$\frac{\mathrm{d}i_2}{\mathrm{d}i_1}=-\frac{\mathrm{d}i_3}{\mathrm{d}i_1} \tag{4-14-8}$$

$$\frac{\mathrm{d}\delta}{\mathrm{d}i_1}=1+\frac{\mathrm{d}i_4}{\mathrm{d}i_1} \tag{4-14-9}$$

$$\cos i_1=n\cos i_2\frac{\mathrm{d}i_2}{\mathrm{d}i_1} \tag{4-14-10}$$

$$n\cos i_3\frac{\mathrm{d}i_3}{\mathrm{d}i_1}=\cos i_4\frac{\mathrm{d}i_4}{\mathrm{d}i_1} \tag{4-14-11}$$

Substituting Equations (4-14-8), (4-14-10) and (4-14-11) into Equation (3-16-9), we obtain

$$\frac{\mathrm{d}\delta}{\mathrm{d}i_1}=1-\frac{\cos i_1\cos i_3}{\cos i_2\cos i_4} \tag{4-14-12}$$

When δ takes an extreme value, both sides of Equation (4-14-12) should be equal to zero, that is

$$\cos i_1 \cos i_3 = \cos i_2 \cos i_4$$

So we have

$$\frac{\cos i_1}{\cos i_2} = \frac{\cos i_4}{\cos i_3},\ \sqrt{\frac{1-n^2\sin^2 i_2}{\cos^2 i_2}} = \sqrt{\frac{1-n^2\sin^2 i_3}{\cos^2 i_3}}$$

which is

$$\sqrt{\sec^2 i_2 - n^2\tan^2 i_2} = \sqrt{\sec^2 i_3 - n^2\tan^2 i_3}$$

$$\sqrt{1-(1-n^2)\tan^2 i_2} = \sqrt{1-(1-n^2)\tan^2 i_3} \tag{4-14-13}$$

Since the function $f(x)=\tan x$ monotonically increases in the interval $(0, \pi/2)$, i_2 and i_3 are in the interval $(0, \pi/2)$. Therefore, the only solution that satisfies Equation (4-14-13) is

$$\tan i_2 = \tan i_3 \Rightarrow i_2 = i_3 = \frac{\alpha}{2}$$

With the law of refraction, we can find that $i_1 = i_4$. Substituting it into Equation (4-14-2), we get $\delta_{\min} = 2i - \alpha$, $i_1 = \frac{\delta_{\min}+\alpha}{2}$, so we have

$$n = \frac{\sin i_1}{\sin i_2} = \frac{\sin\left(\frac{\alpha+\delta_{\min}}{2}\right)}{\sin\frac{\alpha}{2}}$$

4.15 空气比热容比的测定

理想气体的定压比热容 C_p 和定容比热容 C_V 之间满足关系：$C_p - C_V = R$，其中 R 为气体普适常数；两者之比 $\gamma = C_p/C_V$ 称为气体的比热容比，也称气体的绝热指数，它在热力学理论及工程技术的实际应用中起着重要的作用。例如，热机的效率及声波在气体中的传播特性都与空气的比热容比 γ 有关。本实验是用绝热膨胀法测定空气的比热容比。

实验目的

1. 用绝热膨胀法测定空气的比热容比。
2. 观测热力学过程中的状态变化及基本物理规律。
3. 学习空气压力传感器及电流型集成温度传感器的原理和使用方法。

实验原理

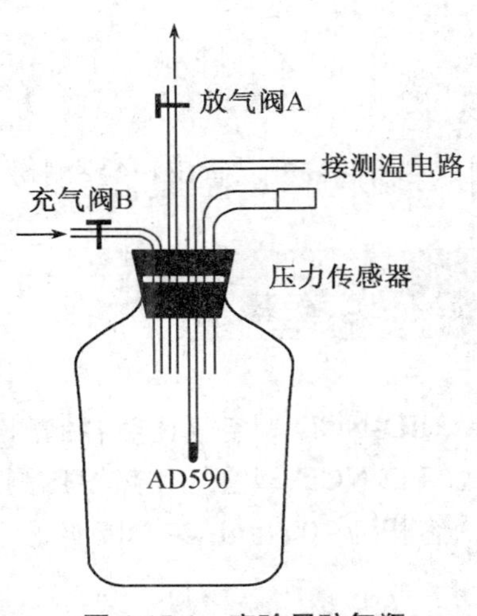

图 4-15-1 实验用贮气瓶

如图 4-15-1 所示，容积为 V_0 贮气瓶内贮有一定量空气，可近似为理想气体。以瓶中气体为研究的热学系统，试进行如下实验过程：

1. 打开放气阀 A，贮气瓶与大气相通，再关闭放气阀 A，瓶内充满与周围空气同温同压的气体，设温度为 T_0，压强为 p_0。

2. 打开充气阀 B，用充气球向瓶内打气，充入一定量的气体后关闭充气阀 B。此时瓶内空气被压缩，压强增大，温度升高。经过一段时间后，由于向瓶外释放热量，瓶内气体温度又恢复到 T_0，达到与周围温度平衡。设此时的气体处于状态Ⅰ(p_1，T_0)。

3. 迅速打开放气阀 A，使瓶内气体与大气相通，当瓶内压强降至 p_0 时，立刻关闭放气阀 A，将有部分气体喷出贮气瓶。由于放气过程较快，瓶内剩余气体来不及与外界进行热交换，可以认为剩余气体经历了一个绝热膨胀的过程，由状态Ⅰ(p_1，V_1，T_0)转变为状态Ⅱ(p_0，V_0，T_1)，其中 V_1 为剩余气体在状态Ⅰ(p_1，T_0)时的体积。此时 $T_1 < T_0$。

4. 由于瓶内气体温度 T_1 低于室温 T_0，所以瓶内气体会缓慢地从外界吸热，直至温度恢复为室温 T_0，此时瓶内气体压强也随之增大为 p_2。设瓶内气体最终稳定后的状态为Ⅲ(p_2，V_0，T_1)。从状态Ⅱ→状态Ⅲ的过程可以看作是一个等容吸热的过程。瓶内剩余气体由状态Ⅰ→状态Ⅱ→状态Ⅲ的过程如图 4-15-2 所示，其对应的 $p-V$ 图如图 4-15-3 所示。

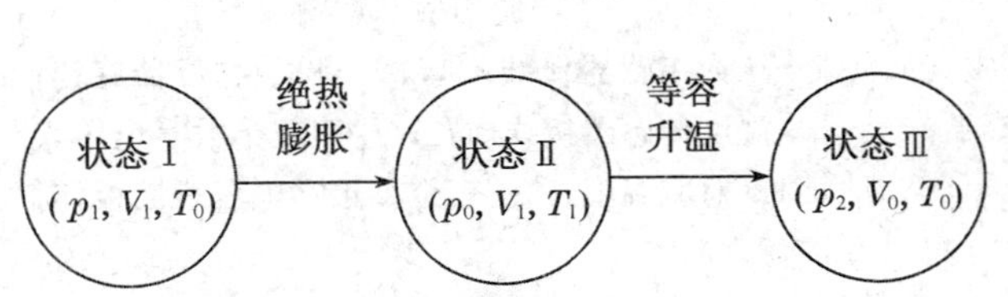

图 4-15-2 瓶内剩余气体的状态变化过程示意图

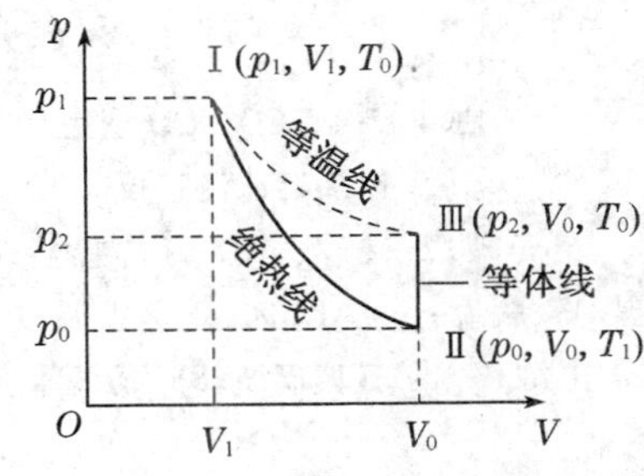

图 4-15-3 瓶内剩余气体状态变化 p-V 图

状态Ⅰ→状态Ⅱ是绝热过程，由绝热过程方程得

$$p_1 V_1^{\gamma} = p_0 V_0^{\gamma} \tag{4-15-1}$$

状态Ⅰ和状态Ⅲ的温度均为 T_0，由气体状态方程得

$$p_1 V_1 = p_2 V_0 \tag{4-15-2}$$

合并式(4-15-1)、式(4-15-2),消去 V_1、V_0 得

$$\gamma = \frac{\ln p_1 - \ln p_0}{\ln p_1 - \ln p_2} = \frac{\ln(p_1/p_0)}{\ln(p_1/p_2)} \tag{4-15-3}$$

或

$$\gamma = \frac{\lg p_1 - \lg p_0}{\lg p_1 - \lg p_2} = \frac{\lg(p_1/p_0)}{\lg(p_1/p_2)}$$

由式(4-15-3)可以看出,只要测得 p_0、p_1、p_2 就可求得空气的绝热指数 γ。

实验器材

FD-NCD 型空气比热容比测定仪、气压计、水银温度计。

FD-NCD 型空气比热容比测定仪由扩散硅压力传感器、AD590 型集成温度传感器、电源、容积为 1000 mL 左右玻璃瓶、充气球及导线等组成,如图 4-15-4 和图 4-15-5 所示。

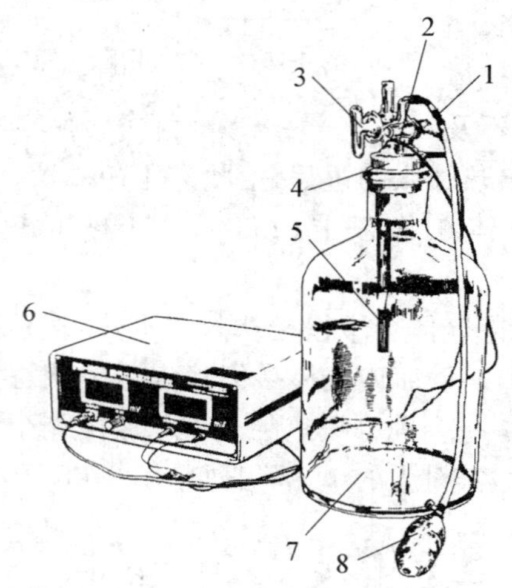

图 4-15-4　FD-NCD 型空气比热容比测定仪

1—充气阀;2—扩散硅压力传感器;3—放气阀;4—瓶塞;5—AD590 型集成温度传感器;6—电源;7—贮气玻璃瓶;8—充气球。

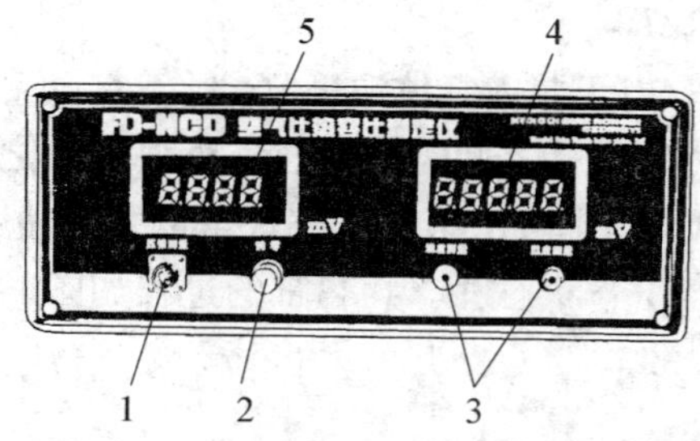

图 4-15-5　FD-NCD 型空气比热容比测定仪电源面板示意图

1—压力传感器接线端口;2—调零电位器旋钮;3—温度传感器接线插孔;4—四位半数字电压表面板(对应温度);5—三位半数字电压表面板(对应压强)。

1. AD590 型集成温度传感器

AD590 是一种新型的半导体温度传感器,测温范围为 −50～150 ℃。当施加 4～30 V 的激励电压时,这种传感器起恒流源的作用,其输出电流与传感器所处的温度呈线性关系。则输出电流为

$$I = Kt + I_0 \tag{4-15-4}$$

式中，$K=1\ \mu A/℃$；I_0 的值从 273～278 μA 略有差异；t 表示温度，℃。AD590 输出的电流 I 可以在距离较远处通过一个适当阻值的电阻 R，转化为电压 U_T，由公式 $I=U_T/R$ 计算出输出的电流，从而计算出温度值。如图 4-15-6 所示，若串联 5 kΩ 电阻后，可产生 5 mV/℃ 的信号电压，接 0～2 V 量程四位半数字电压表，最小可检测到 0.02 ℃温度变化。

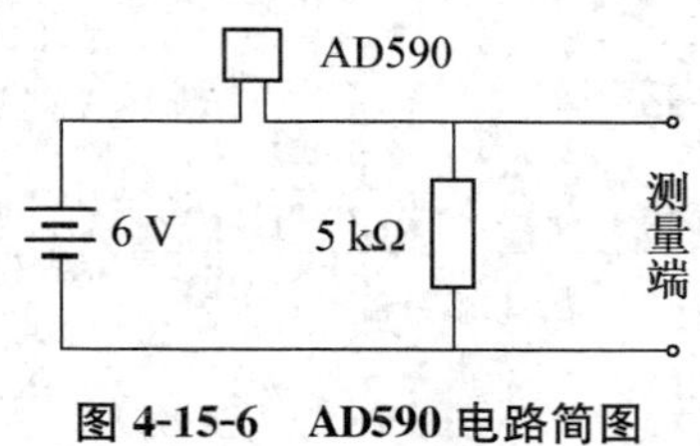

图 4-15-6　AD590 电路简图

2. 扩散硅压力传感器

扩散硅压力传感器是把压强转化为电信号，最终由同轴电缆线输出信号，与仪器内的放大器及三位半数字电压表相接。它显示的是容器内的气体压强大于容器外环境大气压的压强差值。当待测气体压强为 $p_0+10.00$ kPa 时，数字电压表显示为 200 mV，仪器测量气体压强灵敏度为 20 mV/kPa，测量精度为 5 Pa。测量公式为

$$p=p_0+U_p/2000 \qquad (4\text{-}15\text{-}5)$$

式中，电压 U_p 为四位半数字电压表的读数，mV；压强 p、p_0 为 10^5 Pa。

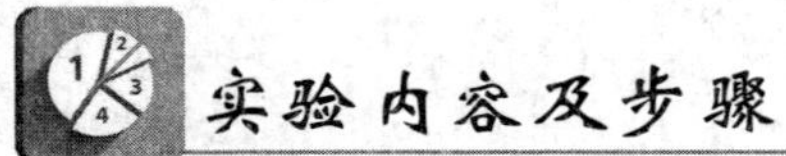

实验内容及步骤

1．连接好仪器的电路，AD590 的正负极请勿接错(红导线为正极、黑导线为负极)。开启电源，预热仪器 20 min，然后用调零电位器旋钮将三位半数字电压表读数调到零。

2．用气压计测定环境大气压强 p_0，用水银温度计测定室温 T_0。

3．关闭放气阀，打开充气阀，用充气球向瓶内打气，使三位半数字电压表示值升高到 100～150 mV，然后关闭充气阀，观察两个数字电压表的变化。经历一段时间后，两个数字电压表示值不变化时，记下三位半电压表的读数 U_p 和四位半电压表的读数 U_T，它们分别对应于瓶内压强和温度。此时瓶内气体近似为状态Ⅰ(p_1，T_0)。

4．迅速打开放气阀，使瓶内气体与大气相通，由于瓶内气压高于大气压，瓶内部分气体将突然喷出，发出“嘶嘶”声。当瓶内空气压强降至环境大气压强 p_0 时(放气声恰好结束)，立刻关闭放气阀，这时瓶内气体温度降低，状态变为Ⅱ(p_0，V_0，T_1)。不用记录数据。

5．等待瓶内空气的温度和压强稳定后，记下 U_p 和 U_T，此时瓶内气体近似为状态 III (p_2，V_0，T_0)。

6．打开放气阀，使贮气瓶与大气相通，重复操作步骤 2～步骤 5。

7．把测得的电压值 U_p 按式(4-15-5)转换成气压值，再按式(4-15-3)计算空气的绝热指数 γ 值，求出平均值，并与理论值比较计算百分误差。

室温时干燥空气中：氧气(O_2)约占 21%，氮气(N_2)约占 78%，氩气(Ar)约占 1%，所以

空气比热容比的理论值近似为

$$\gamma_{理}=\frac{C_p}{C_V}=\frac{\left(\frac{99}{100}\times\frac{5}{2}R+\frac{1}{100}\times\frac{3}{2}R\right)+R}{\frac{99}{100}\times\frac{5}{2}R+\frac{1}{100}\times\frac{3}{2}R}=1.402$$

将计算结果$\overline{\gamma}$与$\gamma_{理}$作比较,按

$$E=\frac{|\overline{r}-r_{理}|}{r_{理}}=\frac{|\overline{\gamma}-1.402|}{1.402}\times 100\% \tag{4-15-6}$$

计算百分误差。

注意事项

1. 转动充气阀和放气阀的活塞时,一定要一只手扶住活塞,另一只手转动活塞,避免活塞损坏。

2. 实验前应检查系统是否漏气,方法是关闭放气阀,打开充气阀,用充气球向瓶内打气,使瓶内压强升高1000～2000 Pa(对应电压值为20～40 mV),关闭充气阀,观察压强是否稳定,若始终下降则说明系统有漏气,须找出原因。

3. 做好本次实验的关键是放气要迅速,即打开放气阀后又关上放气阀的动作要快捷。

思考题

1. 本实验研究的热力学系统是指哪部分气体?
2. 定性分析产生本实验误差的主要原因是什么?应采取什么措施方可减小实验误差?

Experiment 4.15 Determination of Air Specific Heat Capacity Ratio

The relationship between constant pressure specific heat capacity C_p of the ideal gas and its constant volume specific heat capacity C_V can be expressed as: $C_p-C_V=R$, where R is the universal constant of gas. The ratio $\gamma=C_p/C_V$ is referred to as the specific heat capacity of the gas, also known as the gas adiabatic index. It plays an important role in the practical application of thermodynamic theory and engineering technology. For example, both the efficiency of the heat engine and the propagation characteristics of the sound waves in the gas are related to the specific heat capacity ratio of the air. In this experiment, the specific heat capacity ratio of air is determined by adiabatic expansion method.

Experimental Objectives

1. To determine the specific heat capacity ratio of air with adiabatic expansion method.

2. To observe the state changes and basic physics laws in the process of thermodynamics.

3. To learn the principles of air pressure sensors and amperometric integrated temperature sensor and to learn how to use it.

Experimental Principle

As shown in Fig. 4-15-1, a volume of air is stored in a gas cylinder with volume V_0, which is approximately an ideal gas.

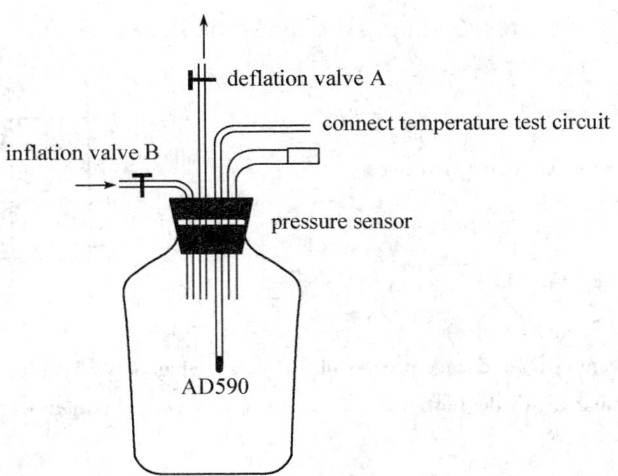

Fig. 4-15-1　Experiment gas cylinders

With the gas in the bottle as a thermal system, the experimental processes are as follows:

(1) Open the deflation valve A in order to make the atmosphere get access to the gas cylinder. Close valve A when the bottle is filled with the gas of the same temperature and pressure as ambient air. Set the temperature as T_0 and pressure as p_0.

(2) Open the inflation valve B and inflate the bottle with a charging balloon. Close the charging valve B after filling a certain amount of gas. When the air inside the bottle is compressed, its pressure increases and the temperature rises. The temperature of the

gas in the bottle returns to T_0 when the gas inside releases heat to the outside of the bottle, the temperature of the gas in the bottle returns to T_0, reaching the equilibrium with the ambient temperature. Set the gas state as state Ⅰ (p_1, T_0)

(3) Quickly open the deflation valve A in order to allow the gas in the bottle circulate with the surrounding atmosphere. When the air pressure in the bottle drops to p_0, immediately close the valve A, and some gas will jet out of the cartridge. Due to rapid deflation, there is not enough time for the residual gas in the bottle to exchange heat with the outside environment. It can be considered that the residual gas has undergone an adiabatic expansion process, from state Ⅰ (p_1, V_1, T_0) to the state Ⅱ (p_0, V_0, T_1), where V_1 is the volume of the residual gas in state Ⅰ (p_1, T_0) and $T_1 < T_0$.

(4) As the gas temperature T_1 is lower than the room temperature T_0, the gas in the bottle will gradually absorb the heat from the outside until the temperature rises to room temperature T_0 with the gas pressure in the bottle increasing to p_2. Set the final steady state of the bottle gas as state Ⅲ (p_2, V_0, T_1). The process from state Ⅱ to state Ⅲ can be regarded as an isometric endothermic process.

The residual gas in the bottle changes from State Ⅰ to state Ⅱ and state Ⅲ as shown in Fig. 4-15-2, and the corresponding pV diagram is shown in Fig. 4-15-3.

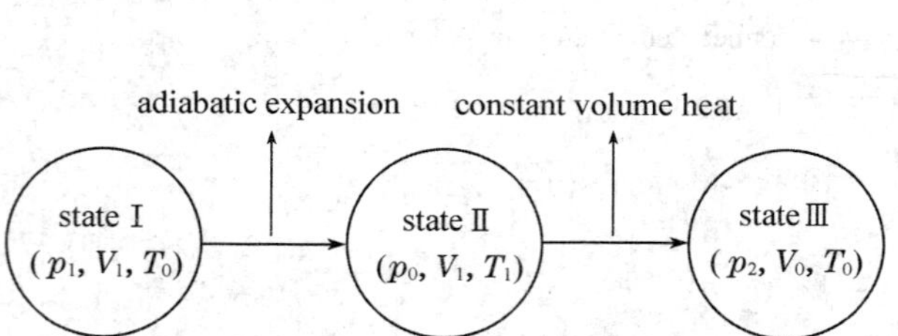

Fig. 4-15-2 Diagram of state change process of the remaining gas in the bottle

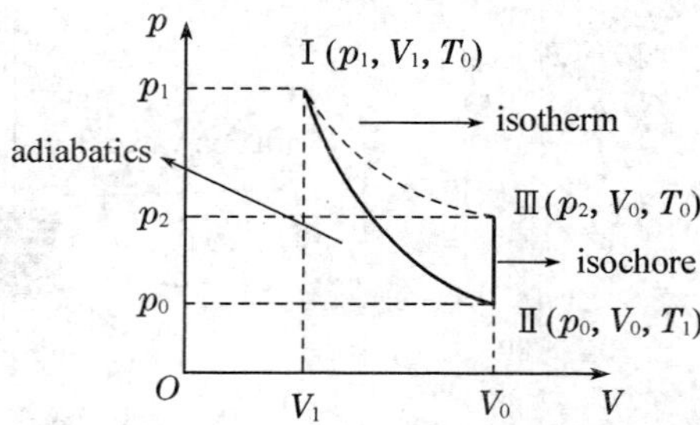

Fig. 4-15-3 Figure of state change of the remaining gas in the bottle

From state Ⅰ to state Ⅱ is the adiabatic process. According to the adiabatic process equation, the following equation can be summarized:

$$p_1 V_1^{\gamma} = p_0 V_0^{\gamma} \tag{4-15-1}$$

The temperatures of both state Ⅰ and state Ⅲ are T_0. From the state equation of gas, the following can be obtained:

$$p_1 V_1 = p_2 V_0 \tag{4-15-2}$$

Combining Equations (4-5-1) with (4-5-2), and eliminating V_1 and V_0, their relationship can be expressed as:

$$\gamma=\frac{\ln p_1-\ln p_0}{\ln p_1-\ln p_2}=\frac{\ln(p_1/p_0)}{\ln(p_1/p_2)} \tag{4-15-3}$$

or

$$\gamma=\frac{\lg p_1-\lg p_0}{\lg p_1-\lg p_2}=\frac{\lg(p_1/p_0)}{\lg(p_1/p_2)}$$

It can be seen from Equation (4-15-3) that the air insulation index γ can be obtained by measuring p_0, p_1 and p_2.

Experimental Instruments

FD NCD air specific heat capacity ratio tester, barometer and mercury thermometer.

FD NCD air specific heat capacity ratio tester, as shown in Fig. 4-15-4, consists of diffusion silicon pressure sensor, AD590 integrated temperature sensor, power supply, a glass bottle with a volume of about 1000 mL, an inflatable ball, wires and etc. Its panel is shown in Fig. 4-15-5.

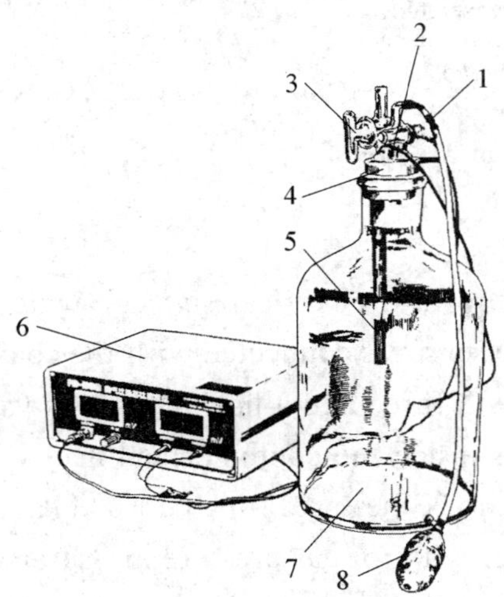

Fig. 4-15-4　FD NCD air specific heat capacity ratio tester

1—Inflation valve; 2—Diffusion silicon pressure sensor; 3—Release valve; 4—Corks; 5—AD590 integrated temperature sensor; 6—Power supply; 7—Gas glass bottles; 8—Inflatable ball.

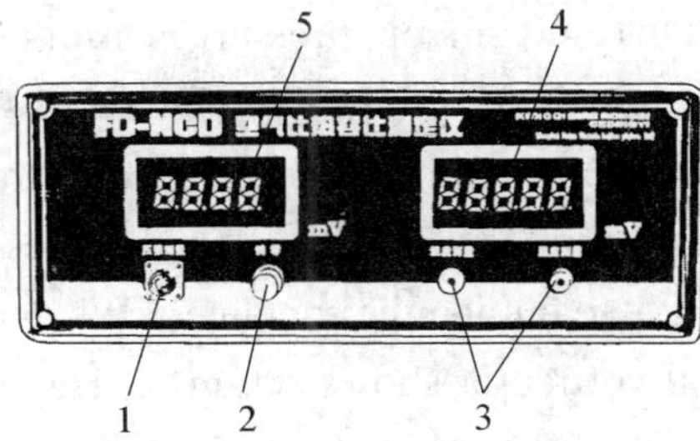

Fig. 4-15-5　FD NCD air specific heat capacity ratio tester power panel

1—Pressure sensor connection port; 2—Zero potentiometer knob; 3—Temperature sensor wiring jack; 4—Digital voltage meter panel (corresponds to temperature); 5—Digital voltage meter panel (corresponds to pressure).

1. AD590 Integrated Temperature Sensor

AD590 is a new type of semiconductor temperature sensor with temperature measurement range of −50 ℃～150 ℃. When the applied excitation voltage is from+4V to+30V, the sensor is a constant current source and its output current is linear with the temperature of the sensor. Set t as Celsius temperature and the output current can be expressed as:

$$I=Kt+I_0 \tag{4-15-4}$$

where $K=1\ \mu A/℃$. I_0 ranges from 273 μA to 278 μA for different AD590 sensors. The output current I of AD590 can be converted to voltage U_T through a resistor R with a suitable resistance at a distance. The output current can be calculated from the equation $I=U_T/R$ and the temperature can be further calculated. As shown in Fig. 4-15-6, if 5 kΩ resistors are connected in series, it produces a signal voltage of 5 mV/℃. Connect to the 0～2V range voltmeter with four-and-a-half digits, and any change over 0.02 ℃ can be detected.

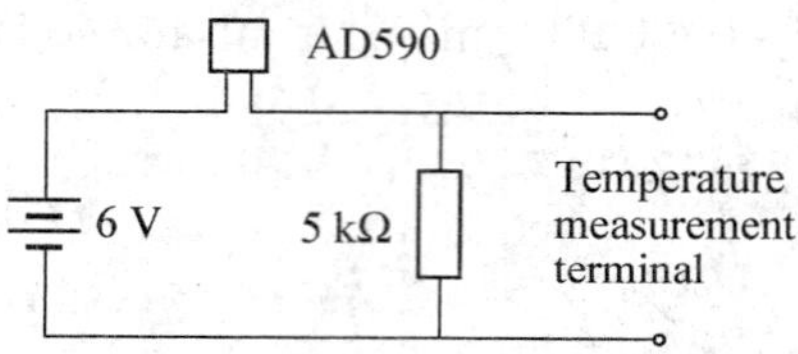

Fig. 4-15-6 Circuit diagram of AD590

2. Diffusion Silicon Pressure Sensor

Diffused silicon pressure sensors convert pressure into electrical signals, which is output by the coaxial cable. Diffused silicon pressure sensor is connected with the coaxial cable, the instrument amplifier and the voltmeter with three-and-a-half digits. Its value shows the difference between the pressure of the gas inside the container and that of the atmosphere outside the container. When the gas pressure to be measured is $p_0+10.00$ kPa, the digital voltmeter shows 200 mV. The instrument sensitivity for gas pressure measurement is 20 mV/kPa, and its measurement accuracy is 5Pa. The measurement formula can be expressed as:

$$p=p_0+U_p/2000 \tag{4-15-5}$$

Where U_p is the reading from the voltmeter with four-and-a-half digits and its unit is mV. p and p_0 are pressures with the unit of 10^5 Pa.

Experimental Contents and Steps

1. Connect the instrument circuit. Be careful not to wrongly connect the positive and negative poles of AD590 (the red lead is positive, and the black lead is negative). Turn on the power and warm up the instrument for 20 minutes. Then use the potentiometer knob to adjust the voltmeter with three-and-a-half digits to zero.

2. Use the barometer to determine the atmospheric pressure p_0 and measure the room temperature T_0 with a mercury thermometer.

3. Close the deflate valve and open the inflatable valve. Inflate the bottle with an inflatable ball until the value of three and a half digital voltage increases to 100～150 mV. Then, close the inflation valve and observe the changes in both digital voltmeters. When the two digital voltage values do not change, record the reading of the three-and-a-half voltmeter as U_p and the reading of the four-and-a-half voltmeter as U_T, corresponding to the pressure and the temperature inside the bottle respectively. And the approximate state of the gas in the bottle can be expressed as Ⅰ(p_1, T_0).

4. Quickly open the deflation valve to make the gas in the bottle exchange with the air outside. As the pressure inside the bottle is higher than the atmospheric pressure, some gas in the bottle will suddenly burst out and make a hissing sound. When the air pressure in the bottle drops to ambient atmospheric pressure p_0 (the sound of the deflation happens to end), close the deflation valve immediately. At this moment, the temperature of the gas in the bottle decreases and its state can be expressed as state Ⅱ (p_0, V_0, T_1). There is no need to record the data.

5. Wait for the temperature and pressure of the air in the bottle to stabilize, and record U_p and U_T. And the approximate state of the gas in the bottle can be expressed as state Ⅲ(p_2, V_0, T_0).

6. Open the deflation valve to make the gas cylinder connected to the atmosphere and repeat steps 2～5 five times.

Based on the Equation (4-15-5), convert the measured voltage value U_p to the atmospheric pressure value, and then calculate the adiabatic index value of the air γ according to the Equation (4-15-3). After obtaining the average, compare it with the theoretical value and then calculate the percentage error.

In the dry air at room temperature, oxygen (O_2) accounts for about 21%, nitrogen (N_2) about 78% and argon (Ar) about 1%, so the theoretical air specific heat capacity ratio is approximately

$$\gamma_{th}=\frac{C_p}{C_V}=\frac{\left(\frac{99}{100}\times\frac{5}{2}R+\frac{1}{100}\times\frac{3}{2}R\right)+R}{\frac{99}{100}\times\frac{5}{2}R+\frac{1}{100}\times\frac{3}{2}R}=1.402$$

Compare the calculated results $\overline{\gamma}$ and γ_{th} and calculate the percentage error according to the following formula:

$$E=\frac{|\overline{\gamma}-1.402|}{1.402}\times 100\% \tag{4-15-6}$$

Precautions

1. When turning the pistons of inflation valve and deflation valve, be sure to hold the piston with one hand and turn the piston with the other hand to avoid damaging the piston.

2. The system should be checked for air leakage before the experiment. Close the deflation valve, open the inflation valve and inflate the bottle with a balloon to increase the pressure in the bottle by about 1000 Pa ~ 2000 Pa (corresponding voltage value is 20 mV~40 mV). Close the inflation valve and observe whether the pressure is stable. If the pressure is always on the decline, it indicates that there is air leakage in the system and the cause must be found out.

3. The key point to do this experiment is the quick deflation, namely the action of opening and closing the vent valve should be fast.

Questions

1. Which part of the gas does the thermodynamic system in this experiment belong to?

2. What are the main reasons for the errors of the qualitative analysis in this experiment? What kind of measures should be taken to reduce the experimental error?

4.16 全息照相

普通的照相术是把从物体表面反射(或漫射)来的光或物体本身发出的光经过物镜成像,并且将光强记录在感光底片上,再在照相纸上显现出物体的平面像。而全息照相术不仅要在感光底片上记录下物光的光强分布,而且还把物光的位相也记录下来,也就是把物光的所有信息全部记录下来,然后通过一定的方法“再现”出物体的立体图像。这种既记录振幅又记录位相的照相就是全息照相(holography)。

全息照相术是20世纪60年代发展起来的。1948年，英国科学家伽伯(D. Gabor)在研究提高电子显微镜的分辨本领时提出了全息原理，并开始了全息照相的研究工作，但是由于缺乏理想的想干光源和高分辨率的记录介质，这方面工作的进展一直相当缓慢。1960年梅曼(T. H. Mainan)研制成世界上第一台红宝石激光器，使全息照相由理论变成了现实，全息照相技术研究才逐渐引起科学界的重视。并在1963年成功地拍摄了第一张全息照片。

全息照相的基本原理是以波的干涉和衍射为基础的，不仅光波可用全息术，还可以有红外的、微波的和声波的全息术，甚至用电子计算机也能制备全息模板，从而使得全息技术在广泛的领域中得到应用：可用于显微技术、信息存储、精密计量、无损探伤等方面，并可发展成立体电影和电视。全息技术的应用涉及家庭娱乐、医疗卫生、军事侦察、工业探伤等众多领域。特别是近年来随着知识经济时代的到来，全息技术将发挥出更大的应用价值。英国科学家伽伯也因发明了全息技术而获得了1971年度的诺贝尔物理学奖。

本实验将通过静态全息照相的拍摄和再现，了解全息照相的主要特征及操作要领。

实验目的

1. 了解全息照像的基本原理和主要特点。
2. 学习拍摄静态全息照片的有关技术和再现观察的方法。

实验原理

一、全息照像的原理

全息照片的拍摄光路如图4-16-1所示，激光束经过分束板 P 后分为强度比例适宜的两束光。其中一束光经 M_1 反射，再经透镜 L_1 扩束后直接投射到照像底片 H 上，形成参考光(R 光)；另一束经 M_2 反射，L_2 扩束后，均匀投在物体表面上，经漫反射后的物光(O 光)照射到底片 H 上。由于激光的高度相干性，两束光在底片上任一点都有确定的位相关系，以干涉叠加形成稳定的干涉图样而被 H 记录下来。

从物体表面任一点出发的物波，都是一个球面波，它与参考光波在 H 相遇，干涉而形成一套干涉图样。从物体上各点发出的物光波与对应点的位置分布有关，不同点发出的物光波其振幅和位相均不相同，对应的干涉图样也不一样。总的物光波可以看作是无数物点发出光波的总和。因此，在全息感光板上形成的，是无数组亮暗程度、疏密、方向各不相同的干涉图样的叠加。曝光以后，经过显影、定影等一系列技术处理过程，就得到了记录物光波全部信息的全息图。

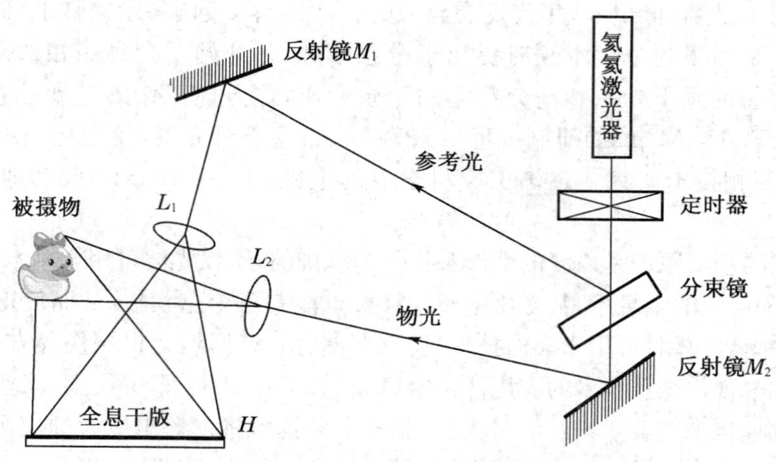

图 4-16-1　全息照相光路图

在全息底片上任一点(x,y)处物光 O 和参考光 R 的光场分布分别为

$$O(x,y,t)=A_o(x,y)\cos[\omega t+\varphi_o(x,y)] \tag{4-16-1}$$

$$R(x,y,t)=A_r(x,y)\cos[\omega t+\varphi_r(x,y)] \tag{4-16-2}$$

将它们用指数函数表示

$$O(x,y,t)=A_o(x,y)\mathrm{e}^{\mathrm{i}[\omega t+\varphi_o(x,y)]} \tag{4-16-3}$$

$$R(x,y,t)=A_r(x,y)\mathrm{e}^{\mathrm{i}[\omega t+\varphi_r(x,y)]} \tag{4-16-4}$$

由于它们是相干光束,所以全息底片上的光强是它们合振幅的平方[略去(x,y)],即:

$$\begin{aligned}I(x,y)&=|O+R|^2=OO^*+O^*R+OR^*+RR^*\\&=A_o^2+A_r^2+A_oA_r\mathrm{e}^{\mathrm{i}}+A_oA_r\mathrm{e}^{\mathrm{i}(\varphi_r-\varphi_o)}\\&=A_o^2+A_r^2+2A_oA_r\cos(\varphi_o-\varphi_r)\end{aligned} \tag{4-16-5}$$

式(4-16-5)中,前两项反映了底片上物光和参考光的强度分布,第三项反映了两束相干光的振幅和相对位相的关系。这样的感光底片上不仅记录了物光束的强度,而且还将物光束的位相信息记录下来了,故称全息照像。全息底片上所记录的不是物体的直观形象,而是一组复杂的干涉条纹,我们要观察全息底片上记录的物像,必须采用一定再现手段。再现观察所用光路图如图 4-16-2 所示。拍摄好的底片经过适当的显影、定影、漂白处理后,底片上各点的振幅透过率与入射光强 $I(x,y)$的关系为

$$t(x,y)=t_0+\beta I(x,y) \tag{4-16-6}$$

式中,t_0 是底片的灰雾度,由底片本身的性质决定;β 是比例系数(对于负片,$\beta<0$)。其绝对值由拍摄技术决定。为了重现物光的波前,必须用一相干光照射全息图。设重现光波的复振幅为 R,透过全息图的复振幅为 $A(x,y)$,所以有

$$
\begin{aligned}
A(x,y) &= t(x,y)R = t_0 R + \beta R\ |O+R|^2 \\
&= t_0 R + \beta R(|O|^2 + |R|^2) + \beta RR^* O + \beta RRO^* \\
&= t_0 R + \beta R(|O|^2 + |R|^2) + \beta |R|^2 O + \beta R^2 O^*
\end{aligned}
\qquad (4\text{-}16\text{-}7)
$$

式(4-16-7)表示经全息图透射后的光包含三个不同的分量。第一、二项代表强度经过衰减的直接透射光；第三项正比于物光 O，是原来物光波的准确再现(振幅发生了改变)，形成物体的虚像；第四项是与物光共轭的光波，它在与虚像相反的一侧会聚成物体的共轭实像，如图 4-16-2 所示。

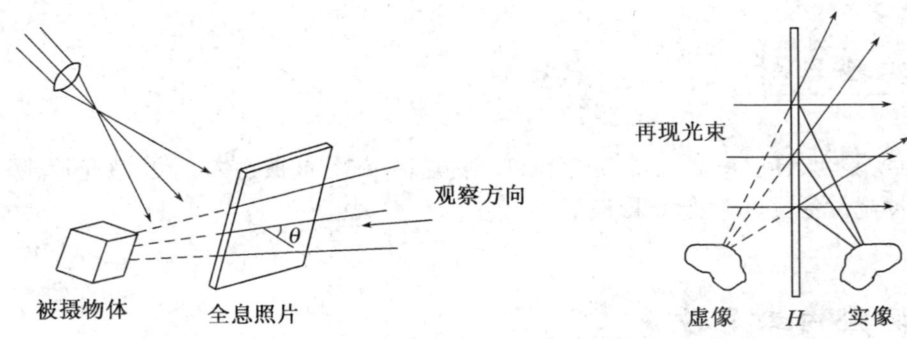

图 4-16-2　全息照相再现光路示意图

二、全息照像的基本条件

要拍摄好一张全息照片，必须具备下列几个基本条件。

1. 相干性好的光源

物光和参考光必须是相干光。He-Ne 激光的相干长度较大，是一般实验室常用的相干光源。激光器的功率在 1～3 mW 即可。当然，功率较大的激光器可使曝光时间相应缩短，减少干扰，效果更好。

2. 光学系统应有足够的机械稳定性

由于全息底片上记录的是精细的干涉条纹，在记录过程中，轻微的振动或其他扰动，只要使光程差发生波长数量级的变化，就会引起干涉条纹的混乱与叠加，使条纹模糊不清，以至看不到物体的衍射像。因此，在曝光的时间内，干涉条纹的移动不得超过条纹间距的1/4。这样全息实验台一般都在远离振源的地方，全息照像各元件全都固定在防振工作台上，以免外界的干扰。

理论指出，全息干涉条纹间距 d 决定于物光与参考光的夹角 θ 和光波长 λ，则

$$d = \frac{\lambda}{2\sin(\theta/2)} \qquad (4\text{-}16\text{-}8)$$

以 He-Ne 激光器作光源，波长 $\lambda = 6328$ nm，$\theta = 30°$，则干涉条纹间距 $d \approx 1.2 \times 10^{-3}$ mm。即每毫米记录下近千条干涉条纹。而且 θ 越大，干涉条纹间距越小。这就要求底片有较高的分辨率。目前国产的超微粒全息底片的分辨率已达到 3000 条/mm。但是，高分辨率的底片感光灵敏度不高，一般全息照像的曝光时间较普通照像时间长。

选择合理的光路是获得优质全息图的关键之一,在安排光路时应注意:

(1) 尽量减小物光与参考光的光程差。实验中要妥善安排光路,使它们的光程差控制在几厘米之内;

(2) 参考光与物光的光强比,一般控制在 3∶1～10∶1 之间,为此应选择合适分光比的分光板和衰减片。检查光强比的简单办法是在准备放全息底片的地方放一白屏,分别挡住参考光和物光,观察它们的照明情况,使参考光强度大于物光强度,但又不非常大;

(3) 投射到感光片上的物光与参考光的夹角 θ,一般选取 30°～60°。

全息实验台、He-Ne 激光器、分束镜、反射镜、扩束镜、被摄物体、全息感光片、曝光定时器、暗室冲洗设备及显影液、定影液等。

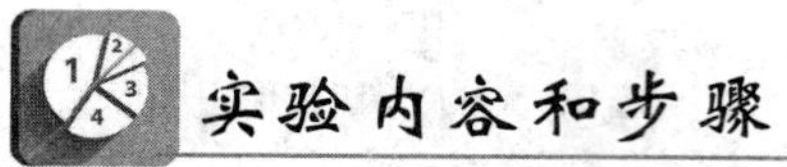

一、全息照片的拍摄

1. 调节分束板、反射镜、物体和屏的位置,使物光和参考光基本等光程。

2. 使物体被均匀照明,物光与参考光能均匀照射在全息底片上。

3. 调整分束板的分光比,使参考光与物光的光强比合适。

4. 设置好曝光定时器的曝光时间(曝光时间由实验室给定)。

5. 打开曝光定时器电源挡住激光束,关闭实验室照明灯。装好底片,注意底片乳胶应面向着光束的方向。

6. 静置数分钟,启动曝光定时器曝光。

7. 将曝光后的感光底片放入 D－19 型显影液中显影,再用 F－5 型定影液定影(显影、定影时间由实验室给定),然后将干板放入漂白液中漂白至全息底片透明;最后用清水冲洗 10 min 后晾干。

二、全息照片的再现

将拍摄冲洗好的全息图放回原位置,遮住物光和被摄物体,只用参考光束照明全息图,即可观察到与原物一模一样的像,仔细观察再现像的特点,特别是立体感,并与原物比较。

三、二次曝光全息照片的拍摄和观察

拍摄物体的一次曝光全息照片,保持光路及装置不变,将待测物加压(或略微转动),再进行一次曝光,经冲洗后就得到二次曝光全息图。也可以对一张全息底片进行多次曝光,即转动底片拍摄多次。再现时,同样转动不同角度出现图像。

四、全息照片观察

数据记录如下表。

拍摄时放置	虚实	大小	正倒	位置
绕 x 轴旋转 180°				
绕 y 轴旋转 180°				
绕 z 轴旋转 180°				

注意：1. "位置"一栏填写与被摄物同侧或异侧；

2. 挡去一半的结果；

3. 观察实像的情况。

注意事项

1. 遵循光学实验的操作规定。
2. 曝光过程中避免人为的振动，保持安静。
3. 遵循暗房操作规程。
4. 注意安全，不要直视激光束，不要随意触摸激光器和电源。

思考题

1. 全息照像有什么特点？
2. 为什么要求光路中物光和参考光的光程差 $\Delta=0$ 是最理想的？
3. 拍摄全息照片必须具备哪些基本条件？

Experiment 4.16 Holography

Ordinary photography takes advantage of the principle of objective lens imaging. When the light reflected (or diffused) from the object surface or emitting by the object passes through the objective lens, the light intensity will be recorded on the photosensitive film, and the plane image of the object will be displayed on the photographic paper. Holography records not only the intensity distribution of object light on the photographic film but also its light phase. In other words, holography collects all the information of light, and then 'reproduce' the three-dimensional image of the object in a certain way. Such a photography records both the amplitude and the phase of the light is called holography.

Holography was developed in the 1960s. In 1948, when studying how to improve the resolution of electron microscopy, the British scientist D. Gabor proposed the holographic principles and began the research work of holography. However, due to the lack of ideal coherent light source and high-resolution recording media, the progress of this work had been quite slow. In 1960, T. H. Maiman developed the world's first ruby laser, making the theory of holography come true. Since then, the research of holographic technology has gradually attracted scientists' attention. In 1963, the first hologram was successfully taken. Based on the principle of wave interference and diffraction as the basic principle, Holographic imaging technology can be classified as light wave holography, infrared holography, microwave holography and acoustic holography, as well as the holographic template made by computer.

Holography technology can be applied in a wide range of fields, including microscopy, information storage, precision measurement, nondestructive testing and so on. Besides, it can also be developed into stereoscopic movies and television. The applications of holographic technology involve family entertainment, health care, military reconnaissance, industrial testing and many other fields. With the advent of knowledge economy, holographic technology will play a greater application role. British scientist Gaber won the Nobel Prize in physics in 1971 thanks to his invention of holographic technology.

This section will discuss the main features of holography technology and its operating rules through the experiment of the static holographic photography and its reproduction.

Experimental Objectives

1. To understand the basic principles and main features of holography.
2. To learn how to shoot static holograms and observe reproduction.

Experimental Principle

1. The Principle of Holography

The light path of producing holographic photo is shown in Fig. 4-16-1.

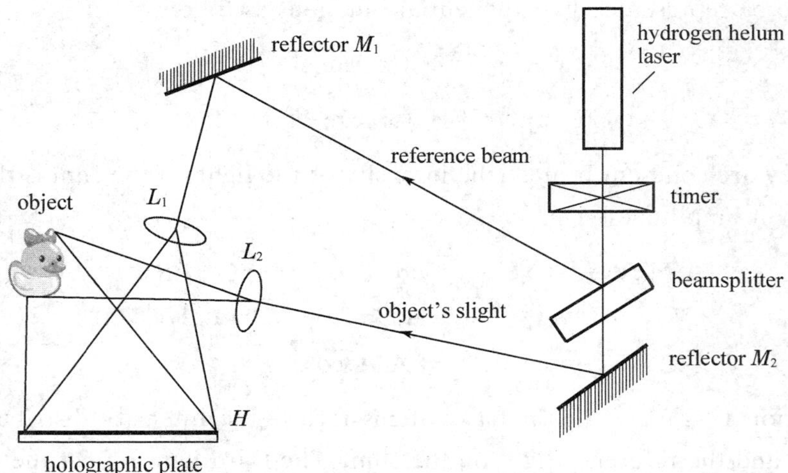

Fig. 4-16-1 Light path of producing holography photo

After passing through the beam splitter plate P, the laser beam is divided into two beams with appropriate intensity ratio.

One beam of light is reflected by M_1 and then expanded by lens L_1 and directly projected onto the photographic film H to form a reference light (light r); the other beam is reflected by M_2, expanded by L_2 and evenly projected on the surface of the object, and then the object light (light o) after diffuse reflection irradiates on the substrate H. Due to the high coherence of the laser, the two beams have a definite phase correlation at any point on the film. The stable interference pattern will be formed by the interference superposition and recorded by film H.

The light waves emitting from any point on the surface of the object is a spherical wave. It meets the reference light wave in the film H, forming a set of interference images. The light waves emitting from different point on the object are related to its position distribution. The amplitude and the phase of the light emitting from different points are different, and thus the corresponding interference images are different.

The total light can be seen as the sum of the waves of light from countless points. Thus the interference pattern formed on the holographic plate is the superposition of numerous groups of interference patterns with different brightness, density and direction. After exposure, the hologram recording all the information of the object light wave can be obtained with some technical processes such as developing and fixing.

On the holographic film, the light field distributions of the object light o and the reference light R at any point (x, y) can be represented respectively as follows:

$$o(x,y,t)=A_o(x,y)\cos[\omega t+\varphi_o(x,y)] \tag{4-16-1}$$

$$r(x,y,t)=A_r(x,y)\cos[\omega t+\varphi_r(x,y)] \tag{4-16-2}$$

They can also be represented by exponential functions as follows:

$$o(x,y,t)=A_o(x,y)e^{i[\omega t+\varphi_o(x,y)]} \tag{4-16-3}$$

$$r(x,y,t)=A_r(x,y)e^{i[\omega t+\varphi_r(x,y)]} \tag{4-16-4}$$

Since they are coherent beams, the intensity of the light on the film is the square of the sum of their amplitudes, i.e.

$$\begin{aligned} I(x,y) &= |o+r|^2 = oo^* + o^*r + or^* + RR^* \\ &= A_o^2 + A_r^2 + A_oA_r e^{i(\varphi_o-\varphi_r)} + A_oA_r e^{i(\varphi_r-\varphi_o)} \\ &= A_o^2 + A_r^2 + 2A_oA_r\cos(\varphi_o-\varphi_r) \end{aligned} \tag{4-16-5}$$

In Equation (4-16-5), the first two items reflect the intensity distribution of the object's light and the reference light on the film. The third item reflect the relationship between the amplitudes and the relative phases of the two beams. Such a film not only records the intensity of the object beam, but also the phase information of the object beam, consequently it is called holography. What is recorded on the hologram is not the visual image of the object, but a group of complex interference fringes, and that is why certain means of reproduction are needed o observe the image recorded on the film. The light path of reproduction is shown in Fig. 4-16-2.

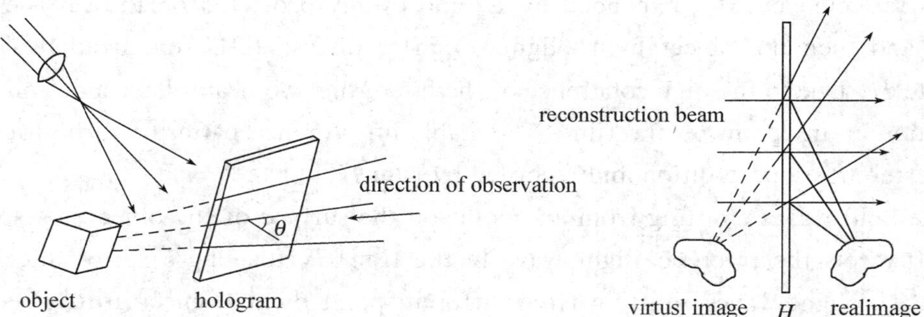

Fig. 4-16-2 Sketch of the optical path of holographic reproduction

After an appropriate process of development, fixing and bleaching, the relationship between the amplitude transmittance at each point on the film and the incident light intensity $I(x,y)$ can be shown as follows:

$$t(x,y)=t_0+\beta I(x,y) \tag{4-16-6}$$

Where t_0 is the film fogging density determined by the nature of the film and β is the ratio factor (for negative film $\beta<0$), whose absolute value is determined by the shooting technique. Thus the holographic image must be illuminated with a coherent light so as to reproduce the wavefront of the light. Set the complex amplitude of the reproduced light as r and the complex amplitude through the hologram as $A(x,y)$, the equation can be summarized as follows:

$$A(x,y)=t(x,y)r=t_0r+\beta r|o+r|^2$$
$$=t_0r+\beta r(|o|^2+|r|^2)+\beta rr^*o+\beta rro^*$$
$$=t_0r+\beta r(|o|^2+|r|^2)+\beta|r|^2o+\beta r^2o^* \quad (4\text{-}16\text{-}7)$$

Equation (4-16-7) shows that the light through the hologram contains three different components. The first and the second items represent the attenuated transmitted light; the third item is the virtual image proportional to the light o and it is the accurate reproduction of the original object's light (amplitude has changed); the fourth item is the light wave conjugated to the object light, which converges into the conjugate real image of the object on the opposite side of the virtual image, as shown in Fig. 4-16-2.

2. Basic Conditions of Holography

To successfully produce a good hologram, the following basic conditions must be met.

(1) The light source with good coherence

Object light and reference light must be coherent light. With a large coherence length, He-Ne laser is commonly used in the experiment as a coherent light source. The acceptable power of the laser ranges from 1mW to 3mW. Generally, the laser with larger power can significantly shorten the exposure time and reduce the interference for a better effect.

(2) The optical system with sufficient mechanical stability

The fine interference fringes are recorded on the holographic film. If any slight vibration or disturbance cause the optical path difference changes in the order of wavelength, the confusion and superposition of interference fringes will be generated, resulting in the blur of fringes and the diffraction pattern of the object can not be seen.

During the exposure time, the movement of interference fringes should not exceed 1/4 of the fringe spacing. In this way, the holographic experimental platform is generally far away from the vibration source, and all the holographic components are fixed on the shockproof working platform to avoid external interference.

The theory points out that the fringe spacing of holographic interference is determined by the angle θ between the object's light and the reference light and the light wavelength λ.

$$d=\frac{\lambda}{2\sin\left(\frac{\theta}{2}\right)} \quad (4\text{-}16\text{-}8)$$

Set the He-Ne laser as the light source and the wavelength λ is 632.8nm, if θ is 30°, the interference fringe spacing d is around 1.2×10^{-3} mm. That is to say, nearly a thousand interference fringes were recorded per millimeter, and the larger the θ is, the smaller the interference fringe spacing will be, which requires a higher resolution of the film. At present, the resolution of domestic ultrafine particle holographic film has reached 3000/mm. However, the photosensitivity of high-resolution films is not high enough, and the general holographic exposure time is longer than that of the general photography.

Choosing a reasonable light path is the key to get high quality holographic images. Attention should be paid to the following points for the arrangement of the light path:

1) Minimize the light path difference between the object's light and the reference light. In the experiment, the light path should be arranged reasonably so that their light path difference is controlled within a few centimeters.

2) The light intensity ratio of the reference light to the object light should be controlled between 3 : 1 and 10 : 1. Accordingly, it is necessary to select a suitable spectroscope and attenuator. A simple way to check the light intensity ratio is to put a white screen in place of the holographic film to block both the reference light and object's light respectively. Observe their lighting conditions and make sure that the reference light intensity is greater than that of object light, yet within certain limits.

3) Angle θ is generally selected to be between 30° and 60°.

Experimental Instruments

Holographic test bench, He-Ne laser, beam splitter, reflector, beam expander lens, subject, holographic film, exposure timer, darkroom processing equipment and developer, etc.

Fig. 4-16-3 The display of holographic light path

Experimental Contents and Steps

1. Shooting of Hologram

(1) Adjust the position of the beam splitter plate, the mirror, the object and the screen until the object's light and reference light have similar light paths.

(2) Make sure that the object is evenly illuminated so that the object light and reference light can be uniformly irradiated on the holographic film.

(3) Adjust the splitting ratio of the splitting plate to make the light intensity ratio of the reference light and the object light suitable.

(4) Set the exposure time of the exposure timer (the exposure time depends on the experiment).

(5) Turn on the power of the exposure timer to block the laser beam and turn off the experiment lights. Install the film and ensure that the latex surface of the film should face the direction of the beam.

(6) Keep the system for a few minutes and start the timer for exposure.

(7) Place the exposed film into the D-19 type developer and then fix it with F-5 type fixing solution (the development and fixing time are given by the experiment). Then get the photographic plate bleached in the bleaching solution until the hologram becomes transparent. Finally, rinse the photographic film with clean water for 10 minutes and make it dry.

2. The Reproduction of Holographic Photos

Place the developed hologram back to the original position so as to block the object and its light. Illuminate the hologram with the reference light so that the image exactly the same as the object can be reproduced. Make a careful observation of the characteristics of the reproduced image, its three-dimensional effect in particular, and compare them with the object.

3. The Second Exposure Holographic Photographing and Observation

Conduct an experiment as shown in Fig. 4-23-1. Take the first exposure hologram and pressurize (or slightly rotate) the object to be measured with the light path and device unchanged. And the second exposure hologram can be obtained by developing the second exposure film. Besides, there can be a multiple exposure of a hologram film, that is, shooting the object while repeatedly rotating the film. Then make the rotation with different angles to reproduce the image.

4. The Observation of Hologram

(1)Data recorded in Table 4-16-1.

Table 4-16-1　Holography date

Original position	Virtual or real	Size	Up or down	Position
Rotate 180°around the x axis				
Rotate 180°around the y axis				
Rotate 180°around the z axis				

Note：(1) Fill the column ‘position’ with “same side or different side of the subject”.

(2) The results of blocking half of the plate.

(3) Observation of the real image.

Precautions

1. Comply with the operating requirements of optical experiments.

2. Avoid the man-made vibration and keep quiet in the exposure process.

3. Abide by the operating procedures of the darkroom.

4. Take caution and do not look at the laser beam directly. Do not touch the laser and power.

Questions

1. What are the characteristics of holography?
2. Why should the optical path difference (between light o and light r) be $\Delta=0$?
3. What are the basic conditions for taking a hologram?

4.17　光电效应普朗克常数测定

1887 年赫兹在用两套电极做电磁波的发射与接收的实验中，发现当紫外光照射到接收电极的负极时，接收电极间更易产生放电，赫兹的发现吸引许多人去做这方面的研究工作。斯托列托夫发现负电极在光的照射下会放出带负电的粒子，形成光电流，光电流的大小与入射光强度成正比，光电流实际是在照射开始时立即产生，无需时间上的积累。1899 年，汤姆逊测定了光电流的荷质比，证明光电流是阴极在光照射下发射出的电子流。赫兹的助手勒纳德从 1889 年就从事光电效应的研究工作，1900 年，他用在阴阳极间加反向电压的方法研究电子逸出金属表面的最大速度，发现光源和阴极材料都对截止电压有影响，但光的强度对截止电压无影响，电子逸出金属表面的最大速度与光强度无关，这是勒纳德的新发现，勒纳德因在这方面的工作获得 1905 年的诺贝尔物理学奖。

光电效应的实验规律与经典的电磁理论是矛盾的，包括勒纳德在内的许多物理学家，提出了种种假设，企图在不违反经典理论的前提下，对上述实验事实作出解释，但都过于牵强附会，经不起推理和实践的检验。1905年，爱因斯坦在其著名论文《关于光的产生和转化的一个试探性观点》中写道："在我看来，如果假定光的能量在空间的分布是不连续的，就可以更好的理解黑体辐射、光致发光、光电效应以及其他有关光的产生和转化的现象的各种观察结果。根据这一假设，从光源发射出来的光能在传播中将不是连续分布在越来越大的空间之中，而是同一个数目有限的局限于空间各点的光量子组成，这些光量子在运动中不再分散，只能整个的被吸收或产生"。作为例证，爱因斯坦由光子假设得出著名的光电效应方程，解释了光电效应的实验结果。密立根从1904年开始做光电效应实验，历经十年，最终用实验证实了爱因斯坦的光量子理论。两位物理大师也因在光电效应等方面的杰出贡献，分别于1921年和1923年获得诺贝尔物理学奖。

光电效应实验对于认识光的本质及早期量子理论的发展，具有里程碑的意义。光量子理论创立后，在固体比热、辐射理论、原子光谱等方面都获得成功，人们逐步认识到光具有波动和粒子二象属性。光子的能量 $E=h\nu$ 与频率有关，当光传播时，显示出光的波动性，产生于涉、衍射、偏振等现象；当光和物体发生作用时，它的粒子性又突出出来。后来科学家发现波粒二象性是一切微观物体的固有属性，并发展了量子力学来描述和解释微观物体的运动规律，使人们对客观世界的认识前进了一大步。

实验目的

1. 通过光电效应实验了解光的量子性。
2. 测量光电管的弱电流特性，找出不同光频率下的截止电压。
3. 验证爱因斯坦方程，并由此求出普朗克常数。

实验原理

在光的照射下，从金属表面释放电子的现象称为光电效应。光电效应的基本实验事实可归纳为：

1. 光电发射率(光电流)与光强度成正比。
2. 光电效应存在一个截止频率(阈频率)，当入射光的频率低于某一阈值时，不论光的强度如何，都没有光电子产生。
3. 光电子的动能与光强度无关，但与入射光的频率成正比。
4. 光电效应是瞬时效应，一经光线照射，立刻产生光电子。

爱因斯坦认为从一点发出的光不是按麦克斯韦电磁学说指出的那样以连续分布的形式把能量传播到空间，而是频率为 ν 的光以 $h\nu$ 为能量单位(光量子)的形式一份一份地向外辐射。至于光电效应，是具有能量 $h\nu$ 的一个光子作用于金属中的一个自由电子，并把它的全部能量都交给这个电子而造成的。如果电子脱离金属表面消耗的能量为 W_S，则由光电效

应打出来的电子动能为

$$E=h\nu=W_S \text{ 或} \frac{1}{2}mv^2=h\nu-W_S \tag{4-17-1}$$

式中，$\frac{1}{2}mv^2$ 是没有受到空间电荷阻止，从金属中逸出的光子的最大初动能。

由式(4-17-1)可见，入射到金属表面的光频越高，逸出来的电子最大初动能必然也越大。正因为逸出的光电子具有一定的初动能，所以即使阳极不加电压也会有部分光电子落入而形成光电流，甚至阳极相对于阴极的电位低时也会有光电子落到阳极，直到阳极电位低于某一数值时，所有光电子都不能到达阳极，光电流才为零。这个相对于阴极为负值的阳极电位 U_S 称为光电效应的遏止电位(或称做遏止电压)。

显然，此时有

$$eU_S-\frac{1}{2}mv^2=0 \tag{4-17-2}$$

将式(4-17-2)代入式(4-17-1)即有

$$eU_S=h\nu-W_S \tag{4-17-3}$$

由于金属材料的逸出功 W_S 是金属的固有属性，对于给定的金属材料 W_S 是一个定值，它与入射光的频率无关。令 $W_S=h\nu_0$，ν_0 为一阈值频率，即具有阈值频率 ν_0 的光子恰恰具有大小为 W_S 的能量，电子吸收了这样的光子不能逸出金属表面。电子只有吸收了频率大于 ν_0 的光子才能逸出金属表面。

将式(4-17-3)改写为

$$U_S=\frac{h}{e}\nu-\frac{W_S}{e}=\frac{h}{e}(\nu-\nu_0) \tag{4-17-4}$$

式(4-17-4)表明，截止电位 U_S 是入射光频率 ν 的线性函数。当入射光的频率 $\nu=\nu_0$ 时，截止电压 $U_S=0$，没有光电子逸出。$U_S \sim \nu$ 曲线斜率 $k=h/e$ 是一个正常数。

$$h=ek \tag{4-17-5}$$

可见，只要用实验方法作出不同频率下的 $U_S \sim \nu$ 曲线，并求出此曲线的斜率 k，就可以通过式(4-17-5)求出普朗克常数 h 的数值。

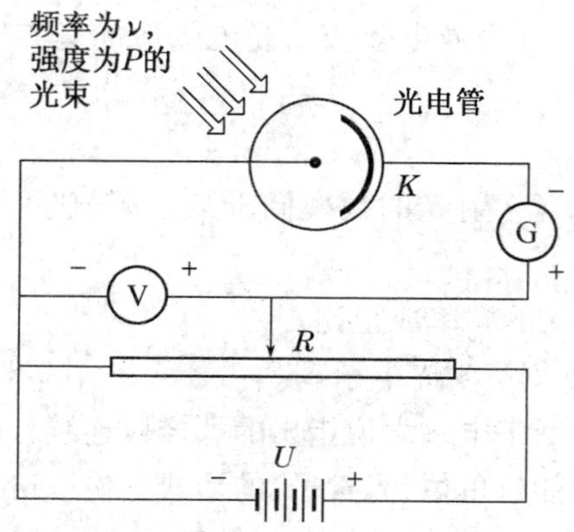

图 4-17-1 光电效应实验原理图

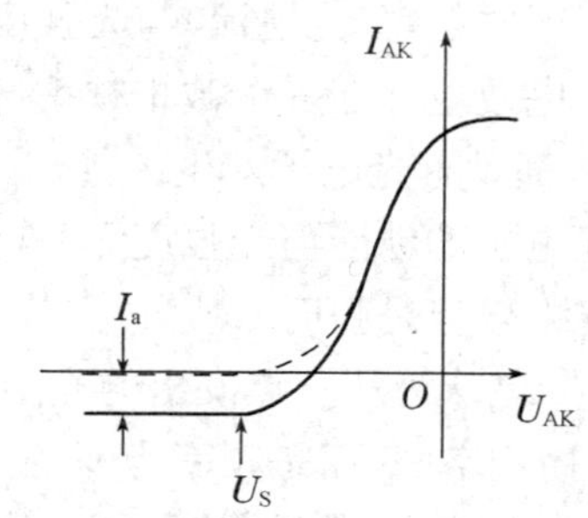

图 4-17-2 光电管的起始伏安特性

分析光电子运动的动力学过程，应用肖特基效应公式，可推导出截止电位 U_S 附近区域的 $I_{AK}-U_{AK}$ 解析关系。再作进一步数学处理，可以得到截止电位 U_S 的解析表示

$$U_S=\frac{U_1 I_a^2-U_0(I_1-I_a)^2}{(I_1-I_a)-I_a^2} \tag{4-17-6}$$

式中，U_0 为 $I_{AK}=0$ 时的 U_{KA} 值（U_{KA} 此时为正值）；I_1 是某个比零稍大的 I_{AK} 值；U_1 为与 I_1 相应的 U_{KA} 值（为正值）；而 I_a 为杂散电流为负值。

实验器材

ZKY-GD－4 型智能光电效应（普朗克常数）实验仪。

实验内容及步骤

1. 连接导线，打开汞灯电源和实验仪电源，并预热 10 min。

2. 电流表调零。

将光电管输入信号端导线断开，“电流量程”旋钮选择“10^{-13}”挡，调节“调零”旋钮，直至“电流指示”屏显示为“000.0”，按下“调零确认/系统清零”键，此时“电压指示”屏显示为“1.998”。

注意，当实验仪开机或变换电流量程时，均需对实验仪调零。调零时，光电管信号输入导线必须与实验仪断开。调零结束后，注意测试前将光电管信号输入导线连接正确，开始测试。

3. 截止电压测试。

按下“截止电压测试”键，取下光电管遮光罩，并装上光阑和 365nm 滤色片，再取下汞灯遮光罩，调节“电压调节”按钮，直至使“电流指示”为 0，此时的“电压指示”值就是 365 滤色片的截止电压。

“◀”/“▶”　当前的修改位将进行循环移动，同时闪烁位随之改变以提示目前修改的位置。

“▲”/“▼”　电压位在当前修改位递增/递减一个增量单位。

盖上汞灯的遮光罩，取下 365 nm 滤色片，依次换上 405 nm、436 nm、546 nm、577 nm 等滤色片，按同样的方法测出其截止电压，并填写下表。注意，每次换滤色片时，都要罩上汞灯的遮光罩，避免汞灯直射光电管而损坏光电管。

波长/nm	365	405	436	546	577	$H\times10^{-34}$ J·s	百分误差/%
频率/$\times10^{14}$ Hz	8.22	7.41	6.88	5.49	5.20		
截止电压 U_S/V							

4. 伏安特性测试。将“电流量程”旋钮调到“10^{-10}”挡,断开光电管信号输入导线,重新调零并按下“系统清零”后,按下“伏安特性测试”键,此时“电压指示”显示为“-1 V”,测试时,接上光电管信号输入导线,依次放上各滤色片,将电压为-1 V时光电流的值依次填入下表。然后将电压值按下表格增大为0、1、5、10、15……,并将所对应各个滤色片的光电流的值对应的填入下表。

U_{AK}/V		-1	0	5	10	15	20	30	40	50
365 nm	$I_{AK}/\times10^{-10}$ A									
405 nm	$I_{AK}/\times10^{-10}$ A									
436 nm	$I_{AK}/\times10^{-10}$ A									
546 nm	$I_{AK}/\times10^{-10}$ A									
577 nm	$I_{AK}/\times10^{-10}$ A									

5. 作图。根据步骤3测量的结果,作出截止电压$U_S \sim \nu$图。如果光电效应遵循爱因斯坦方程,则$U_S=f(\nu)$关系曲线应该是一条直线。求出直线的斜率k,求出普朗克常数$h=ek$。并计算所测值与公认值之间的误差。

根据伏安特性测试中的表格内容作出$U_{AK}-I_{AK}$图。

思考题

1. 从截止电压U_S与入射光频率ν的关系曲线图,能确定阴极材料的逸出功吗?

2. 如果某种材料的逸出功为2.0 eV,用它做成光电管阴极时能探测的截止波长是多少?

3. 实验中,电流为零的点所对应的电压是否为要测遏止电压?为什么?

Experiment 4.17 Determination of Planck Constant of Photoelectric Effect

Experimental Objectives

1. To understand the quantum nature of light by photoelectric effect experiment.

2. To measure the weak current characteristics of the photoelectric tube and to find the cut-off voltage at different optical frequencies.

3. To verify the Einstein equation and calculate Planck's constant.

Experimental Principle

The phenomenon of releasing electrons from the metal surface under light irradiation is called the photoelectric effect. The basic experimental facts of the photoelectric effect can be summarized as follows:

1. The photoelectric emissivity is proportional to the light intensity.

2. There is a cut-off frequency (threshold frequency) for the photoelectric effect. When the frequency of the incident light is lower than a certain threshold, no photoelectron is produced regardless of the light intensity.

3. The kinetic energy of photoelectron is independent of light intensity, yet it is proportional to the frequency of incident light.

4. Photoelectric effect is an instantaneous effect. Photoelectrons are generated immediately after a light exposure.

Einstein argues that light from a point does not propagate energy to space in the form of continuous distribution as pointed out by Maxwell's electromagnetic theory, but radiates outward in the form of $h\nu$ (light quantum) as energy unit. The photoelectric effect is caused by a photon with energy $h\nu$ acting on a free electron in the metal, and its entire energy is given to the electron. If the energy dissipated from the metal surface is W_S, the kinetic energy of the electrons generated by the photoelectric effect can be expressed as:

$$\frac{1}{2}mv^2 = h\nu - W_S \tag{4-17-1}$$

where the $\frac{1}{2}mv^2$ is the maximum initial kinetic energy of photoelectrons that are not trapped by the space charge. It can be seen from the Equation (4-17-1) that the higher the frequency of light incident on the metal surface is, the greater the maximum kinetic energy of the escaping electrons will be. The escaping photoelecton has a certain initial kinetic energy, thus some photoelectrons will fall into the anode and form photocurrent even with no voltage application to the anode. Even when the potential of the anode is lower than that of the cathode, some photoelectrons will fall into the anode until the anode potential is lower than a certain value. Only when all photoelectrons cannot reach the anode, the photocurrent is zero.

Such an anode potential U_S negatively relative to the cathode is called the suppression potential (or suppression voltage) of the photoelectric effect. It can be expressed as

$$eU_S - \frac{1}{2}mv^2 = 0 \tag{4-17-2}$$

Combining Equations (4-17-1) and (4-17-2), the formula can be converted into

$$eU_S = h\nu - W_S \tag{4-17-3}$$

The work function W_S of the metallic material is an intrinsic property of the metal. For a given metallic material, W_S is constant and is independent of the frequency of the incident light. Let $W_S = h\nu_0$, where ν_0 represents the threshold frequency of the metal. This equation means that the photon with threshold frequency ν_0 exactly has the amount of energy WS and electrons absorbing such photons cannot escape from the metal surface. The electron can escape from the metal surface only when it absorbs photons with frequency greater than ν_0.

After divided by e, the Equation (4-17-3) can be rewritten as

$$U_S = \frac{h}{e}\nu - \frac{W_S}{e} = \frac{h}{e}(\nu - \nu_0) \tag{4-17-4}$$

Equation (4-17-4) shows that the stopping potential U_S is a linear function of the incident light frequency ν. According to Equation (4-17-4), when the incident light frequency is $\nu = \nu_0$, the stopping potential will be $U_S = 0$ and no electrons will escape. The slope of $U_S \sim \nu$ line is expressed as h/e, and it can be converted into

$$h = ek \tag{4-17-5}$$

According to the Equation (4-17-5), the Planck constant h can be obtained if the the value of slope k of the curve is given. Thus, it is necessary to draw the $U_S \sim \nu$ curves with different frequencies so as to obtain the value of k.

Fig. 4-17-1 is the schematic diagram of Planck constant measurement by using photoelectric effect. The dotted line in Fig. 4-17-2 is a negative part of the volt-ampere characteristic curve in the U_{AK} for the photoelectric tube. Due to the presence of stray current (reverse current of the photoelectric tube), the volt-ampere characteristic curve of the photoelectric tube is shifted down. The measured photocurrent is shown by the solid lines in Fig. 4-17-2. The stopping potential moves from U_S to U'_S, The stray current is so small that I_a is approximately equal in the vicinity of U_S. Consequently, it can be considered that U'_S coincides with U_S.

By analyzing the dynamics of photoelectron's motion and applying the Schottky effect formula, the $I_{AK} - U_{AK}$ analytic relation in the vicinity of U_S can be deduced. After the further mathematical processing, the analytical expression of the stopping potential U_S can be obtained as

$$h = ek \tag{4-17-6}$$

$$U_S=\frac{U_1 I_a^2-U_0(I_1-I_a)^2}{(I_1-I_a)-I_a^2} \tag{4-17-7}$$

where U_0 is the value of U_{AK} (positive value) under the condition of $I_{AK}=0$, and the value of I_1 is slightly larger than that of I_{AK}. U_1 is the value of U_{AK} (positive) corresponding to I_1 and I_a is the stray current (negative).

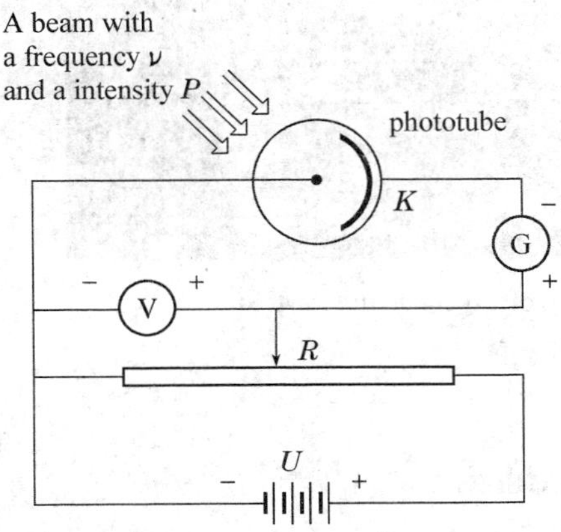

Fig. 4-17-1　The schematic diagram of the photoelectric effect

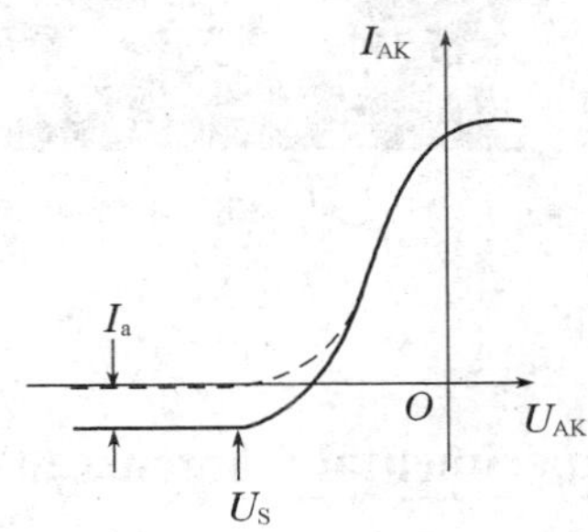

Fig. 4-17-2　The volt-ampere characteristic curve

Experimental Instruments

ZKY-GD-4 photoelectric effect (planck constant) tester. The panel is shown in Fig. 4-17-3.

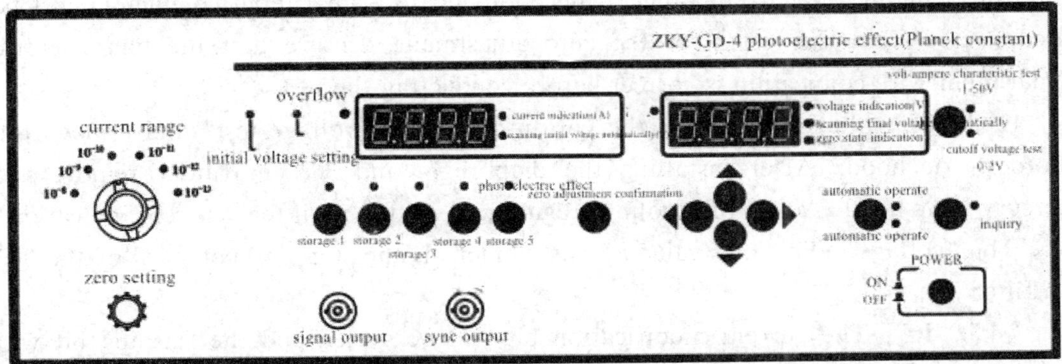

Fig. 4-17-3　ZKY-GD-4 photoelectric effect tester

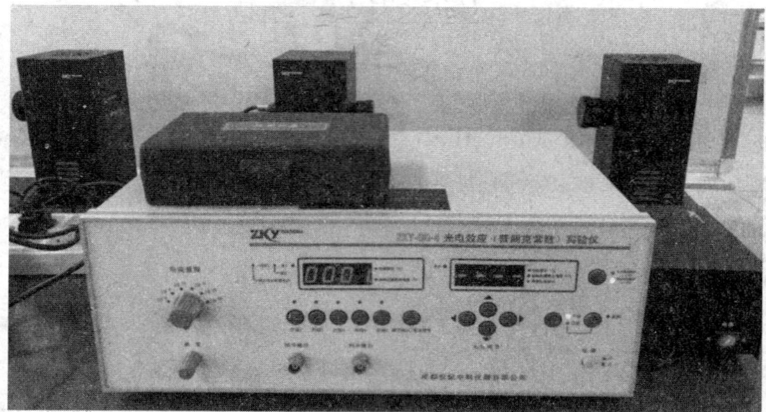

Fig. 4-17-4 The display of experimental devices

 Experimental Contents and Procedures

1. Connect the wires. Turn on the mercury lamp power and the experimental device power to pre-heat for about 10 minutes.

2. Adjust the ammeter to zero. Disconnect the input of the photoelectric tube into the signal terminal. Turn the "Current Range" knob to select "10^{-13}". Adjust the "Zero" knob until the "Current Indicator" screen displays "000.0". Press the "Zero Confirm/System Clear" key and the "voltage indication" screen shows "1.998".

Precautions: it is necessary to make the zero adjustment before switching on the tester or changing the current range. The photoelectric signal input conductor must be disconnected from the tester for the zero adjustment. Ensure that the photoelectric signal input wire connection is correct before conducting the test.

4. Cut-off voltage test. Press the button of "cut off voltage test" and remove the photoelectric hood. After installing the diaphragm and 365 nm filter, remove the mercury lamp hood. Adjust the "voltage regulation" button until the "current indication" is 0. The "voltage indication" value at this moment is the stopping potential of the 365 nm filter.

"◀"/"▶": The current modification bit will be cycled and the flashing bit will change to indicate the currently modified position.

"▲"/"▲": The voltage bit is incremented/decremented by one unit in the current modification.

Cover the hood of the mercury lamp and remove the 365 nm filter to replace the filters of 405 nm, 436 nm, 546 nm and 577 nm respectively. Measure the cut-off voltage in the same way and fill the Table 4-17-1. When changing the color filter, the operator must cover the mercury lamp hood to avoid direct exposure of the mercury lamp to the phototube.

Table 4-17-1

λ/nm	365	405	436	546	577	$h/\times10^{-34}$ J·s	percentage error/%
$\nu/\times10^{14}$ Hz							
U_S/V							

Volt-ampere characteristic curve test. Turn the "Current Range" knob to "10^{-10}" and disconnect the photoelectric signal input conductor. After zeroing and pressing "System cleared", press the "Volt-ampere characteristic test" key and the "Voltage indication" displays "-1 V". Connect to the photoelectric signal input wire and place the filters in turn. Fill respectively the values of the photocurrent at a voltage of -1 V in Table 4-17-1. Then increase the voltage value to 0, 1, 5, 10, 15, … and fill the corresponding photocurrent values of each color filter in Table 4-17-2.

Table 4-17-2

λ/nm	U_{AK}/V	-1	0	5	10	15	20	30	40	50
365	$I_{AK}/\times10^{-10}$ A									
405	$I_{AK}/\times10^{-10}$ A									
436	$I_{AK}/\times10^{-10}$ A									
546	$I_{AK}/\times10^{-10}$ A									
577	$I_{AK}/\times10^{-10}$ A									

5. Mapping. Make the stopping potential diagram "$U_S-\nu$" with the data in Table 4-17-1. If the photoelectric effect follows the Einstein equation, the $U_S=f(\nu)$ relation curve should be a straight line. After obtaining the slope k of the curve and the Planck constant $h=ek$, calculate the errors between the measured value and the accepted value.

Make the diagram I_{AK}-U_{AK} with the data in Table 4-17-2.

Questions

1. How to determine the work function of the cathode material based on the $U_S-\nu$ curve?

2. If a photocell cathode is made of the material with the work function of 2.0 eV, what will the cut-off wavelength this photocell cathode can detect?

3. Will the voltage corresponding to the zero current in the experiment be the

stopping potential to be measured? And why?

4.18 热敏电阻的温度特性

热敏电阻是由半导体材料制成的,是对温度变化表现非常敏感的电阻元件。热敏电阻能测量温度的微小变化,并且体积小,工作稳定,结构简单。因此,热敏电阻在测温技术、无线电技术、自动化和遥控等领域都有广泛应用。

实验目的

1. 了解热敏电阻的温度特性。
2. 学习用直流电桥测量热敏的温度特性。

实验原理

热敏电阻是用半导体材料制成的热敏器件,根据其电阻率随温度变化的特性,大致可分为三种类型:(1) NTC(负温度系数)型热敏电阻;(2) PTC(正温度系数)型热敏电阻;(3) CTC(临界温度系数)型热敏电阻。其中,PTC 和 CTR 型热敏电阻在某些温度范围内,其电阻值会产生急剧变化,适用于某些狭窄温度范围内的一些特殊应用;而 NTC 热敏电阻可用于较宽温度范围的测量。在温度测量中使用较多的是 NTC 型热敏电阻,本实验将测量其电阻温度特性。热敏电阻的电阻-温度特性曲线如图 4-18-1 所示。

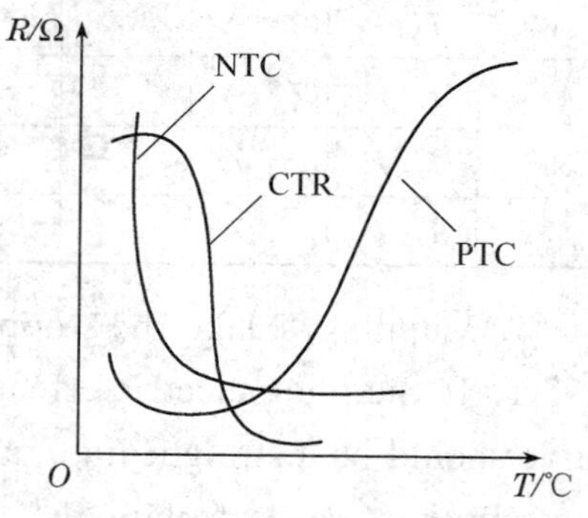

图 4-18-1 热敏电阻的电阻-温度特性曲线

NTC 型热敏电阻是由一些金属氧化物,如钴、锰、镍、铜等过渡金属的氧化物,按照不同比例的配方,经高温烧结,并采用不同的封装形式制成珠状、片状、杠状、垫圈状等各种形状。与金属导热电阻比较,NTC 型热敏电阻具有以下特点:

(1) 具有很大的负电阻温度系数,其温度测量的灵敏度比较高。

(2) 体积小,目前最小的珠状热敏电阻的尺寸可达 0.2 mm,热容量很小可作为点温或表面温度以及快速变化温度的测量。

(3) 具有很大的电阻值($10^2 \sim 10^5$ Ω),可以忽略线路中导线的电阻和接触电阻等的影响,特别适用于远距离的温度测量和控制。

(4) 制造工艺比较简单,价格便宜。

半导体热敏电阻的缺点是温度测量范围较窄。

NTC 型热敏电阻具有负温度系数，其电阻值随温度的升高而减小，电阻与温度的关系可以用下面的经验公式表示

$$R_T = A\exp(B/T) \tag{4-18-1}$$

式中，R_T 为在温度为 T 时的电阻值；T 为绝对温度；K；A 和 B 分别为具有电阻量纲和温度量纲，并且与热敏电阻的材料和结构有关的常数。

由式(4-18-1)可得到当温度为 T_0 时的电阻值 R_0，即

$$R_0 = A\exp(B/T_0) \tag{4-18-2}$$

比较式(4-18-1)和式(4-18-2)，可得

$$R_T = R_0\exp\left[B\left(\frac{1}{T} - \frac{1}{T_0}\right)\right] \tag{4-18-3}$$

由式(4-18-3)可以看出，只要知道常数 B 和在温度为 T_0 时的电阻值 R_0，就可以利用式(4-18-3)计算任意温度 T 时的 R_T 值。常数 B 可以通过实验来确定。将式(4-18-3)两边取对数，则有

$$\ln R_T = \ln R_0 + B\left(\frac{1}{T} - \frac{1}{T_0}\right) \tag{4-18-4}$$

由式(4-18-4)可以看出，$\ln R_T$ 与 $1/T$ 呈线性关系，直线的斜率就是常数 B，热敏电阻的材料常数 B 的值一般在 2000～6000 K 范围内。

热敏电阻的温度系数 α_T 定义如下

$$\alpha_T = \frac{1}{R_T} \times \frac{dR_T}{dT} = -\frac{B}{T^2} \tag{4-18-5}$$

显然，半导体热敏电阻的温度系数是负的，并与温度有关，且 α_T 是随温度降低而迅速增大。确定半导体材料常数 B 后，便可计算出这种材料的激活能 $E = Bk$（k 为玻耳兹曼常数，其值见附录）以及它的电阻温度系数。α_T 决定热敏电阻在全部工作范围内的温度灵敏度。热敏电阻的测温灵敏度比金属热电阻的高很多。例如，B 值为 4000 K，当 $T = 293.15$ K (20 ℃) 时，热敏电阻的 $\alpha_T = 4.7\%$(℃$^{-1}$)，约为铂电阻的 12 倍。

热敏电阻在不同温度时的电阻值，可采用惠斯通电桥测得。

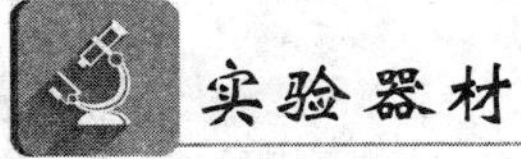

实验器材

热敏电阻、控温仪、烧杯（或恒温箱）、箱式惠斯通电桥、直流电稳压电源等。

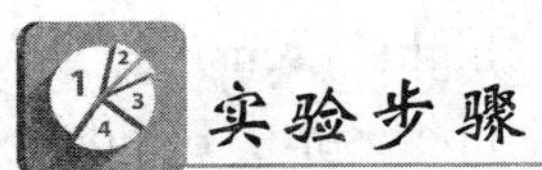

实验步骤

1. 按图 4-18-2 所示的实验装置图接好电路，安置好仪器。

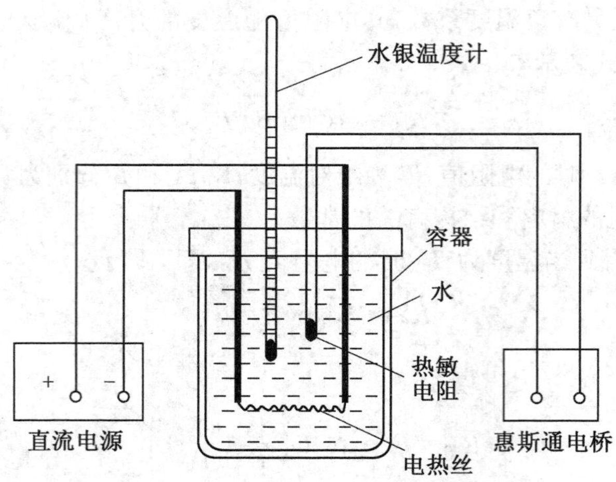

图 4-18-2　热敏电阻实验装置图

2. 在容器内盛入水，开启直流电源开关，在电热丝中通以 2.5～3.0 A 的电流，对水加热，使水温逐渐上升，温度由水银温度计读出。热敏电阻的两条引出线连接到惠斯通电桥的待测电阻“R_x”接线柱上。

3. 升温的过程，测试温度从 20 ℃开始，每增加 5 ℃，测量一次 R_x 的值，直到温度为 80 ℃为止。

4. 降温的过程，测试温度从 80 ℃开始，每降低 5 ℃，测量一次 R_x 的值，直到温度为 20 ℃为止。

5. 求升温降温过程的平均值，作 $\ln R_T \sim 1/T$（$T=273+t$）的直线，求此直线的斜率 B，计算出半导体热敏电阻的激活能 E 和温度系数 α_T。

思考题

1. 半导体热敏电阻与金属导热电阻比较，具有什么特点？
2. 数据记录及处理中为什么要验证 $\ln R_T$ 与 $1/T$ 是否呈线性关系？
3. 当温度变化时，NTC 热敏电阻与 P_t100 的电阻值分别做什么变化，变化的趋势各有什么特点？

Experiment 4.18　Temperature Characteristics of Thermistors

A thermistor is made of semiconductor materials. Being highly sensitive to temperature changes, the thermistor can measure tiny change in temperature. Besides, the thermistor has the advantages of small volume, stable operation and simple structure. Therefore, it has been widely applied in temperature measurement, radio technology, automation and remote control.

Experimental Objectives

1. To understand the temperature characteristics of a thermistor.

2. To learn how to use DC bridge to measure the temperature characteristics of a thermistor.

Experimental Principle

Thermistor is a thermosensitive device made of semiconductor materials. According to its different resistivity to temperature change, Thermistor can be roughly classified into three types: ① NTC (negative temperature coefficient) thermistors; ② PTC (positive temperature coefficient) thermistors; and ③ CTC (critical temperature coefficient) thermistors. The resistance of PTC and CTR thermistors change sharply within some temperature ranges, making them suitable for special applications in some narrow temperature ranges. Being suitable for the measurement with a wider temperature range, NTC thermistor is most frequently used in temperature measurement. This experiment is designed to measure the resistance-temperature characteristics of thermistor and its resistance-temperature characteristic curve is shown in Fig. 4-18-1.

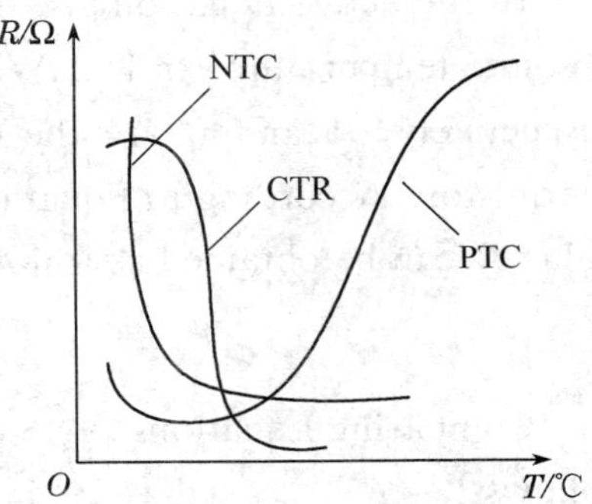

Fig. 4-18-1 Characteristic curve of thermistors

A NTC semiconductor thermistor is made of various metal oxides, including cobalt, manganese, nickel, copper and other transition metal oxides. These metal oxides are a blend at a high temperature with different proportion, and then shaped into packaging forms, like beads, flakes, bars and gaskets etc.. Compared with metal thermal resistance, the NTC semiconductor thermistor has the following characteristics:

(1) NTC semiconductor thermistor has a large negative resistance temperature coefficient, and its temperature measurement sensitivity is relatively high.

(2) NTC semiconductor thermistor has a small volume and the size of the smallest bead thermistor is about 0.2mm. As a result, its small thermal capacity can be used for the measurement of point temperature, surface temperature, and the rapid temperature changing.

(3) NTC semiconductor thermistor has a large resistance value ($10^2 \sim 10^5$ Ω), thus the influence of conductor resistance and contact resistance in the line can be ignored, making it especially suitable for long-distance temperature measurement and control.

(4) NTC semiconductor thermistor has relatively simple manufacturing process and a lower price.

The disadvantage of semiconductor thermistors lies in its narrow range of temperature measurement.

A NTC semiconductor thermistor has a negative temperature coefficient, and its resistance value decreases with the increase of temperature. The relationship between its resistance and temperature can be expressed by the following empirical formula:

$$R_T = A\exp(B/T) \tag{4-18-1}$$

In the above Equation, R_T is the resistance value at the temperature of T; T is the absolute temperature (in K). With the resistance dimension and temperature dimension respectively, A and B are the constants related to the material and structure of the thermistor. According to Equation (4-18-1), when the temperature is T_0, the resistance value R can be obtained as follows:

$$R_0 = A\exp[B/T_0] \tag{4-18-2}$$

Comparing Equations (4-18-1) and (4-18-2), their relationship can be expressed as follows:

$$R_T = R_0\exp\left[B\left(\frac{1}{T} - \frac{1}{T_0}\right)\right] \tag{4-18-3}$$

According to Equation (4-18-3), it can be seen that if the constant B and the resistance value R_0 at the temperature T_0 are given, the R_T value at any temperature T can be calculated. The constant B can be set by experiment and if the logarithm is taken from both sides of Equation (4-18-3), the equation can be converted as follows:

$$\ln R_T = \ln R_0 + B\left(\frac{1}{T} - \frac{1}{T_0}\right) \tag{4-18-4}$$

It can be seen from Equation (4-18-4) that ln R_T is in linear relationship with $1/T$ and the slope of the straight line is constant B. Given that the material constant B of the thermistor is generally in the range of 2000～6000 K, the temperature coefficient α_T of the thermistor can be defined as follows:

$$\alpha_T = \frac{1}{R_T}\frac{dR_T}{dT} = -\frac{B}{T^2} \tag{4-18-5}$$

According to Equation (4-18-5), it is easy to find that the temperature coefficient of a semiconductor thermistor is negative and increases rapidly with the decrease of temperature. If the constant B of semiconductor material is set, the activation energy $E=Bk$ (k is the Boltzmann constant; its value is shown in the Appendix) and its resistance temperature coefficient can be calculated. α_T can determine the temperature sensitivity of thermistors within the whole working range, thus the temperature measurement sensitivity of the thermistor is much higher than that of the metal thermal resistor. For example, when the B value is 4000 K and $T=293.15$ K (20 ℃), α_T is 4.7% (k^{-1}), which is about 11 times higher than that of the platinum's resistance.

The resistance of thermistors at different temperatures can be measured by Wheatstone bridges.

Experimental Instruments

Thermistor, temperature controller, beaker (or incubator), box-type Wheatstone bridge, and DC stabilized power supply, etc.

Experimental Contents and Steps

1. Connect the circuit and install the instrument according to Fig. 4-18-2.

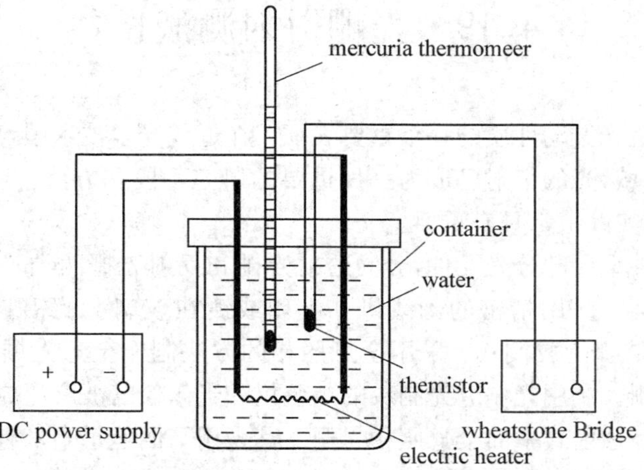

Fig. 4-18-2 Experimental setup diagram of thermistors

2. Fill the container with water, turn on the DC power switch and heat the water with a current of 2.5～3.0 A in the electric heating wire. The mercury thermometer reads the temperature while the water temperature gradually rises. Two outgoing lines of the thermistor are connected to the terminal of the resistance to be measured, that is "R_x" of the Wheatstone bridge.

3. During the heating process, the test temperature rises from 20 ℃, and the Rx value is measured for every increase of 5 ℃ until it reaches 80 ℃.

4. During the cooling process, the test temperature starts from 80 ℃, and the value of R_x is measured once every 5 ℃ until it drops to 20 ℃.

5. Calculate the average value of the temperature rise and fall and make the line ln $R_T - 1/T$ ($T = 273 + t$), then calculate the slope B of the line. Based on the collected data, the activation energy E and temperature coefficient α_T of this semiconductor thermistor can be calculated.

Questions

1. What are the distinctive characteristics of the semiconductor thermistor if compared with the metal thermal resistance?

2. Why is it necessary to verify the linear relationship between ln R_T and $1/T$ in data recording and processing?

3. During the process of temperature changing, what happens to the resistance value of the NTC thermistor and P_t 100? What are the characteristics of their changing trends?

4.19 光栅衍射测波长

光的衍射现象是光波动性质的一个重要表征。在近代光学技术中,如光谱分析、晶体分析、光信息处理等领域,光的衍射已成为一种重要的研究手段和方法。所以,研究衍射现象及其规律,在理论和实践上都具有重要意义。

衍射光栅是一种重要的分光元件,分为透射光栅和反射光栅两类。目前使用的光栅主要通过以下方法获得:① 用精密的刻线机在玻璃或镀在玻璃上的铝膜上直接划刻得到;② 用树脂在优质母光栅上复制;③ 采用全息照相的方法制作全息光栅。实验室通常使用复制光栅或全息光栅。利用光栅分光制成的单色仪和光谱仪已被广泛应用。

本实验使用的是透射式平面全息照相光栅。其光栅平面上每毫米有 300 条(或 600 条)平行的直痕迹,痕迹处由于散射不易透光,光线只能从痕迹间的狭缝中通过,因此光栅实际上是一排密集、均匀而又平行的狭缝。

实验目的

1. 进一步熟悉分光计的调节和使用。

2．了解平面光栅的构造与特性。

3．掌握用光栅测量单色光波波长的原理与方法。

4．加深理解光栅衍射公式及其成立条件。

实验原理

如果平行单色光垂直地照射到光栅平面上，则透过各狭缝的光线因衍射将向各个方向传播，经透镜会聚后相互干涉，并在透镜焦平面上形成一系列被相当宽的暗区隔开的间距不同的明条纹。

按照光栅衍射理论，衍射光谱中明条纹的位置由光栅方程决定

$$d\sin\theta_k = \pm k\lambda \ (k=0,1,2,3,\cdots) \tag{4-19-1}$$

式中，d 为光栅常数；λ 为入射光线的波长；k 为明条纹的级数；θ_k 为 k 级明条纹的衍射角，如图 4-19-1 所示。$k=0$ 为中央明条纹，k 的正负两组值所对应的两组光谱线，对称地分布在零级光谱的两侧。因此，若光栅常数 d 为已知，在实验中测定了某谱线的衍射角 θ_k 和对应的级数 k，则可根据式(4-19-1)求出该谱线的波长 λ；反之如果波长 λ 为已知，则可求出光栅常数 d。

如果入射光是复色光，由于波长不同，其衍射角 θ_k 也各不相同，于是复色光被分解，在中央 $k=0$ 时，任何波长的光均满足上式，也即在 $\theta=0$ 的方向上，各种波长的光重叠在一起，形成中央零级明条纹，仍为白色。两侧对称地分布着按波长大小的顺序依次排列的彩色谱线。这样就把复色光分解为单色光。

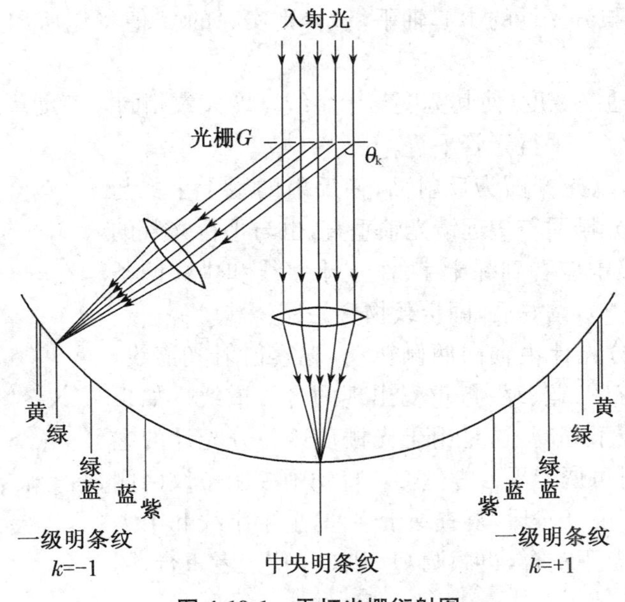

图 4-19-1　汞灯光栅衍射图

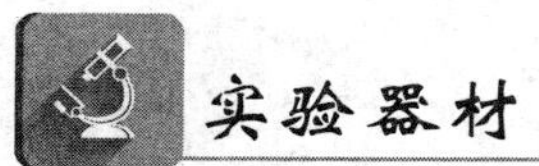

实验器材

分光计、平面反射镜、汞灯、全息照相光栅。

实验内容和步骤

一、调整分光计

光栅按图 4-19-2 所示位置放置，光栅面经过螺丝 G_1，与螺丝 G_2、G_3 连线垂直。将光栅作平面镜，调节分光计达到测量要求。

1. 用自准直法调节望远镜，使其聚焦于无穷远。
2. 调节望远镜光轴与分光计主轴垂直。
3. 调节平行光管，使其聚焦于无穷远。
4. 调节平行光管光轴与分光计主轴垂直。

具体步骤见“4.15 分光计的调整和使用”内容。

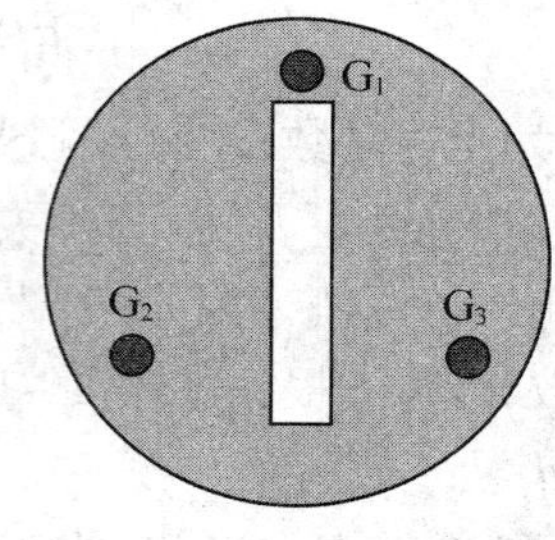

图 4-19-2　光栅在分光计载物台的放置

二、调节光栅

入射的平行光要垂直照射光栅表面，否则左右两侧的衍射角将不对称，$d\sin\theta_k=\pm k\lambda$ 不适用；光栅痕迹要与分光计的中心轴平行，否则实际的 d 值将比所用光栅的光栅常数大(为什么?)。

1. 调节狭缝至适当宽度，使其宽度与十字线宽度大致相同。转动望远镜使叉线的竖线对准狭缝，此时望远镜与平行光管平行。

2. 转动载物台，以光栅面为反射面，使反射的绿十字像与上十字线重合，此时光栅面与望远镜光轴垂直，也与平行光管的光轴垂直。在望远镜中应看到绿十字像、上十字线和中央明条纹三者仍重合(三合一)，调好后，固定载物台。

3. 将望远镜绕分光计主轴向两侧转动，观察衍射光谱线，在中央明条纹两侧有紫、蓝、绿、黄重复出现的各级谱线。如果中央明条纹两侧谱线不等高，这是由于光栅痕迹与分光计的主轴不平行所致。此时可调节平台螺丝 G_1，直到中央明条纹两侧的衍射谱线在同一高度。然后，复查绿十字像，上十字线和中央明条纹三者仍重合，若不重合，再微调 G_2 和 G_3，使三者重合(图 4-19-3)。

图 4-19-3　狭缝像、绿十字像和叉丝竖线“三合一”

三、测量汞灯谱线的衍射角

1．使望远镜叉丝的竖线对准中央明条纹，记录标志其角坐标的两窗口刻度盘读数。

2．逐次对准中央明条纹两侧的蓝、绿、黄1、黄2的二级谱线，记录标志其角坐标的两窗口刻度盘读数。

3．计算各谱线的衍射角，用测得的衍射角，根据光栅方程式(4-19-1)计算相应光波的波长。

注意事项

1．望远镜、平行光管上的镜头、平面镜镜面、光栅表面均不能用手摸拭。有尘埃等时，应用擦镜纸轻轻揩。但光栅的药膜面不能揩擦，必要时用清水缓缓冲洗；

2．望远镜和游标盘在止动螺钉旋紧的情况下不能强行扳转它们，以免损伤转轴。为此，每次转动望远镜和游标盘前，先检查一下止动螺钉是否放松；

3．在调整分光计的过程中一定要耐心按正确步骤进行调整；

4．平面镜、光栅等要放置好，以免摔破；

5．测量时，刻度盘的零刻线经过游标零刻线，需加上360°再计算。

思考题

1．光栅分光与棱镜分光有何异同？

2．测量时光栅光谱的级次是如何确定的？

3．望远镜对准平面全息透射光栅观察时，发现不止一个自准直分划反射像应如何解释？

4．利用本实验装置如何测定光栅常数？

5．怎样利用光栅方程 $d\sin\theta_k=\pm k\lambda$ 推导间接测量的误差公式？

Experiment 4.19　Wavelength Measurement with Diffraction Grating

Experiment Motivation

The diffraction of light is an important characterization of the wave properties of light. As an important research method, light diffraction has been widely applied in the spectral analysis, crystal analysis and optical information processing, etc.. The study of diffraction phenomenon and its relevant principles are of great significance both in the theory and practice.

Diffraction grating is an important spectroscopic component and it can be divided into two types: transmission grating and reflection grating. The grating used at present is mainly obtained by the following methods: 1. Directly engraving on glasses or aluminum film coated on glass with delicate marking-Line machine. 2. Multiple engraving on high quality master grating with resin. 3. Holographic grating can be produced with holographic method. Replica grating and holographic grating are commonly used in the laboratory. Monochromater and spectrometer made by grating spectrometry have been widely applied in various experiments. Transmissive planar holographic grating is adopted in this experiment. There are 300 (or 600) parallel straight traces per millimeter on the grating plane. It is not easy for light to transmit at trace due to light scattering and thus light can only pass through the slit between the traces. In short, the grating is actually a row of dense, uniform and parallel slits.

Experimental Objectives

1. To comprehend the key points about the adjustment and operation of the spectrometer.

2. To understand the configuration and properties of diffraction grating.

3. To master the principles and methods of wavelength measurement for monochromatic light.

4. To have a better understanding about the formula of Diffraction Grating and its application criteria.

Experimental Principle

If the parallel monochromatic light irradiates vertically on the grating plane, the light passing through each slit will propagate in all directions due to diffraction. After converging by the lens, the light will interfere with each other and form a series of bright fringes with different spacing separated by quite wide dark areas on the focal plane of the lens.

According to the grating diffraction theory, the position of the bright stripes in the diffraction spectrum is determined by the following equation:

$$d\sin\theta_k = \pm k\lambda \quad (k=0,1,2,3,\cdots) \tag{4-19-1}$$

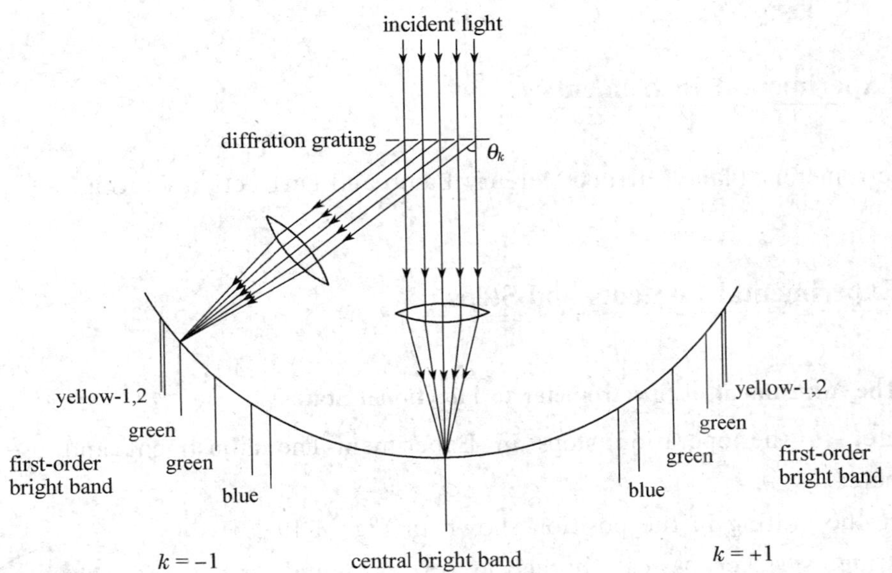

Fig. 4-19-1　Mercury light diffraction grating

The above Equation (4-19-1) is the formula of diffraction grating, where d is diffraction grating constant, λ is the wavelength of incident light, k is the orders of bright stripes, and θ_k is the diffraction angle of bright fringes with order K as is shown in Fig. 4-19-1. The stripe with $k=0$ is the central bright stripe. Two groups of spectral lines corresponding to the two groups of positive and negative values of k are symmetrically distributed on each side of the zero-level spectrum.

Consequently, if the grating constant d is given, and the diffraction angle θ_k and its corresponding order k of a certain line have been determined in the experiment, the wavelength of the line can be obtained according to the above formula. On the other hand, if the wavelength λ is known, the grating constant d can also be calculated based on the above formula.

If the incident light is a polychromatic light, the diffraction angle θ_k is different due to the different wavelength.

When the polychromatic light is decomposed, the light with different wavelength can satisfy the above formula under the condition of the center $k=0$. To put it in another way, in the direction of $\theta=0$, the light with various wavelengths overlaps to form a white central zero-order bright stripe. The color lines arranged in the order of wavelength are symmetrically distributed on both sides. In this way, the polychromatic light is decomposed into monochromatic light.

Experimental Instruments

Spectrometer, Plane Mirror, Mercury Lamp and Diffraction Grating.

Experimental Contents and Steps

1. The Adjustment of Spectrometer to Functional Status

(Refer to the operation steps in Experiment-The adjustment and use of the spectrometer)

Place the grating in the position shown in Fig. 4-19-2. The grating surface passes through screw G_1 and is perpendicular to the connection line between screws G_2 and G_3. Set the grating as a planar mirror and adjust the spectrometer to meet the measurement requirements.

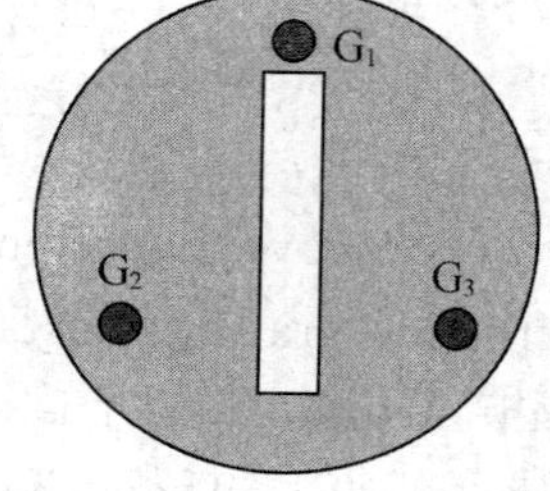

Fig. 4-19-2 Placement of diffraction grating on the objective table

(1) Adjust the telescope by self-alignment and keep it focused on infinity.

(2) Adjust the optical axis of the telescope perpendicular to the main axis of the spectrometer.

(3) Adjust the parallel light tubes and keep it focused on infinity.

(4) Adjust the parallel optical tube axis perpendicular to the spectrometer axis.

2. Adjustment of Diffraction Grating

The incident parallel light should irradiate the grating surface vertically, otherwise the diffraction angles on the left and right sides will be asymmetric, resulting in the inapplicability of the formula $d\sin\theta k = \pm k\lambda$. The grating trace should be parallel to the central axis of the spectrometer, or the actual value of d will be larger than the grating constant (why?)

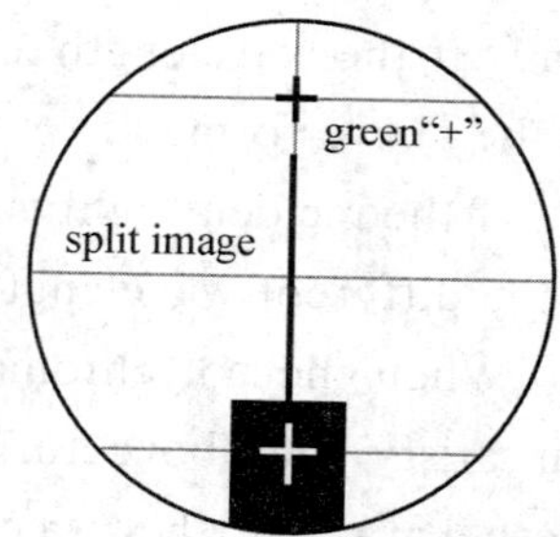

Fig. 4-19-3 Merging of slit image, green cross and the vertical line of the upward cross line

(1) Adjust the slit in the parallel tube to a suitable width so that its width is roughly the same as that of the "cross hairs". Rotate the telescope to make the vertical lines of the "cross hairs" align with the slit and the telescope parallel to the collimator.

(2) Tune the stage and use the grating surface as the reflecting surface to make the

reflected green cross image coincide with the upper cross line. Under such a condition, the grating surface is perpendicular to the optical axis of the telescope and the collimator. Through the telescope, it should be seen that the green cross image, the upper cross line and the central bright stripe are overlapped with each other (three in one). Fix the carrier stage after the adjustment.

(3) Rotate the telescope around the principal axis of the spectrometer to both sides and observe the diffraction spectral lines. The spectral lines with purple, blue, green and yellow on both sides of the central bright stripe keep emerging. The unparallelism between the grating trace and the main axis of the spectrometer may cause the unequal height between the spectral lines on both sides of the central bright stripe. To address this problem, adjust the platform screw G_1 until the diffraction lines on both sides of the central bright stripe reach the same height. Thereafter, recheck the position of green cross image, the upper cross line and the central bright stripe. Then tune the G_2 and G_3 until the three of them coincide with each other (see Fig. 4-19-3).

3. Measure the Diffraction Angle for the Spectrum of Mercury Lamp

(1) Align the vertical line of the cross of the telescope with the central bright band. Record the angular coordinates from the twin window dail readings.

(2) Align the secondary spectral lines of blue line, green line and yellow 1 and yellow 2 lines on both sides of the central bright band gradually. Record the corresponding angular coordinates from the twin window dail readings.

(3) Calculate the diffraction angles for the four spectral lines in step (2) and calculate the corresponding wavelength from the equation of diffraction grating.

Precautions

(1) Do not touch the surfaces ofthe telescope, plane mirror or diffraction grating. Please use special lens cleaning tissue (not for diffraction grating) to clean the dust. Rinse slowly the diffraction grating with clear water if necessary.

(2) Do not force the rotation of the telescope when it is locked (fastened) with a stabilizing screw to avoid damaging the axle. Check whether the stabilizing screw is loose before rotating.

(3) Follow the steps for spectrometer adjustment carefully and patiently (refer to the operation steps in experiment: The adjustment and use of the spectrometer)

(4) Carefully place the plane mirror and diffraction grating to avoid breaking.

(5) If the zero scale passes the Vernier zero line, the measured value should be added 360° before the final calculation.

Questions

(1) What are the similarities and differences between the diffraction grating and prismatic decomposition?

(2) How to determine the order of the diffracted spectral lines during measurement?

(3) Align the telescope with the surface of the diffraction grating, only to find more than one autocollimation graticule reflection image. How to account for this phenomenon?

(4) How to measure the constant of the diffraction grating by using the current experiment apparatus?

(5) How to deduce the error formula for indirect measurement with the diffraction grating equation $d\sin\theta_k = \pm k\lambda$?

4.20 用牛顿环测量透镜的曲率半径

在光学发展史上,光的干涉实验验证了光的波动性。当薄膜层的上、下表面有一很小的倾角时,由同一光源发出的光,经薄膜的上、下表面反射后在上表面附近相遇时产生干涉,并且厚度相同的地方形成同一干涉条纹,这种干涉就叫等厚干涉。光的等厚干涉原理在生产实践中具有广泛的应用,它可用于检测透镜的曲率,测量光波波长,精确地测量微小长度、厚度和角度,检验物体表面的光洁度、平整度等。

实验目的

1. 观察光的等厚干涉现象,了解等厚干涉的特点。
2. 学习用干涉方法测量平凸透镜的曲率半径和微小待测物的厚度。
3. 掌握读数显微镜的原理和使用。

实验原理

牛顿环是由一块曲率半径很大的平凸透镜的凸面放在一块光学平板玻璃上构成的,如图 4-20-1 所示,在平凸透镜和平板玻璃的上表面之间形成一层空气薄膜,其厚度由中心到边缘逐渐增加,当平行单色光垂直照射到牛顿环上时,经空气薄膜层上、下表面反射的光在凸面附近相遇产生干涉,其干涉图样是以玻璃接触点为中心的一组明暗相间的圆环,如图

4-20-2 所示。

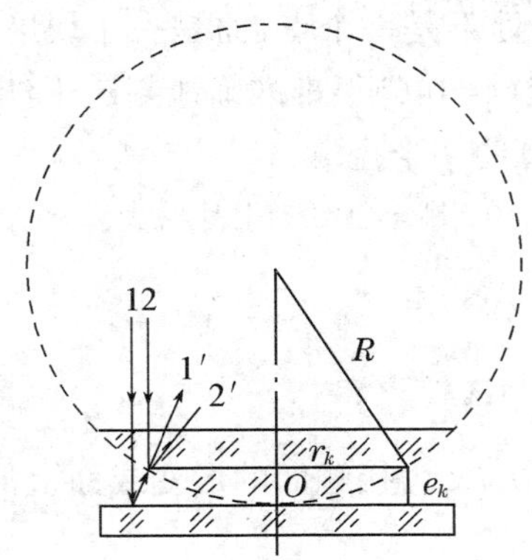

图 4-20-1　牛顿环装置示意图

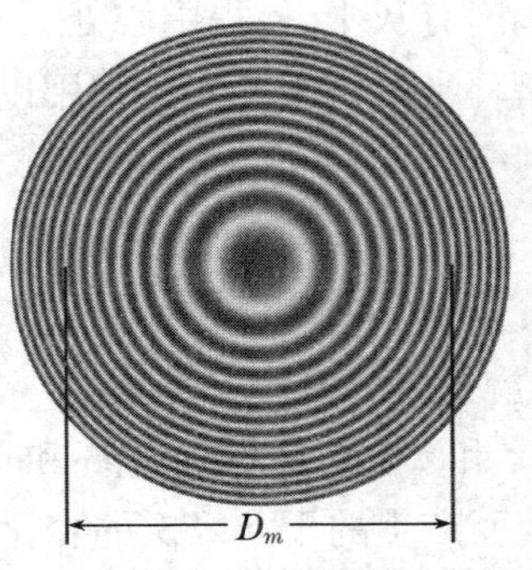

图 4-20-2　牛顿环干涉图样

设平凸透镜的曲率半径为 R，与接触点 O 相距为 r_k 处的空气薄层厚度为 e_k，那么由几何关系

$$R^2=(R-e_k)^2+r_k^2=R^2-2Re_k+r_k^2-e_k^2 \tag{4-20-1}$$

因 $R\gg e_k$，所以 e_k^2 项可以被忽略，则有

$$e_k=\frac{r_k^2}{2R} \tag{4-20-2}$$

现在考虑垂直入射到 r_k 处的一束光，它经薄膜层上、下表面反射后在凸面处相遇时其光程差

$$\delta=2e_k+\lambda/2 \tag{4-20-3}$$

式中，$\lambda/2$ 为光从平板玻璃表面向空气层反射时的半波损失（从折射率大的介质向折射率小的介质反射时有半小损失）。

将式(4-20-2)代入式(4-20-3)得

$$\delta=\frac{r_k^2}{R}+\frac{\lambda}{2} \tag{4-20-4}$$

由干涉理论，产生暗环的条件为

$$\delta=(2k+1)\frac{\lambda}{2}\ (k=0,1,2,3,\cdots) \tag{4-20-5}$$

由式(4-20-4)和式(4-20-5)可以得出，第 k 级暗纹的半径

$$r_k^2=kR\lambda \tag{4-20-6}$$

所以只要测出 r_k，如果已知光波波长 λ，即可求出曲率半径 R；反之，已知曲率半径 R 也可由式(4-20-6)求出波长 λ。

式(4-20-6)是在透镜与平玻璃面相切于一点（$e_0=0$）时的情况，但实际测量时需要考虑

各种因素的影响。实际测量时可能观测到的牛顿环中心是一个或明或暗的小圆斑,这是因为接触面间或有弹性形变,使得透镜与平玻璃接触处不是一个点,而是一小块区域;或因透镜与平玻璃间有灰尘,使得中心处 $e_0>0$,所以用式(4-20-6)很难准确地判定干涉级次 k,也不易测准暗环半径。因此实验中用以下方法来计算曲率半径 R。

由式(4-20-6),第 m 环暗纹和第 n 环暗纹的直径(图 4-20-2)可表示为:

$$D_m^2=4(m+x)R\lambda \tag{4-20-7}$$

$$D_n^2=4(n+x)R\lambda \tag{4-20-8}$$

式中,$m+x$ 和 $n+x$ 分别为 m 环和 n 环的干涉级次;x 为接触面的形变或面上的灰尘所引起光程改变而产生的干涉级次的变化量。

将式(4-20-7)和式(4-20-8)式相减得到

$$D_m^2-D_n^2=4(m-n)R\lambda \tag{4-20-9}$$

则曲率半径

$$R=\frac{D_m^2-D_n^2}{4(m-n)\lambda} \tag{4-20-10}$$

从式(4-20-10)可知,只要测出第 m 环直径和第 n 环直径以及数出环数差 $(m-n)$,就无需确定各环的级数和圆心的位置。

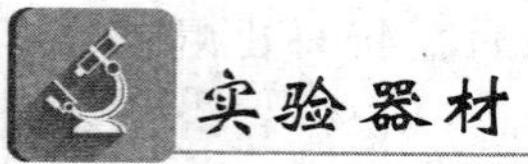

实验器材

读数显微镜、钠光灯、牛顿环仪。

实验步骤

一、观察牛顿环的干涉图样

1. 调整牛顿环仪的 3 个调节螺丝,在自然光照射下能观察到牛顿环的干涉图样,并将干涉条纹的中心移到牛顿环仪的中心附近;

2. 把牛顿环测试仪置于显微镜的正下方,使单色光源与读数显微镜上 45°角的反射透明玻璃片等高。旋转反射透明玻璃,直至从目镜中能看到明亮均匀的光照;

3. 调节读数显微镜的目镜,使十字叉丝清晰;自下而上调节物镜直至观察到清晰的干涉图样。移动牛顿环仪,使中心暗斑(或亮斑)位于视域中心,调节目镜系统,使叉丝横丝与读数显微镜的标尺平行,消除视差。平移读数显微镜,观察待测的各环左右是否都在读数显微镜的读数范围之内。

二、测量牛顿环的直径

1．松开读数显微镜球体固定螺丝，转动目镜球体，使叉丝的水平线与牛顿环沿 x 轴移动的方向一致，即转动 x 轴向测微器鼓轮时，叉丝的水平线始终与某一移动的牛顿环相切，若不相切，则继续调节球体，直至相切为止。同一暗环在左右两边的坐标值差即为该环的直径；

2．选取要测量的 m 和 n（各 5 环），如取为 20，19，18，…，11，从中央暗斑开始数起，向右移动，当数到 25 环时，将鼓轮反转，使 y 方向的叉丝与暗环的切点左移至 20 环时，开始记下各环的位置，记到 11 环后，再继续左移，让叉丝过中央暗斑到左边，记下左边的 11，12，…，20 环时的位置值，每隔 5 环组成一对数。

数据处理

1．将实验数据记入下表，根据式(4-20-10)计算半径的平均值。

环的级别	m	20	19	18	17	16
环的位置/mm	右（x_m）					
	左（x_m）					
环的直径/mm	D_m					
环的级别	n	15	14	13	12	11
环的位置/mm	右（x_n）					
	左（x_n）					
环的直径/mm	D_n					
$D_m^2-D_n^2$						
$D_m^2-D_n^2$						
$u_A(D_m^2-D_n^2)$						

2．计算 $D_m^2-D_n^2$ 的 A 类不确定度，其 B 类不确定度由实验室给出；

3．计算 $D_m^2-D_n^2$ 的合成标准不确定度及相对不确定度 $u_R(D_m^2-D_n^2)$；

4．由式(4-20-10)可知 $u_R(D_m^2-D_n^2)=u_R(R)$，故可得由下式求出 R 的合成标准不确定度；

$$u_R=\overline{R}\times u_R(R)=\overline{R}\times u_R(D_m^2-D_n^2)$$

用不确定度表示测量结果，即

$$R=\overline{R}+u_R$$

注意事项

1. 调节牛顿环仪时,调节螺丝不能太紧,以免中心暗斑太大,甚至损坏牛顿环仪;

2. 注意在一次测量过程中,测微鼓轮应沿一个方向旋转,中途不得反转,以免引起回程差。

1. 实验中为什么通过公式 $R=\frac{D_m^2-D_n^2}{4(m-n)\lambda}$ 求出 R,而不用更简单的函数关系式 $R=\frac{r_k^2}{k\lambda}$ 求出 R 的值?

2. 在实验中若遇到下列情况,对实验结果是否有影响?为什么?

(1) 牛顿环中心是亮斑而非暗斑。

(2) 测各个 D_m 时,叉丝交点未通过圆环的中心,因而测量的是弦长而非真正的直径。

(3) 在测量过程中,读数显微镜为什么只准单方向前进,而不准后退?

Experiment 4.20 Measuring the Curvature Radius of Lens with Newton's Ring

Throughout the history of optical development, the interference experiment of light confirms that light has the wave property.

With a small inclination angle between the upper and lower surfaces of the film layer, the light with the same source will meet near the upper surface and form the mutual interference after being reflected by the upper and lower surfaces of the film. At the places with same thickness, the same interference fringes will be formed and generally called as equal thickness interference. The principle of equal thickness interference of light can be widely applied in practice, such as detecting the curvature of a lens, measuring the wavelengths of a light wave, the accurate measurement of tiny lengths, thicknesses and angles, and the inspection of the surface smoothness and flatness.

Experimental Objectives

1. To observe the equal thickness interference of light and understand its characteristics.

2. To learn how to measure the curvature radius of a plano convex lens and the thickness of a small object by using an interference method.

3. To master the principles and effective use of the reading microscope.

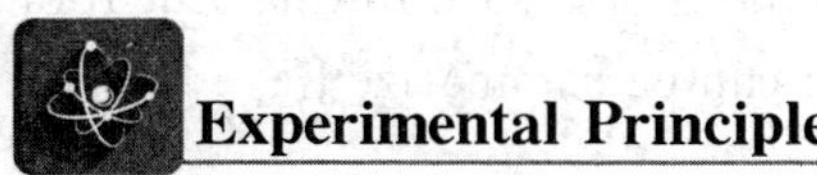

Experimental Principle

Newton's ring is composed of a plano convex lens with a large radius of curvature and an optical flat glass. As shown in Fig. 4-20-1, the convex surface of the plano convex lens is placed over the flat glass, between which lies an air film and its thickness gradually increases from the center to the edge. When the parallel monochromatic light irradiates vertically on Newton's ring, the light reflected from the upper and lower surfaces of the air film layer meets near the convex surface, resulting in the interference. The interference pattern is a group of light and dark rings centered on the glass contact point, as shown in Fig. 4-20-2.

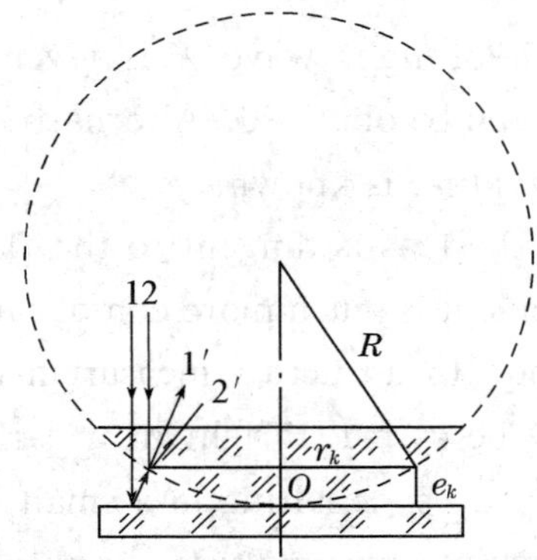

Fig. 4-20-1 Schematic diagram of Newton's rings

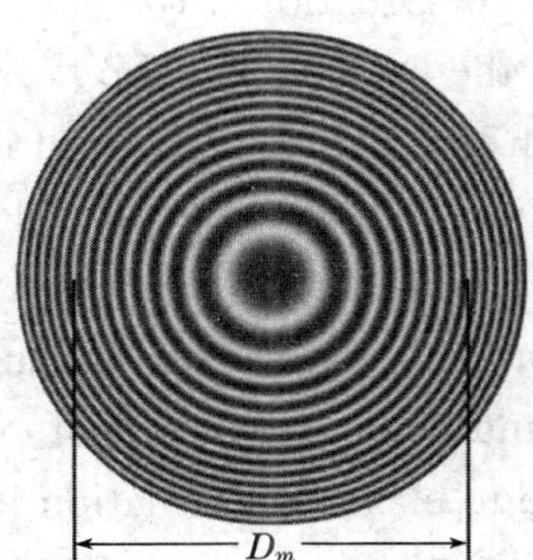

Fig. 4-20-2 Interference pattern of Newton's rings

Set the curvature radius of the plano convex lens as R, the distance from the contact point O to the air layer as r_k, and the thickness of thin air layer as e_k, and their geometric relationship can be expressed as:

$$R^2=(R-e_k)^2+r_k^2=R^2-2Re_k+r_k^2-e_k^2 \tag{4-20-1}$$

ignoring e_k^2 term as $R \gg e_k$, the above equation can be converted as:

$$e_k=\frac{r_k^2}{2R} \tag{4-20-2}$$

Suppose there is a beam of light perpendicular to r_k. When it is reflected by the upper and lower surfaces of the film layer and meets at the convex surface, its optical path difference can be expressed as:

$$\delta=2e_k+\lambda/2 \tag{4-20-3}$$

where $\lambda/2$ is the half wave loss of light reflected from the surface of flat glass to the air layer, (there is a half wave loss when the light is reflected from the medium with high refractive index to the medium with low refractive index). By substituting Equation (4-20-2) into Equation (4-20-3), the following equation can be summarized:

$$\delta=\frac{r_k^2}{R}+\frac{\lambda}{2} \tag{4-20-4}$$

According to interference theory, the conditions for generating dark rings are as follows:

$$\delta=(2k+1)\frac{\lambda}{2}\ (k=0,1,2,3,\cdots) \tag{4-20-5}$$

The radius of the k-th dark lines can be summarized from Equation (4-20-4) and Equation (4-20-5) as follows:

$$r_k^2=kR\lambda \tag{4-20-6}$$

According to Equation (4-20-6), if the wavelength of light wave λ is given, after measuring the value of r_k, the radius of curvature R can be obtained. Alternatively, the wavelength can also be calculated if the radius of curvature is known.

Equation (4-20-6) is applicable to the case that the lens is tangent to the flat glass surface at a point ($e_0=0$). Yet the practical measurement is much more complicated and the influence of various factors should be considered. In an actual measurement, the center of interference pattern of Newton's ring may be a small bright or a dark spot. This is due to elastic deformation between contact surfaces, resulting in a small contact area between the lens and the flat glass rather than a point. In some case, there might be some dust between the lens and the flat glass, causing $e_0>0$ at the center. As a result, it is difficult to accurately determine the interference order k or accurately measure the dark ring radius by using Equation (4-20-6).

In the experiment, the curvature radius R is often calculated with the following method:

With Equation (4-20-6), the diameters of the m-th and n-th dark rings (see Fig. 4-20-2) can be expressed as:

$$D_m^2=4(m+x)R\lambda \tag{4-20-7}$$

$$D_n^2=4(n+x)R\lambda \tag{4-20-8}$$

In the above equations, $m+x$ and $n+x$ are the interference orders of the m-th and n-th rings. x refers to the change of interference order due to the change of optical path caused by the deformation of the contact surface or the dust on the surface.

Subtract Equation (4-20-7) and Equation (4-20-8) to obtain the following equation:

$$D_m^2 - D_n^2 = 4(m-n)R\lambda \tag{4-20-9}$$

Then the radius of the curvature can be expressed as:

$$R = \frac{D_m^2 - D_n^2}{4(m-n)\lambda} \tag{4-20-10}$$

According to Equation (4-20-10), to calculate the radius of curvature, if the diameters of the m-th ring and the n-th ring are measured, and the ring number difference $m-n$ is obtained, it is unnecessary to measure the orders of each ring and the position of the center of the circle.

Experimental Instruments

Reading microscope, sodium light, and Newton's ring instrument

Experimental Contents and Steps

1. Observing the Interference Pattern of Newton's Ring

(1) While adjusting the three adjustment screws of the Newton's ring instrument, observe the interference pattern of the Newton's ring under natural light, and move the center of the interference fringe to the vicinity of the center of the Newton's ring instrument.

(2) Place the Newton's ring instrument directly under the microscope to make the monochromatic light source situated at the same height as the reflection transparent glass plate with an angle of 45° on the reading microscope. Rotate the reflective transparent glass until a bright and uniform light can be seen from the eyepiece.

(3) Adjust the eyepiece of the reading microscope to make the crosshair clear. Make the adjustment of the objective lens from the bottom to the top until a clear interference pattern can be observed. Move the Newton's ring instrument so that the central dark spot (or bright spot) is in the center of the field of vision. Adjust the eyepiece system to make the horizontal crosshair parallel to the scale of the reading microscope for the elimination of the parallax. Move the reading microscope to observe whether the measurements of the left and right rings are within the reading range of the reading microscope.

2. Measuring the Diameter of Newton's Ring

(1) Loosen the fixing screw of the reading microscope sphere and rotate the

eyepiece sphere, so that the horizontal line of the fork wire is consistent with the moving direction of Newton's ring along the x-axis. That is, when rotating the drum of the x-axis micrometer, the horizontal line of the fork wire is always tangent to a moving Newton's ring. If not, continue to adjust the sphere until it is tangent. The diameter of the same dark ring is determined by the difference of the coordinate values between the left and right sides.

(2) Select m and n (5 rings each) to be measured, such as twenty, nineteen, eighteen, …, eleven. Start counting from the central dark spot and move to the right. When counting to the twenty-five rings, reverse the drum to make the tangent point of the y-direction fork wire and the dark ring move to the left. Start to record the position of each ring when the tangent point reaches the twentieth ring. After recording the eleventh ring, continue to move to the left. Make the fork wire pass through the central dark spot to the left, and record the values of the eleventh, twelfth, …, and twentieth ring's position. The position values of every 5 rings form a logarithm.

Data Recording and Processing

1. Record the experimental data in the table below and calculate the average radius according to the Equation(4-20-10).

Ring's order	m	20	19	18	17	16
Ring's positions/mm	Right (x_m)					
	Left (x_m)					
Ring's diameters/mm)	D_m					
Ring's order	n	15	14	13	12	11
Ring's position/(mm	Right (x_n)					
	Left (x_n)					
Ring's diameter/mm	D_n					
$D_m^2-D_n^2$						
$D_m^2-D_n^2$						
$u_A(D_m^2-D_n^2)$						

2. Calculate the uncertainty of type A $D_m^2-D_n^2$, and its uncertainty of type B is provided by the laboratory.

3. Calculate the combined standard uncertainty of $D_m^2-D_n^2$ and its relative uncertainty of $u_R(D_m^2-D_n^2)$.

4. According to Equation (4-20-10), the combined standard uncertainty of R can be obtained with the following Equation:

$$u_R = \overline{R} \times u_R(R) = \overline{R} \times u_R(D_m^2 - D_n^2)$$

By using uncertainty, the final measurement result can be expressed as follows:

$$R = \overline{R} + u_R$$

Precautions

1. When adjusting the Newton's ring instrument, the adjusting screw should not be too tight so as to avoid too large dark spot in the center or even damaging the Newton's ring instrument.

2. During the process of measurement, the operator should always rotate the micrometer drum in one direction rather than reversing the direction, so as to avoid a backhaul difference。

Questions

1. Why is R calculated with the formula $R = \dfrac{D_m^2 - D_n^2}{4(m-n)\lambda}$ instead of a simpler functional relation $R = \dfrac{r_k^2}{k\lambda}$?

2. In this experiment, what kind of the effects the following situations will exert on the final results of the experiment? And why?

(1) The center of Newton's ring is a bright spot rather than a dark spot.

(2) In measuring each D_m, the intersection point of the fork wire does not pass through the center of the ring. As a result, what has been measured is the chord length instead of the real diameter.

3. In the process of measurement, why must the reading microscope only move forward in one direction rather than moving backward?

4.21 刚体转动惯量的测定

转动惯量是刚体转动中惯性大小的量度。它取决于刚体的总质量、质量分布、形状大小和转轴位置。对于形状简单、质量均匀分布的刚体,可以通过数学方法计算出它绕特定转轴的转动惯量,但对于形状比较复杂或质量分布不均匀的刚体,用数学方法计算其转动惯量是非常困难的,因而大多采用实验方法来测定。

转动惯量的测定，在涉及刚体转动的机电制造、航空、航天、航海、军工等工程技术和科学研究中具有十分重要的意义。测定转动惯量常采用扭摆法或恒力矩转动法，本实验采用恒力矩转动法测定转动惯量。

实验目的

1. 学习用恒力矩转动法测定刚体转动惯量的原理和方法。
2. 观测刚体的转动惯量随其质量，质量分布及转轴不同而改变的情况，验证平行轴定理。

实验原理

一、恒力矩转动法测定转动惯量的原理

根据刚体的定轴转动定律：

$$M=J\beta \tag{2-21-1}$$

只要测定刚体转动时所受的总合外力矩 M 及该力矩作用下刚体转动的角加速度 β，则可计算出该刚体的转动惯量 J。

设以某初始角速度转动的空实验台转动惯量为 J_1，未加砝码时，在摩擦阻力矩 M_μ 的作用下，实验台将以角加速度 β_1 做匀减速运动，即：

$$-M_\mu=J_1\beta_1 \tag{4-21-2}$$

将质量为 m 的砝码用细线绕在半径为 R 的实验台塔轮上，并让砝码下落，系统在恒外力作用下将做匀加速运动。若砝码的加速度为 a，则细线所受张力为 $T=m(g-a)$。若此时实验台的角加速度为 β_2，则有 $a=R\beta_2$。细线施加给实验台的力矩为 $TR=m(g-R\beta_2)R$，此时有：

$$m(g-R\beta_2)R-M_\mu=J_1\beta_2 \tag{4-21-3}$$

将(4-21-2)、(4-21-3)两式联立消去 M_μ 后，可得：

$$J_1=\frac{mR(g-R\beta_2)}{\beta_2-\beta_1} \tag{4-21-4}$$

同理，若在实验台上加上待测物体后系统的转动惯量为 J_2，加砝码前后的角加速度分别为 β_3 与 β_4，则有：

$$J_2=\frac{mR(g-R\beta_4)}{\beta_4-\beta_3} \tag{4-21-5}$$

由转动惯量的迭加原理可知，待测物体的转动惯量 J_3 为：

$$J_3 = J_2 - J_1 \tag{4-21-6}$$

测得 R、m 及 β_1、β_2、β_3、β_4，由(4-21-4)，(4-21-5)，(4-21-6)式即可计算待测物体的转动惯量。

二、β的测量

实验中采用通用计时计数记录遮挡次数和相应的时间。固定在载物台圆周边缘相差 π 角的两遮光细棒，每转动半圈遮挡一次固定在底座上的光电门，即产生一个计数光电脉冲，计数器记下遮挡次数 k 和相应的时间 t。若从第一次挡光($k=0$，$t=0$)开始计次、计时，且初始角速度为 ω_0，则对于匀变速运动中测量得到的任意两组数据(k_m，t_m)、(k_n，t_n)，相应的角位移 θ_m、θ_n 分别为：

$$\theta_m = k_m \pi = \omega_0 t_m + \frac{1}{2}\beta t_m^2 \tag{4-21-7}$$

$$\theta_n = k_n \pi = \omega_0 t_n + \frac{1}{2}\beta t_n^2 \tag{4-21-8}$$

从(4-21-7)、(4-21-8)两式中消去 ω_0，可得：

$$\beta = \frac{2\pi(k_n t_m - k_m t_n)}{t_n^2 t_m - t_m^2 t_n} \tag{4-21-9}$$

由(4-21-9)式即可计算角加速度 β。

三、平行轴定理

理论分析表明，质量为 m 的物体围绕通过质心 O 的转轴转动时的转动惯量 J_0 最小。当转轴平行移动距离 d 后，绕新转轴转动的转动惯量为：

$$J = J_0 + md^2 \tag{4-21-10}$$

四、转动惯量实验组合仪简介

刚体转动惯量实验仪如图 4-21-1 所示，绕线塔轮通过特制的轴承安装在主轴上，使转动时的摩擦力矩很小。塔轮半径为 15、20、25、30 和 35 mm 共 5 挡，可与大约 5 g 的砝码托及 3 个 5 g，3 个 10 g 的砝码组合，产生大小不同的力矩。载物台用螺钉与塔轮连接在一起，随塔轮转动。组合仪中含待测物体有 1 个圆盘、1 个圆环、两个圆柱；待测圆柱可插入载物台上的不同孔，这些孔离中心的距离分别为 45、60、75、90 和 105 mm，便于验证平行轴定理。铝制小滑轮的转动惯量与实验台相比可忽略不计。实验台两侧各有一个光电门，可根据实际情况随意选取一个接口与通用计时计数器相连。

图 4-21-1 转动惯量实验组合仪

实验中所用的通用计时计数器前面板如图 4-21-2 所示，连接电源和光电门，通过切换开关设置光电门被遮挡单次或双次为一周期，选择好计数模式后再开启电源(开启电源后不能再切换计数模式)，将光标移至“总周期”位置，按下“确定”，可通过上下按钮对测量的总周期数进行设置，设置完成后按下“确定”按钮进行保存。

NMS-JS 通用计时计数器

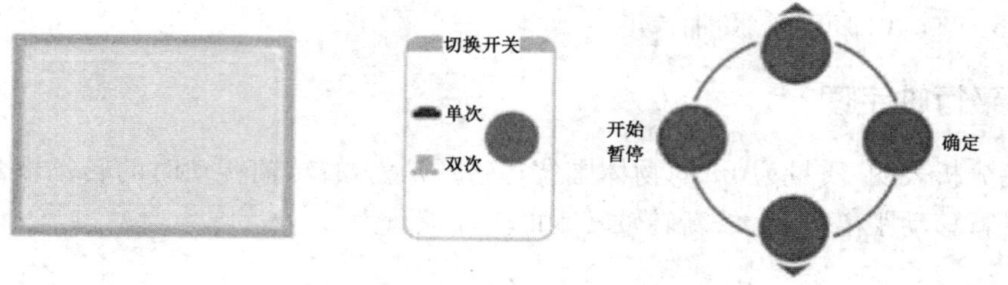

图 4-21-2 通用计时计数器

按下“开始/暂停”按钮后，仪器进入计时准备状态，当接收到光电门发来的第一个信号后开始计时，每接收一个信号为一周期，当接收的信号数达到设定的总周期数后，计时自动停止。

移动光标至“第×周期”，按下“确定”后，可使用上下键对记录的各周期数的总时间进行回查。

查看结束后，按下“确定”按钮使光标回到上一级位置，此时按下“开始/暂停”按钮，上一次记录的数据清零，仪器进入计时准备状态。

实验器材

刚体转动惯量实验组合仪、电子天平、游标卡尺。

实验内容及步骤

1．实验准备。在桌面上放置转动惯量实验仪，利用基座上的三颗调平螺钉，将仪器调平。将滑轮支架固定在实验台面边缘，调整滑轮高度及方位，使滑轮槽与选取的绕线塔轮槽等高，且其方位相互垂直，并且用数据线将通用计时计数器接口与转动惯量实验仪上的一个光电门相连。

2．利用天平测量砝码和样品的质量，游标卡尺测量样品的尺寸。

3．测量并计算实验台的转动惯量 J_1。

（1）测量 β_1

① 打开通用计时计数器为单次计数，打开电源，设置总周期数为 10；

② 用手轻轻拨动载物台，使实验台有一初始转速并在摩擦阻力矩作用下做匀减速运动；

③ 迅速按下“开始/暂停”按钮，通用计时计数器会在下一次接收到挡光信号时开始计时和计数，并在记录完预设的 15 个信号后停止计时和计数；

④ 将光标移至“第×周期”栏，查看各计时周期的时间，并将数据记入表 4-21-1 中；

⑤ 采用逐差法处理数据，用(4-21-9)式计算对应各组的 β_1 值，然后求其平均值作为 β_1 的测量值。

（2）测量 β_2

① 选择塔轮半径 R 及砝码质量，将一端打结的细线沿塔轮上开的细缝塞入，并且不重叠地密绕于所选定半径的轮上，细线另一端通过滑轮后连接砝码托上的挂钩，用手将载物台稳住；

② 按下通用计时计数器的“开始/暂停”按钮，进入计时准备状态；

③ 释放载物台，砝码重力产生的恒力矩使实验台产生匀加速转动；

④ 记录 8 组数据后停止测量，查阅、记录数据于表 4-21-1 中并计算 β_2 的测量值；

⑤ 由(4-21-4)式即可算出 J_1 的值。

（3）测量并计算实验台放上待测物体后的转动惯量 J_2，计算待测物体的转动惯量 J_3 并与理论值比较。

将待测物体放在载物台上，并使其几何中心轴与塔台转轴中心重合，按与测量 J_1 同样的方法可分别测量未加砝码的角加速度 β_3 与加砝码后的角加速度 β_4。由(4-21-5)式可计算 J_2 的值，已知 J_1、J_2，由(4-21-6)式可计算待测物体的转动惯量 J_3。

已知圆盘、圆柱绕几何中心轴转动的转动惯量理论值为：

$$J=\frac{1}{2}mR^2 \tag{4-21-11}$$

圆环绕几何中心轴的转动惯量理论值为：

$$J=\frac{1}{2}m(R_{外}^2+R_{内}^2) \tag{4-21-12}$$

(4) 验证平行轴定理

将两圆柱体对称插入载物台上与中心距离为 d 的圆孔中，测量并计算两圆柱体在此位置的转动惯量。将测量值与理论值比较，验证平行轴定理。

数据记录与处理

1. 测量实验台空载时的角加速度。

表 4-21-1 实验台空载时的角加速度

匀减速						匀加速 $R_{塔轮}=$ mm $m_{砝码}=$ g					
k	1	2	3	4	平均	k	1	2	3	4	平均
$t(s)$						$t(s)$					
k	5	6	7	8		k	5	6	7	8	
$t(s)$						$t(s)$					
$\beta_1(1/s^2)$						$\beta_2(1/s^2)$					

2. 测量实验台加载待测圆环后的角加速度，由(4-21-6)式计算待测圆环的转动惯量测量值，由(4-21-2)式得到待测圆环转动惯量的理论值，并求出相对误差。

表 4-21-2 实验台加载待测圆环后的角加速度

$R_{外}=$ mm $R_{内}=$ mm $m_{圆环}=$ g

匀减速						匀加速 $R_{塔轮}=$ $m_{砝码}=$ g					
k	1	2	3	4	平均	k	1	2	3	4	平均
$t(s)$						$t(s)$					
k	5	6	7	8		k	5	6	7	8	
$t(s)$						$t(s)$					
$\beta_3(1/s^2)$						$\beta_4(1/s^2)$					

3. 测量实验台加载待测圆柱后的角加速度，由(4-21-6)式计算其转动惯量测量值，与理论值相比较，求出相对误差。同时验证平行轴定理。

表 4-21-3 实验台加载待测圆柱后的角加速度

d= mm $R_{圆柱}$ = mm $m_{圆柱}\times 2$ = g

匀减速						匀加速 $R_{塔轮}$ = mm $m_{砝码}$ = g					
k	1	2	3	4	平均	k	1	2	3	4	平均
$t(s)$						$t(s)$					
k	5	6	7	8		k	5	6	7	8	
$t(s)$						$t(s)$					
$\beta_3(1/s^2)$						$\beta_4(1/s^2)$					

1. 分析影响实验精度的各种因素，如何减少这些因素的影响？
2. 是否可以通过实验和作图，既求出转动惯量，又求出摩擦力矩？

Experiment 4.21 Measurement of Moment of Inertia

Moment of inertia is a measure of the inertia in the rotational motion of a rigid body. It depends on the total mass of the body, the distribution of mass, the size and shape of the body, and the position of the axis of rotation. For rigid bodies with simple shapes and uniform mass distribution, the moment of inertia about a specific axis of rotation can be calculated using mathematical methods. However, for bodies with complex shapes or non-uniform mass distribution, it is very difficult to calculate the moment of inertia using mathematical methods, and experimental methods are mostly used to measure it.

The measurement of moment of inertia has great significance in engineering technology and scientific research related to mechanical and electrical manufacturing, aviation, aerospace, navigation, military industry, and so on. The moment of inertia is often measured using either the torsion pendulum method or the constant torque rotation method. In this experiment, the constant torque rotation method is used to measure the moment of inertia.

Experimental Objectives

1. To learn the principles and methods of measuring the moment of inertia of a rigid

body using the constant torque rotation method.

2. To observe how the moment of inertia of a rigid body changes with its mass, mass distribution, and different axes of rotation, and to verify the parallel axis theorem.

3. To learn how to use a general-purpose timer counter to measure time.

Experimental Principle

1. Principles of Measuring Moment of Inertia Using the Constant Torque Rotation Method

According to the law of rotational motion of a rigid body about a fixed axis:

$$M=J\beta \tag{4-21-1}$$

The moment of inertia J of the body can be calculated if the total external torque M acting on the body and the angular acceleration β of the body are measured.

Assume that the moment of inertia of an empty experimental platform rotating at an initial angular velocity is J_1. When no weights are added, the experimental platform will undergo uniform deceleration with angular acceleration β_1 under the action of the frictional torque M_μ:

$$-M_\mu=J_1\beta_1 \tag{4-21-2}$$

A weight of mass m is hung on the tower wheel of the experimental platform with a thin wire, and the system undergoes uniform acceleration under the action of a constant external force. If the acceleration of the weight is a, the tension in the wire is $T=m(g-a)$. If the angular acceleration of the experimental platform is β_2, the force moment applied to the experimental platform by the wire is $TR=m(g-R\beta_2)R$, where R is the radius of the tower wheel. Therefore, we have:

$$m(g-R\beta_2)R-M_\mu=J_1\beta_2 \tag{4-21-3}$$

Eliminating M_μ by combining equations (4-21-2) and (4-21-3), we obtain:

$$J_1=\frac{mR(g-R\beta_2)}{\beta_2-\beta_1} \tag{4-21-4}$$

Similarly, if the system has a moment of inertia J_2 when the object to be tested is added to the experimental platform, and the angular accelerations before and after the weight is added are β_3 and β_4, respectively, then we have:

$$J_2=\frac{mR(g-R\beta_4)}{\beta_4-\beta_3} \tag{4-21-5}$$

According to the principle of superposition of moment of inertia, the moment of inertia J_3 of the object being tested is:

$$J_3 = J_2 - J_1 \tag{4-21-6}$$

By measuring R, m, β_1, β_2, β_3 and β_4, the moment of inertia of the object being tested can be calculated using Eqs. (4-21-4), (4-21-5), and (4-21-6).

2. Measurement of β

In the experiment, a general-purpose timer counter is used to record the number of times the light is blocked and the corresponding time. Two light-blocking rods are fixed on the circumference of the load platform at a π angle difference, and the light-blocking gates fixed on the base are blocked once every half circle, generating a counting photocurrent pulse. If we start counting from the first light blocking ($k=0$, $t=0$) and the initial angular velocity is ω_0, then for any two sets of data (k_m, t_m) and (k_n, t_n) obtained in uniform acceleration motion, their corresponding angular displacements θ_m and θ_n are:

$$\theta_m = k_m \pi = \omega_0 t_m + \frac{1}{2}\beta t_m^2 \tag{4-21-7}$$

$$\theta_n = k_n \pi = \omega_0 t_n + \frac{1}{2}\beta t_n^2 \tag{4-21-8}$$

Eliminating ω_0 from Eqs. (4-21-7) and (4-21-8), we get:

$$\beta = \frac{2\pi(k_n t_m - k_m t_n)}{t_n^2 t_m - t_m^2 t_n} \tag{4-21-9}$$

Eq. (4-21-9) can be used to calculate the angular acceleration β.

3. Parallel Axis Theorem

Theoretical analysis shows that the moment of inertia J_0 of an object with mass m rotating about an axis passing through its center of mass O is minimum. When the axis of rotation is moved parallel to a distance of d, the moment of inertia of the object rotating about the new axis is:

$$J = J_0 + md^2 \tag{4-21-10}$$

Experimental Instruments

1. Rigid Body Moment of Inertia Experiment Instrument

The rigid body moment of inertia experiment instrument is shown in Fig. 4-21-1. The wire tower wheel is installed on the main spindle through a special bearing to minimize the frictional torque during rotation. The tower wheel has five gears with different radii of 15, 20, 25, 30, and 35mm, which can be combined with a support for weights of about 5g and three weights of 5g and 10g to generate torques of different magnitudes. The load platform is connected to the tower wheel by screws and rotates with the tower wheel. The included test samples are one disk, one ring, and two cylinders, with their geometric dimensions and mass marked to facilitate comparison of the measured values of moment of inertia with theoretical calculations. The cylinders can be inserted into different holes on the load platform, which are located at distances of 45, 60, 75, 90, and 105mm from the center, to verify the parallel axis theorem. The moment of inertia of the small aluminum pulley can be ignored compared to the experimental platform. There are two photoelectric gates on both sides of the experimental platform, and one of the interfaces can be connected to a general-purpose timer counter according to the actual situation.

Fig. 4-21-1 Rigid body moment of inertia experiment instrument

2. General-Purpose Timer Counter

NMS-JS Universal timer counter

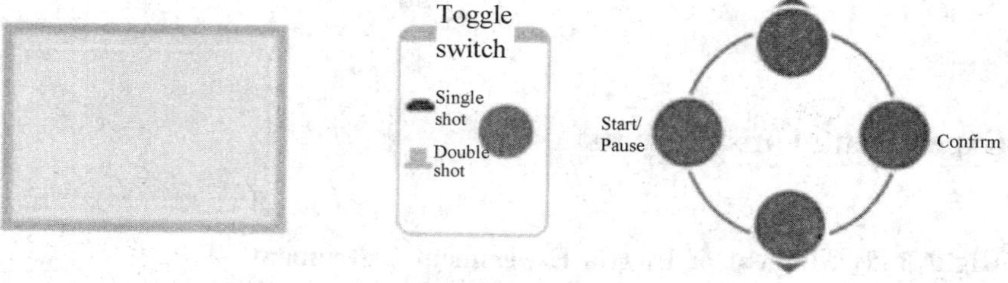

CHENG DU NI MU Digital Technology Co., Ltd.

Fig. 4-21-2 General-purpose timer counter

The front panel of the general-purpose timer counter is shown in Fig. 4-21-2. It is connected to the power supply and photoelectric gates. The switching switch is used to set whether a single or double blocking of the photoelectric gate counts as one cycle. After selecting the counting mode, the power supply is turned on (the counting mode cannot be changed after the power is turned on), and the cursor is moved to the "total cycles" position. Press the "OK" button, and use the up and down buttons to set the total number of cycles measured. After setting, press the "OK" button to save the setting.

After pressing the "start/pause" button, the instrument enters the timing preparation state. When the first signal is received from the photoelectric gate, the timer starts counting. Each signal received is counted as one cycle. When the total number of cycles set is reached, the timer automatically stops. Move the cursor to the "cycle #" position, press the "OK" button, and use the up and down keys to review the total time of each recorded cycle.

After reviewing, press the "OK" button to return the cursor to the previous level, and press the "start/pause" button to clear the data from the previous recording and enter the timing preparation state.

Experimental Contents and Steps

1. Experimental Preparation

Place the moment of inertia experiment instrument on the desktop and use the three leveling screws on the base to level the instrument. Fix the pulley bracket on the edge of the experimental platform, adjust the height and orientation of the pulley to make the pulley groove at the same height as the selected wire tower wheel groove, and their orientations are perpendicular to each other. Connect one of the photoelectric gates of the general-purpose timer counter to the interface of the moment of inertia experiment instrument using a data cable.

2. Measurement and Calculation of the Moment of Inertia J_1 of the Experimental Platform

(1) Measurement of β_1

(a) Turn on the general-purpose timer counter for single counting, turn on the power supply, and set the total number of cycles to 10.

(b) Gently shake the load platform by hand to give the experimental platform an initial rotational speed, and it will undergo uniform deceleration motion under the action of frictional torque.

(c) Quickly press the "start/pause" button. The general-purpose timer counter will start timing and counting at the next blocking signal and stop timing and counting after recording the preset 15 signals.

(d) Move the cursor to the "cycle #" column, check the time of each timing cycle, and record the data in Table 4-21-1.

(e) Use the method of differences to process the data. Form four groups by pairing the first and fifth groups, the second and sixth groups, etc. Calculate the corresponding β_1 value for each group using Eq. (4-21-9), then calculate the average value as the measured value of β_1.

(2) Measurement of β_2

(a) Select the radius R of the tower wheel and weight mass. Insert one end of a thin wire tied in a knot into the slot on the tower wheel, and wind the wire evenly around the wheel without overlapping at the selected radius. The other end of the wire passes through the pulley and is connected to the hook on the weight support. Hold the load platform steady by hand.

(b) Press the "start/pause" button of the general-purpose timer counter to enter the timing preparation state.

(c) Release the load platform. The constant torque generated by the weight's gravity causes the experimental platform to rotate uniformly with acceleration.

(d) Stop the measurement after recording 8 sets of data. Check and record the data in Table 4-21-1 and calculate the measured value of β_2.

(e) The value of J_1 can be calculated by using Eq. (4-21-4).

3. Measure and Calculate the Moment of Inertia J_2 of the Experimental Platform with the Sample and Compare the Theoretical Value of the Moment of Inertia J_3

Place the test sample on the load platform and align the geometric center axis of the sample with the axis of rotation. Measure the angular acceleration β_3 without adding weights and β_4 with weights added using the same method as measuring J_1. The value of J_2 can be calculated using Eq. (4-21-5). Given J_1 and J_2, J_3 can be calculated using Eq. (4-21-6).

The theoretical values of the moment of inertia for the disk and cylinder rotating around the geometric center axis are given by:

$$J=\frac{1}{2}mR^2 \tag{4-21-11}$$

The theoretical value of the moment of inertia for the ring rotating around the geometric center axis is given by:

$$J=\frac{m}{2}(R_{外}^2+R_{内}^2) \tag{4-21-12}$$

Calculate the theoretical value of the moment of inertia J_3 of the sample and compare it with the measured value. Calculate the relative error of the measured value:

$$E=\frac{J_3-J}{J}\times 100\% \tag{4-21-13}$$

4. Verification of the Parallel Axis Theorem

Insert two symmetrical cylinders into the circular hole on the load platform with a center distance of d, measure and calculate the moment of inertia of the two cylinders in this position. Compare the measured value with the calculated value obtained from Eqs. (4-21-11) and (4-21-10). If they are consistent, the parallel axis theorem is verified.

Data Recording and Processing

Table 4-21-1　Measurement of angular acceleration of experimental platform

Uniform Deceleration						Uniform Acceleration　R_tower wheel=　mm　m_weight=　g					
k	1	2	3	4	Average	k	1	2	3	4	Average
t(s)						t(s)					
k	5	6	7	8		k	5	6	7	8	
t(s)						t(s)					
$\beta_1(1/s^2)$						$\beta_2(1/s^2)$					

Table 4-21-2　Measurement of angular acceleration of experimental platform with ring sample

Uniform Deceleration						Uniform Acceleration　R_tower wheel =25 mm　m_weight=50.4 g					
k	1	2	3	4	Average	k	1	2	3	4	Average
t(s)						t(s)					
k	5	6	7	8		k	5	6	7	8	
t(s)						t(s)					
$\beta_3(1/s^2)$						$\beta_4(1/s^2)$					

Table 4-21-3　Measurement of angular acceleration of two cylinders with center-to-axis distance

d=100 mm, R_cylinder=　　mm, m_cylinder×2=　　g

Uniform Deceleration						Uniform Acceleration　R_tower wheel=　mm　m_weight=　g					
k	1	2	3	4	Average	k	1	2	3	4	Average
t(s)						t(s)					
k	5	6	7	8		k	5	6	7	8	
t(s)						t(s)					
$\beta_3(1/s^2)$						$\beta_4(1/s^2)$					

(1) Substitute the data from Table 4-21-1 into Eq. (4-21-4) to calculate the moment of inertia of the empty experimental platform $J_1=$_______kgm^2

(2) Substitute the data from Table 4-21-2 into Eq. (4-21-5) to calculate the moment of inertia of the experimental platform with the ring sample $J_2=$_______kgm^2

(a) Calculate the measured moment of inertia of the ring sample $J_3=$_______kgm^2

(b) Calculate the theoretical moment of inertia of the ring sample $J=$_______kgm^2

(c) Calculate the relative error of the measured value using $E=$_______%

(3) Substitute the data from Table 4-21-3 into Eq. (4-21-5) to calculate the moment of inertia of the experimental platform with two cylinders $J_2=$_______kgm^2

(a) Calculate the measured moment of inertia of the two cylinders $J_3=$ _______ kgm^2

(b) Calculate the theoretical moment of inertia of the two cylinders using Eqs. (4-21-11) and (4-21-10).

(c) $J=$_______kgm^2

(d) Calculate the relative error of the measured value using $E=$_______%

Precautions

1. The moment of inertia of the sample is indirectly measured using the formula $J_3=J_2-J_1$, and the transfer formula for standard error is $\Delta J_3=(\Delta J_2^2+\Delta J_1^2)^{1/2}$. When the moment of inertia of the sample is much smaller than that of the experimental platform, the transfer of errors may increase the relative error of the measurement.

2. Theoretically, the moment of inertia of the same sample does not change with the variation of the torque. By changing the radius of the tower wheel or the mass of the weight (five tower wheels and five weights), 25 combinations can be obtained to form different torques. The experimental conditions can be changed for measurement and data analysis to explore the rules, identify the causes of errors, and seek the optimal measurement conditions.

4.22 亥姆霍兹线圈的磁场测量

实验目的

1. 用霍尔传感器测量载流圆线圈和亥姆霍兹线圈轴线上的磁感应强度,验证磁场的叠加原理。

2. 研究载流圆线圈轴线平面上的磁场分布规律。

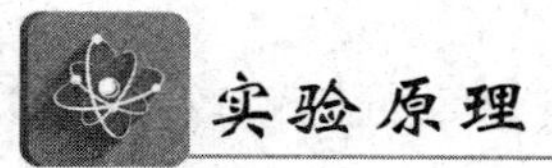

一、载流圆线圈的磁场

设有一平均半径为 R 通有电流 I 的圆线圈，如图 4-22-1(a)所示，根据毕奥-萨伐尔定律，载流线圈在轴线(通过圆心并与线圈平面垂直的直线)上某点的磁感应强度为

$$B=\frac{\mu_0 NR^2}{2(R^2+x^2)^{3/2}}I \tag{4-22-1}$$

式中，x 为轴线上某点到圆心的距离，$\mu_0=4\pi\times10^{-7}\ \mathrm{T\cdot m\cdot A^{-1}}$ 为真空磁导率，N 为线圈的匝数。线圈轴线上的磁场分布如图 4-22-1(b)所示，圆心处的磁感应强度 B_0 为

$$B_0=\frac{\mu_0 N}{2R}I \tag{4-22-2}$$

二、亥姆霍兹线圈的磁场

如图 4-22-2(a)所示，一对载流圆线圈彼此平行且共轴，两线圈的匝数均为 N，当线圈内通有大小相同、方向一致的电流，且线圈之间距离 d 等于圆形线圈的平均半径 R 时，则在两圆线圈间轴线中点附近的较大范围内为均匀磁场，这对线圈称为亥姆霍兹线圈，其轴线上的磁场分布如图 4-22-2(b)所示。由于亥姆霍兹线圈能较容易地提供范围较大而又相当均匀的磁场，因此，在生产和科学实验中有较大的实用价值，也常用于弱磁场的计量标准。

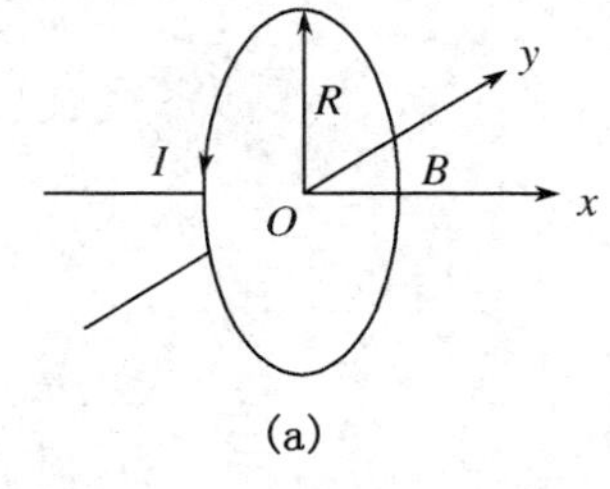

(a)

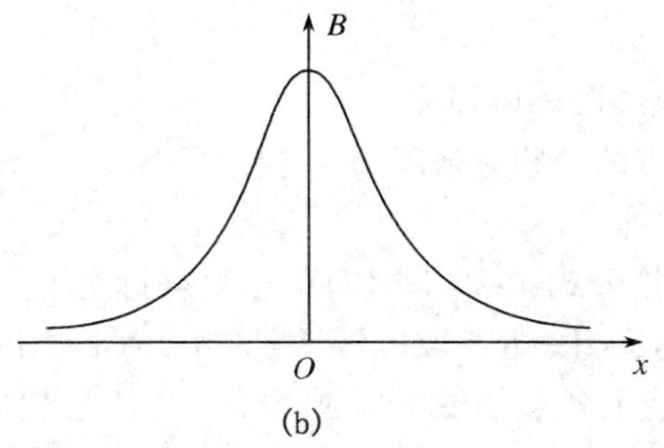

(b)

图 4-22-1　载流圆线圈轴线上的磁场

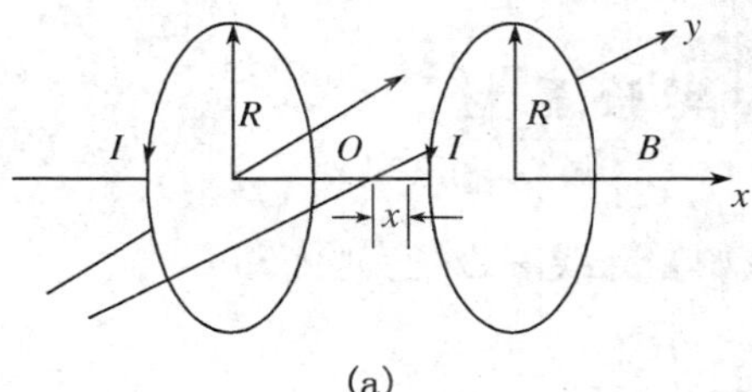

(a)

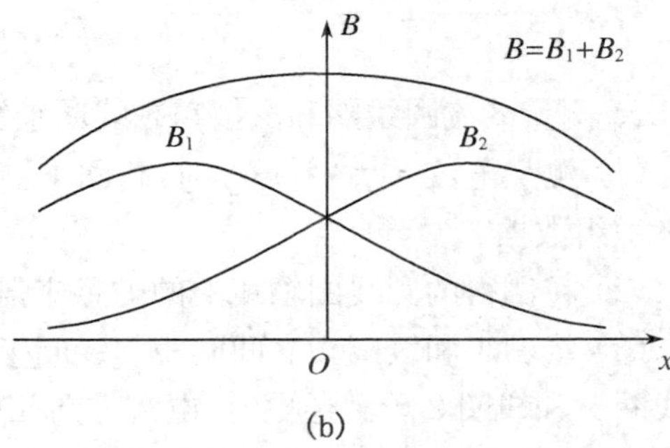

(b)

图 4-22-2　亥姆霍兹线圈轴线上的磁场

两线圈在距中心 O 点 x 处所产生的磁场分别为

$$B_1=\frac{\mu_0 NI}{2}\cdot\frac{R^2}{\left[R^2+\left(\frac{R}{2}+x\right)^2\right]^{3/2}} \tag{4-22-3}$$

$$B_2=\frac{\mu_0 NI}{2}\cdot\frac{R^2}{\left[R^2+\left(\frac{R}{2}-x\right)^2\right]^{3/2}} \tag{4-22-4}$$

则亥姆霍兹线圈轴线上任一点的磁感应强度为

$$\begin{aligned}B&=B_1+B_2\\&=\frac{\mu_0 NI}{2}\cdot\frac{R^2}{\left[R^2+\left(\frac{R}{2}+x\right)^2\right]^{3/2}}+\frac{\mu_0 NI}{2}\cdot\frac{R^2}{\left[R^2+\left(\frac{R}{2}-x\right)^2\right]^{3/2}}\end{aligned} \tag{4-22-5}$$

亥姆霍兹线圈轴线中点 O 处($x=0$)的磁感应强度为

$$B_0=\frac{8}{5\sqrt{5}}\frac{\mu_0 NI}{R} \tag{4-22-6}$$

通过对式(4-22-5)在 $x=0$ 处进行泰勒级数展开可以证明,在轴线中心区磁场基本与 B_0 相同,一般可以认为是均匀的。

利用霍尔元件的霍尔效应测量亥姆霍兹线圈的磁场分布的原理请参考本书《霍尔效应测磁场》一节的内容。此处不再赘述。

实验器材

亥姆霍兹线圈磁场实验仪。

实验内容及步骤

1. 测量载流圆线圈和亥姆霍兹线圈轴线上各点的磁感应强度。

(1) 测量电流 $I=300$ mA 时,线圈 1 轴线上各点的磁感应强度 B_1。要求每隔 1.00 cm 测量一组数据。

(2) 将测得的圆线圈轴线上的磁感应强度与理论公式(4-22-1)计算的结果进行比较。

(3) 线圈 1 和线圈 2 之间间距与线圈半径相等,即 $d=R$。取电流 $I=300$ mA,分别测量线圈 1 和线圈 2 单独通电时(电流方向相同),轴线上各点的磁感应强度 B_1 和 B_2。然后在两线圈内通有大小相等、方向相同的电流 $I=300$ mA,测量亥姆霍兹线圈轴线上各点的磁感应强度 B_{1+2}。

2. 描绘载流圆线圈轴线平面上的磁感应线分布图。

(1) 亥姆霍兹线圈通有电流 $I=300$ mA，分别测量与轴线平行的几条直线上各点的磁感应强度。

(2) 当两个圆线圈通以大小相等($I=300$ mA)、方向相反的电流时，测量其轴线上的磁场分布。

数据处理

1. 圆线圈轴线上磁场分布测量数据记录在如表 4-22-1 所示的表格中。坐标原点设在圆心处，要求在同一坐标系内画出实验曲线与理论曲线。

表 4-22-1　圆线圈轴线上磁场分布测量数据

$R=$　　m，　$N=$　　匝

轴向距离 x/cm	−7.00	…	−1.00	0	1.00	…	7.00
B/mT		…				…	
$B_{理}=\frac{\mu_0 NIR^2}{2(R^2+x^2)^{3/2}}$/mT		…				…	

2. 亥姆霍兹线圈轴线上磁场分布测量数据记录在如表 4-22-2 所示的表格中。坐标原点设在两个线圈圆心连线的中点处。在同一坐标系里用坐标纸或计算机作出 B_1-x、B_2-x、$B_{1+2}-x$、$(B_1+B_2)-x$ 四条曲线。考察 $B_{1+2}-x$ 与 $(B_1+B_2)-x$ 曲线，验证磁场叠加原理，即亥姆霍兹线圈轴线上任一点磁感应强度 B_{1+2} 是两个单线圈分别在该点上产生的磁感应强度之和 B_1+B_2。

表 4-22-2　亥姆霍兹线圈轴线上磁场分布测量数据

轴向距离 x/cm	−10.00	…	−1.00	0	1.00	…	10.00
B_1/mT							
B_2/mT							
B_1+B_2/mT							
B_{1+2}/mT							

(1) 根据测量数据，近似画出亥姆霍兹线圈轴线平面上的磁感应线分布图。

(2) 与亥姆霍兹线圈磁场比较，分析当两个圆线圈通大小相等方向相反电流时磁场分布的特点。

注意事项

1. 仪器使用时，应避开周围有强磁场源的地方。

2. 开机后，预热 10 分钟左右，方可进行实验；测量前，应断开线圈电路，在电流为零时调零，然后接通线圈电路，进行测量和读数。

思考题

1. 霍尔效应法测量磁场的原理是什么？与电磁感应法相比有何优点？

2. 亥姆霍兹线圈是怎样组成的？其基本条件是什么？它的磁场分布有什么特点？

3. 分析用霍尔效应法测量磁场时，当流过线圈中的电流为零时，显示的磁感应强度值为什么不为零？

Experiment 4.22 Helmholtz Coil

Experimental Objectives

1. Mastering the basic principles and methods of measuring magnetic fields using the Hall effect.

2. Measuring the magnetic induction intensity on the axis of a current-carrying circular coil and a Helmholtz coil using a Hall sensor to validate the principle of magnetic field superposition.

3. Studying the distribution law of magnetic field on the axis plane of a current-carrying circular coil.

Experimental Instruments

Helmholtz coil magnetic field experimental apparatus, mainly consists of Helmholtz coil device and magnetic field measuring instrument.

Experimental Principle

1. Magnetic field of a current-carrying circular coil

Assuming a circular coil with an average radius of R and carrying current I, as shown in Fig. 4-22-1(a), according to the Biot-Savart law, the magnetic induction intensity at a certain point on the axis (a straight line passing through the center of the coil and perpendicular to the coil plane) of the current-carrying coil is

$$B=\frac{\mu_0 NR^2}{2(R^2+x^2)^{3/2}}I \tag{4-22-1}$$

where x represents the distance from a certain point on the axis to the center of the coil, $\mu_0=4\pi\times10^{-7}\ \text{T}\cdot\text{m}\cdot\text{A}^{-1}$ represents the permeability of vacuum, and N represents the number of turns of the coil.

The distribution of the magnetic field along the axis of the coil is shown in Fig. 4-22-1(b), and the magnetic induction intensity at the center is

$$B_0=\frac{\mu_0 N}{2R}I \tag{4-22-2}$$

2. Magnetic field of a Helmholtz coil

As shown in Fig. 4-22-2(a), a pair of parallel and coaxial current-carrying circular coils with the same number of turns N are placed. When the coils carry currents of the same magnitude and direction, and the distance d between the coils is equal to the average radius R of the circular coil, a uniform magnetic field is generated in a larger region near the midpoint of the axis between the two coils. This pair of coils is called a Helmholtz coil, and the distribution of the magnetic field along the axis is shown in Fig. 4-22-2(b). Due to its ability to provide a relatively large and uniform magnetic field easily, Helmholtz coils have great practical value in production and scientific experiments. They are also commonly used for metrological standards of weak magnetic fields.

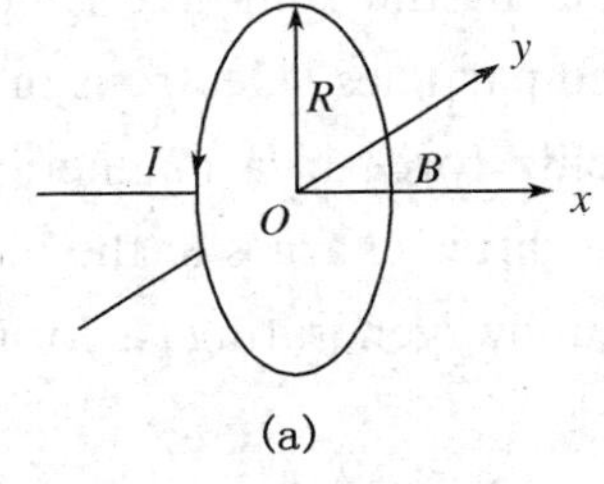

(a)

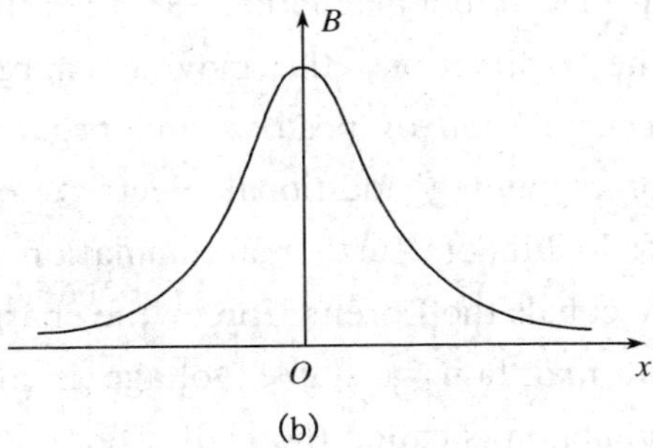

(b)

Fig. 4-22-1 Magnetic field along the axis of a current-carrying circular coil

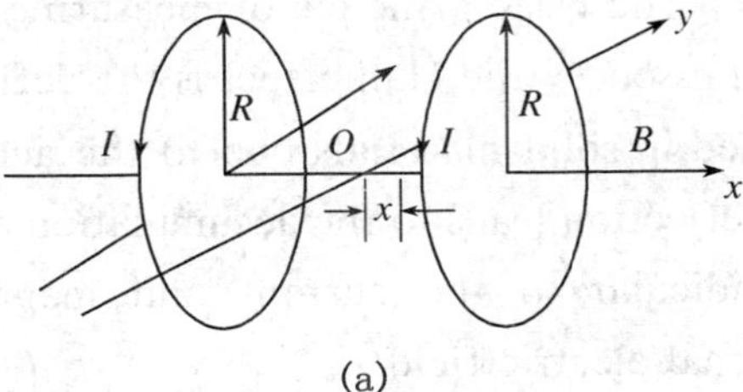

(a)

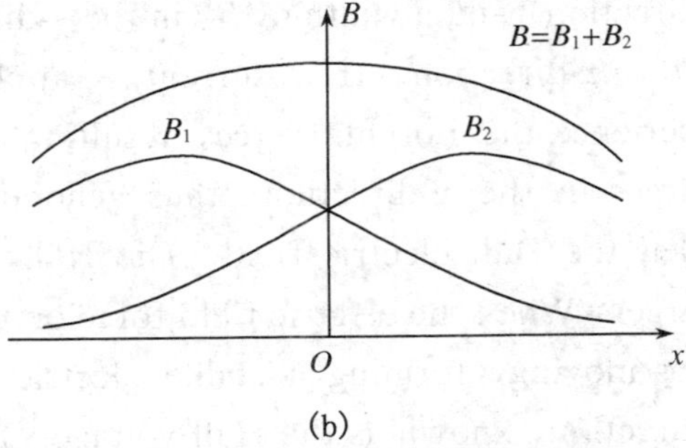

(b)

Fig. 4-22-2 Magnetic field along the axis of a Helmholtz coil

The magnetic fields produced by the two coils at a distance x from the center point O are

$$B_1=\frac{\mu_0 NI}{2}\cdot\frac{R^2}{\left[R^2+\left(\frac{R}{2}+x\right)^2\right]^{3/2}} \tag{4-22-3}$$

$$B_2=\frac{\mu_0 NI}{2}\cdot\frac{R^2}{\left[R^2+\left(\frac{R}{2}-x\right)^2\right]^{3/2}} \tag{4-22-4}$$

The magnetic induction intensity at any point on the axis of the Helmholtz coil is

$$B=B_1+B_2$$

$$=\frac{\mu_0 NI}{2}\cdot\frac{R^2}{\left[R^2+\left(\frac{R}{2}+x\right)^2\right]^{3/2}}+\frac{\mu_0 NI}{2}\cdot\frac{R^2}{\left[R^2+\left(\frac{R}{2}-x\right)^2\right]^{3/2}} \tag{4-22-5}$$

The magnetic induction intensity at the midpoint O ($x=0$) of the axis of the Helmholtz coil is

$$B_0=\frac{8}{5\sqrt{5}}\frac{\mu_0 NI}{R} \tag{4-22-6}$$

By performing a Taylor series expansion of Eq. (4-22-5) at $x=0$, it can be proven that the magnetic field in the central region of the axis is approximately the same as B_0 and can generally be considered uniform.

(1) The basic principle of measuring magnetic fields using the Hall effect method

In essence, the Hall effect is the deflection of charged particles (electrons or holes) confined in solid materials due to the action of the Lorentz force in a magnetic field. This deflection leads to the accumulation of positive and negative charges in the direction perpendicular to the current and magnetic field, thereby generating a transverse additional electric field.

As shown in Fig. 4-22-3, a semiconductor thin slice with a thickness of d (in the z-direction) and a width of b (in the y-direction) is placed in a magnetic field B parallel to the z-direction. If a current is applied in the x-direction, the moving charges experience the Lorentz force, resulting in the accumulation of positive and negative charges in the y-direction, thus generating a corresponding additional electric field called the Hall electric field. The Hall electric field hinders further accumulation of charges. When the electric field force on the charge equals the Lorentz force, the charge stops moving, forming a stable electric field that maintains a stable voltage in the y-direction, known as the Hall voltage. This phenomenon is called the Hall effect.

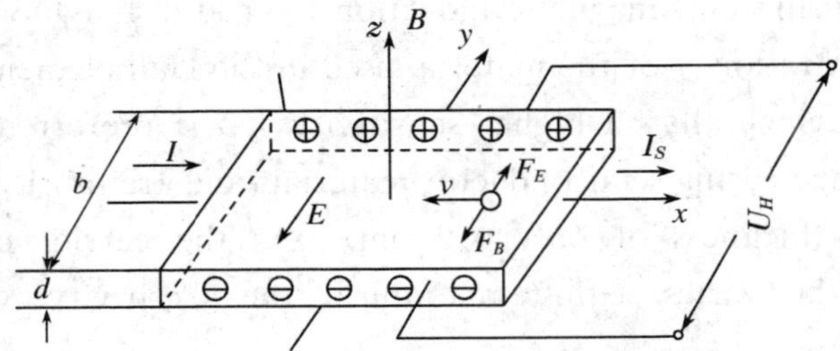

Fig. 4-22-3 Schematic diagram of the Hall effect principle

Let's assume the concentration of charge carriers in the semiconductor thin slice is represented by n, and the average drift velocity in the direction of current is represented by v. The relationship between the current intensity, I, and n and v can be expressed as follows:

$$I_S = envbd \tag{4-22-7}$$

where e represents the elementary charge of an electron.

The relationship between the stable Hall electric field E_H and the magnetic induction intensity B is expressed as

$$eE_H = evB \tag{4-22-8}$$

The generated Hall voltage is

$$U_H = E_H b \tag{4-22-9}$$

By Eqs. (4-22-7) to (4-22-9), we can obtain

$$U_H = \frac{1}{ne}\frac{I_S B}{d} = R_H \frac{I_S B}{d} \tag{4-22-10}$$

This means that the Hall voltage U_H is inversely proportional to the thickness d and directly proportional to the current I_S and the magnetic induction intensity B. Therefore, the Hall effect can be used to convert the magnetic induction intensity B into the Hall voltage U_H for measurement. In Eq. (4-22-10), $R_H = 1/ne$ is called the Hall coefficient, which is an important parameter that reflects the strength of the Hall effect in a material.

A Hall element is a magnetic-electric conversion device made using the Hall effect, which is also known as a Hall sensor. For a finished Hall element, both the Hall coefficient of the material and the thickness d are known. Therefore, in practical applications, Eq. (4-22-10) is generally written as:

$$U_H = K_H I_S B \tag{4-22-11}$$

where K_H represents the sensitivity of the Hall element, and its commonly used unit is mV/(mA·T). The sensitivity of the Hall element indicates the Hall voltage output per unit working current and unit magnetic induction intensity. It is inversely proportional to the carrier concentration n of the material used in the Hall element and the thickness d of the thin film. Generally, a higher sensitivity K_H is preferred. Therefore, Hall elements are often made using semiconductor materials and are fabricated to be very thin (commonly with a thickness of only 0.2 mm). As the carrier concentration of a semiconductor material varies with temperature, the sensitivity K_H is temperature-dependent.

From Eq. (4-22-11), it is known that given the sensitivity K_H of the Hall element, the magnitude of the magnetic induction intensity B can be calculated by measuring the working current I and the Hall voltage U_H, using the formula:

$$B=\frac{U_H}{K_H I_S} \tag{4-22-12}$$

This is the basic principle of using the Hall effect to measure magnetic fields.

Experimental Contents and Steps

1. The experimental Content

(1) Measure the magnetic induction intensity at various points along the axis of both the current-carrying circular coil and the Helmholtz coil.

(2) Draw a distribution map of magnetic field lines on the plane along the axis of the current-carrying circular coil.

2. The Experimental Procedure

(1) Measure the magnetic induction intensity at various points along the axis of the current-carrying circular coil and the Helmholtz coil.

a. Measure the magnetic induction intensity at various points along the axis of coil while varying the current is $I=300$ mA. Take measurements at intervals of 1.00 cm.

b. Compare the measured magnetic field intensity on the axis of the circular coil with the results calculated using the theoretical Eq. (4-22-1).

c. The distance between coil 1 and coil 2 is equal to the radius of the coil, i. e. $d=R$. With a current of $I=300$ mA, measure the magnetic induction intensity B_1 and B_2 at various points along the axis when coil 1 and coil 2 are individually energized (with the current flowing in the same direction). Then, with an equal and parallel current passing through both coils $I=300$ mA, measure the magnetic induction intensity at various points along the axis of the Helmholtz coil B_{1+2}.

(2) Illustrate the distribution of magnetic field lines on the plane of the axis of the current-carrying circular coil.

a. With a current $I=300$ mA passing through the Helmholtz coil, measure the magnetic induction intensity at various points on several straight lines parallel to the axis.

b. When two circular coils carry equal and opposite currents ($I=300$ mA), measure the magnetic field distribution along their axis.

Data Recording and Processing

a. The measurement data of the magnetic field distribution along the axis of the circular coil is recorded in the table shown as Table 4-22-1. The coordinate origin is set at the center of the coil, with $R=0.105$m and $N=500$ turns. It is required to plot the experimental curve and the theoretical curve in the same coordinate system.

Table 4-22-1 Measurement of the magnetic field along the axis of the circular coil

x/cm	-7.00	…	-1.00	0	1.00	…	7.00
B/mT		…				…	
$B_{理}=\dfrac{\mu_0 NIR^2}{2(R^2+x^2)^{3/2}}$/mT		…				…	

b. The measurement data of the magnetic field distribution along the axis of the Helmholtz coil is recorded in the table shown as Table 4-22-2. The coordinate origin is set at the midpoint of the line connecting the centers of the two coils. Using graph paper or a computer, plot four curves: B_1-x, B_2-x, $B_{1+2}-x$ and $(B_1+B_2)-x$. Examine the relationship between the curves $B_{1+2}-x$ and $(B_1+B_2)-x$ to verify the principle of magnetic field superposition. According to this principle, the magnetic induction intensity at any point along the axis of the Helmholtz coil is the sum of the magnetic induction intensities produced by each individual coil at that point.

Table 4-22-2 Measurement of the magnetic field along the axis of the Helmholtz coil

x/cm	-10.00	…	-1.00	0	1.00	…	10.00
B_1/mT							
B_2/mT							
B_1+B_2/mT							
B_{1+2}/mT							

c. According to the measurement data, please briefly describe the distribution of the magnetic field along the axis of the Helmholtz coil.

d. Based on the measurement data, please sketch an approximate diagram illustrating the distribution of magnetic field lines along the plane of the axis of the Helmholtz coil. Compare the magnetic field of the Helmholtz coil and analyze the characteristics of the magnetic field distribution when equal currents in opposite directions flow through the two circular coils.

Precautions

1. When using the instrument, it is important to avoid areas with strong magnetic field sources in the vicinity.

2. After powering on, please allow approximately 10 minutes for the instrument to warm up before conducting any experiments.

3. Before taking measurements, the coil circuit should be disconnected. Zero adjustment should be performed when the current is zero. After that, the coil circuit should be connected to carry out measurements and readings.

Questions

1. What is the principle of the Hall effect method to measure magnetic fields? What are the advantages compared to electromagnetic induction?

2. How are Helmholtz coils composed? What are the basic conditions? What are the characteristics of its magnetic field distribution?

3. When analyzing magnetic fields measured with the Hall-effect method, when the current flowing through the coil is zero, why is the magnetic induction intensity value displayed zero?

4.23 迈克尔逊干涉仪的调节和使用

19 世纪末,美国物理学家迈克尔逊(A. A. Michelson)为测量光速,依据分振幅产生双光束实现干涉的原理,设计制造了迈克尔逊干涉仪这一精密光学仪器。迈克尔逊与莫雷用这仪器完成了相对论研究中具有重要意义的“以太”漂移实验,实验结果否定了“以太”的存在,为爱因斯坦建立狭义相对论奠定了基础。

在近代物理学和近代计量科学中,迈克尔逊干涉仪不仅可以观察光的等厚、等倾干涉现象,精密地测定光波波长、微小长度、光源的相干长度等,还可以测量气体、液体的折射率等,因此,迈克尔逊于 1907 年获诺贝尔物理学奖。迈克尔逊干涉仪的基本原理已经被推广到许多方面,研制成各种形式的精密仪器,广泛地应用于生产和科学研究领域。

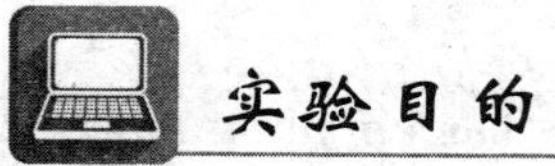

1. 了解迈克尔逊干涉仪的特点,学会其调节和使用方法。

2. 调节和观察迈克尔逊干涉仪产生的干涉图，加深对各种干涉条纹特点的理解。

3. 应用迈克尔逊干涉仪测定钠D双线的平均波长和波长差。

实验原理

实验室中最常用的迈克尔逊干涉仪，其原理和结构如图4-23-1和图4-23-2所示。M_1和M_2是相互垂直的两臂上放置的两个平面反射镜，其背面各有三个调节螺旋，用来调节镜面的俯仰；M_2是固定的，M_1由精密丝杆控制，可沿臂轴前后移动，其移动距离由转盘读出。仪器前方粗动手轮最小分格值为10^{-2} mm，右侧微动手轮的最小分格值为10^{-4} mm，可估读至10^{-5} mm，两个读数手轮属于蜗轮蜗杆传动系统。在两臂轴相交处，有一与两臂轴各成45°的平行平面玻璃板P_1，且在P_1的第二平面上镀以半透（半反射）膜，以便将入射光分成振幅近乎相等的反射光1和透射光2，故P_1板又称为分光板。P_2也是一平行平面玻璃板，与P_1平行放置，厚度和折射率均与P_1相同。补偿板P_2的作用是使光束2也两次透过玻璃板，以“补偿”光束1在P_1板中往返两次所多走的光程，使干涉仪对不同波长的光能同时满足等光程要求。从扩展光源S射来的光，到达分光板P_1后被分成两部分。反射光1在P_1处反射后向着M_1前进，透射光2透过P_1后向着M_2前进。这两列光波分别在M_1、M_2上反射后逆着各自的入射方向返回，最后都到达E处。这两列光波来自光源上同一点O，因而是相干光，在E处的观察者能看到干涉图样。

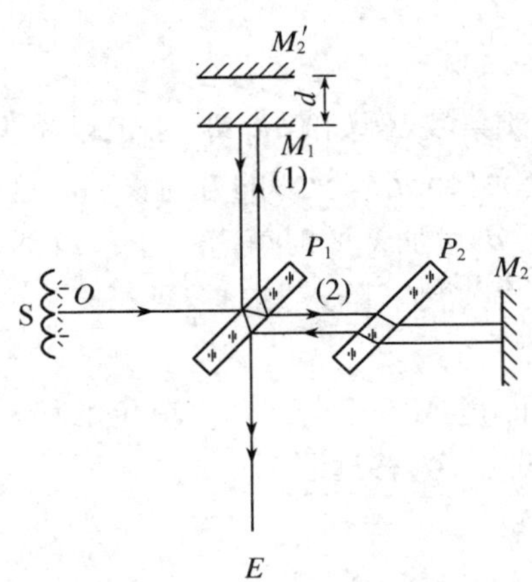

图4-23-1　迈克尔逊干涉仪原理图

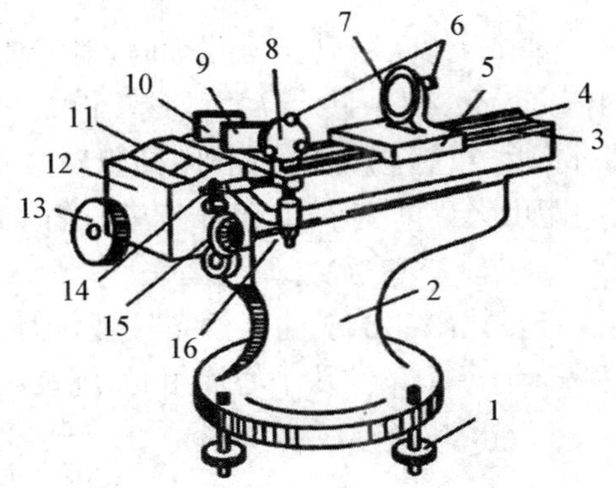

图4-23-2　迈克尔逊干涉仪结构图

1—水平调节螺丝；2—底座；3—导轨；4—精密丝杆；5—托板；6—反射镜调节螺丝；7—可动反射镜M_1；8—固定反射镜M_2；9—补偿板P_2；10—分光板P_1；11—读数窗口；12—传动系统罩；13—粗动手轮；14—水平拉簧螺丝；15—微动手轮；16—垂直拉簧螺丝。

由于从 M_2 返回的光线在分光板 P_1 的第二面上反射,使 M_2 在 M_1 附近形成一平行于 M_1 的虚像 M_2',因而光在迈克尔逊干涉仪中自 M_1 和 M_2 的反射,相当于自 M_1 和 M_2' 的反射。由此可见,在迈克尔逊干涉仪中所产生的干涉与厚度为 d 的空气膜所产生的干涉是等效的。

一、扩展光源照明产生的干涉图

(1) 当 M_1 和 M_2' 严格平行时,所得的干涉为等倾干涉。所有倾角为 i 的入射光束,由 M_1 和 M_2' 反射光线的光程差 Δ 均为

$$\Delta=2d\cos i \tag{4-23-1}$$

式中 i 为光线在 M_1 镜面的入射角,d 为空气薄膜的厚度,它们将处于同一级干涉条纹,并定位于无限远。这时,在图 4-23-1 中的 E 处用眼睛正对 P_1 观察(或在 E 处放一会聚透镜,在其焦平面上),便可观察到一组明暗相间的同心圆纹。这些条纹的特点是:

① 干涉条纹的级次以中心为最高

在干涉纹中心,因 $i=0$,如果不计反射光线之间相位突变,由圆纹中心出现亮点的条件为

$$\Delta=2d=k\lambda \tag{4-23-2}$$

得圆心处干涉条纹的级次为

$$k=\frac{2d}{\lambda} \tag{4-23-3}$$

当 M_1 和 M_2' 的间距 d 逐渐增大时,对于任一级干涉条纹,例如第 k 级,必定以减少其 $\cos i_k$ 的值来满足 $2d\cos i_k=k\lambda$,故该干涉条纹向 i_k 变大($\cos i_k$ 变小)的方向移动。这时,观察者将看到条纹好像从中心向外"涌出",且每当间距 d 增加 $\lambda/2$ 时就有一个条纹涌出;反之,当间距由大逐渐变小时,最靠近中心的条纹将一个一个地"陷入"中心,且每陷入一个条纹,间距 d 的改变亦为 $\lambda/2$。

因此,只要数出涌出或陷入的条纹数,即可得到平面镜 M_1 以波长 λ 为单位的移动距离。显然,若有 N 个条纹从中心涌出时,则表明 M_1 相对于 M_2' 移远了

$$\Delta d=N\frac{\lambda}{2} \tag{4-23-4}$$

反之,若有 N 个条纹陷入时,则表明 M_1 向 M_2' 移近了同样的距离。根据式(4-23-4),如果已知光波的波长 λ,便可由条纹变动的数目,计算出 M_1 移动的距离,这就是长度的干涉计量原理;反之,如果已知 M_1 移动的距离和干涉条纹变动数目,便可算出光波的波长。

② 干涉条纹的分布是中心宽边缘窄

对于相邻的 k 级和 $k-1$ 级干涉条纹,有

$$2d\cos i_k=k\lambda$$

$$2d\cos i_{k-1}=(k-1)\lambda$$

将两式相减，当 i 较小时，利用 $\cos i=1-\frac{i^2}{2}$，可得相邻条纹的角距离 Δi_k 为

$$\Delta i_k=i_k-i_{k-1}\approx\frac{\lambda}{2di_k} \tag{4-23-5}$$

上式表明：d 一定时，视场里干涉条纹的分布是中心较宽（i_k 小，Δi_k 大），边缘较窄（i_k 大，Δi_k 小）；i_k 一定时，d 越小，Δi_k 越大，即条纹随着薄膜厚度 d 的减小而变宽。所以在调节和测量时，应选择 d 为较小值，即调节 M_1 和 M_2 到分光板 P_1 上镀膜面的距离大致相同。

（2）当 M_1 和 M_2' 有一很小的夹角 α，且入射角 i 也较小时，所得的干涉一般为等厚干涉，其条纹定位于空气薄膜表面附近。此时，由 M_1 和 M_2' 反射光线的光程差仍近似为

$$\Delta=2d\cos i=2d\left(1-\frac{i^2}{2}\right) \tag{4-23-6}$$

① 在两镜面的交线附近处，因厚度 d 较小，$d\cdot i^2$ 的影响可略去，相干的光程差主要由膜厚 d 决定，因而在空气膜厚度相同的地方光程差均相同，即干涉条纹是一组平行于 M_1 和 M_2' 交线的等间隔的直线条纹。

② 在离 M_1 和 M_2' 的交线较远处，因 d 较大，干涉条纹变成弧形，而且条纹弯曲的方向是背向两镜面的交线。

这是由于式(4-23-6)中的 $d\cdot i^2$ 作用已不容忽略。由于同一 k 级干涉条纹是等光程差的光点的轨迹，为满足 $\Delta=k\lambda$，因此用扩展光源照明时，当 i 逐渐增大，必须相应增大 d 值，以补偿由 i 增大时引起的光程差的减小。所以干涉条纹在 i 增大的地方要向 d 增加的方向移动，使条纹成为弧形，如图 4-23-3 所示。随着 d 的增大，条纹弯曲越厉害。

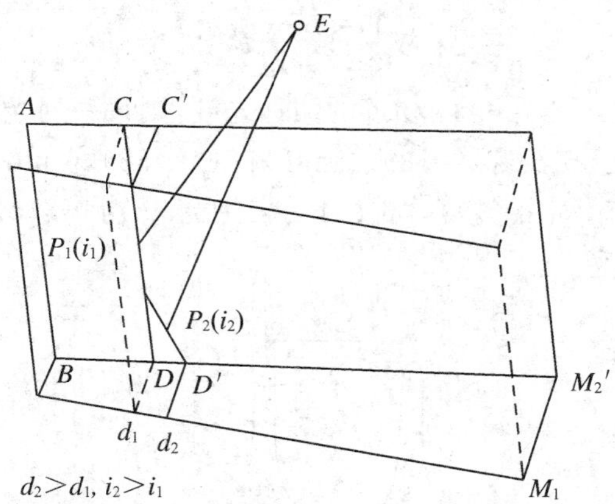

图 4-23-3　等厚干涉 d 与 i 关系示意图

（3）白光照射下看到彩色干涉条纹的条件为：

对于等倾干涉，在 d 接近于零时可以看到；对于等厚干涉，在 M_1 和 M_2' 的交线附近可以看到。因为在 $d=0$ 时，所有波长的干涉情况相同，不显彩色。当 d 较大时因不同波长干涉条纹互相重叠，使照明均匀，彩色消失。只有当 d 接近零时才可看到数目不多的彩色干涉条纹。

二、点光源照明产生的非定域干涉图样

点光源 S 经 M_1 和 M_2'的反射产生的干涉现象，等效于沿轴向分布的两个虚光源 S_1、S_2 所产生的干涉。因而从 S_1 和 S_2 发出的球面波在相遇的空间处处相干，故为非定域干涉。如图 4-23-4 观察屏 E 放在不同位置上，则可看到不同形状的干涉条纹。

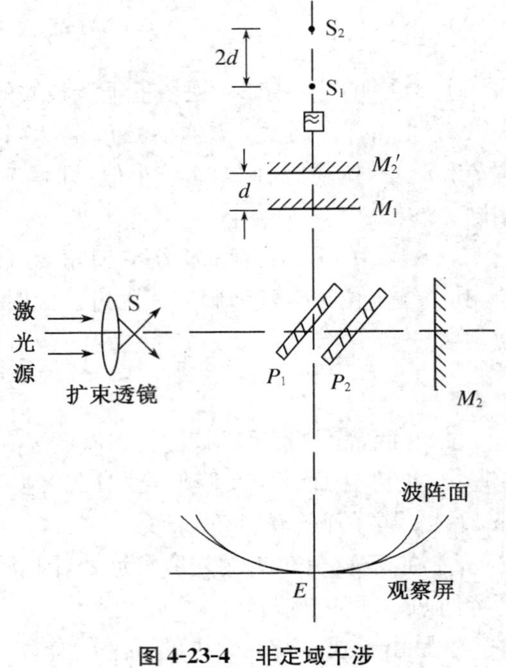

图 4-23-4　非定域干涉

当观察屏 E 垂直于 S_1、S_2 连线时，屏上呈现出圆形的干涉条纹。同等倾条纹相似，在圆环中心处，光程差最大，$\Delta=2d$，级次最高；当移动 M_1 使 d 增加时，圆环一个个地从中心"涌出"，当 d 减小时，圆环一个个地向中心"陷入"。每变动一个条纹，M_1 移动的距离为 $\lambda/2$。因此，也可用于计量长度或测量波长。

当 M_1 与 M_2'互相平行时，得到明暗相间的圆形干涉条纹。如果光源是绝对单色的，则当 M_1 镜缓慢地移动时，虽然视场中条纹不断涌出或陷入，但条纹的对比度应当不变。对比度描述的是条纹清晰的程度。

设亮条纹光强为 I_1，相邻暗条纹光强为 I_2，则对比度 V 可表示为

$$V=\frac{I_1-I_2}{I_1+I_2} \tag{4-23-7}$$

如果光源中包含有波长 λ_1 和 λ_2 相近的两种光波，则每一列光波均不是绝对单色光，以钠黄色为例，它是由中心波长 $\lambda_1=589.0$ nm 和 $\lambda_2=589.6$ nm 的双线组成，波长差为 0.6 nm。每一条谱线又有一定的宽度，如图 4-23-5 所示。由于双线波长差 $\Delta\lambda$ 与中心波长相比甚小，故称之为准单色光。

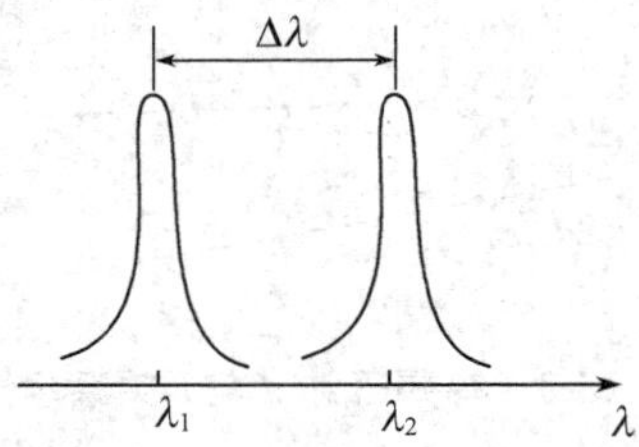

图 4-23-5　钠黄色光波长示意图

用这种光源照明迈克尔逊干涉仪，它们将各自产生一套干涉图。干涉场中的强度分布则是两组干涉条纹的非相干叠加，由于 λ_1 和 λ_2 有微小差异，对应 λ_1 的亮环的位置和对应 λ_2 的亮环的位置，将随 d 的变化而呈周期性的重合和错开。因此 d 变化时，视场中所见叠

加后的干涉条纹交替出现“清晰”和“模糊甚至消失”。设在 d 值为 d_1 时，λ_1 和 λ_2 均为亮条纹，对比度最佳，则有

$$d_1 = m\frac{\lambda_1}{2}, d_1 = n\frac{\lambda_2}{2} \quad (m \text{ 和 } n \text{ 为整数})$$

如果 $\lambda_1 > \lambda_2$，当 d 值增加到 d_2，如果满足

$$d_2 = (m+k)\frac{\lambda_1}{2}, d_2 = (n+k+0.5)\frac{\lambda_2}{2} \quad (k \text{ 为整数})$$

此时对 λ_1 是亮条纹，对 λ_2 则为暗条纹，对比度最差（可能分不清条纹）。从对比度最佳到最差，M_1 移动的距离为

$$d_2 - d_1 = k\frac{\lambda_1}{2} = (k+0.5)\frac{\lambda_2}{2}$$

由 $d_2 - d_1 = k\frac{\lambda_1}{2}$ 和 $k\frac{\lambda_1}{2} = (k+0.5)\frac{\lambda_2}{2}$，消去 k 可得两波长差为

$$\lambda_1 - \lambda_2 = \frac{\lambda_1\lambda_2}{4(d_2 - d_1)} \approx \frac{\overline{\lambda_{12}^2}}{4(d_2 - d_1)} \tag{4-23-8}$$

式中 $\overline{\lambda_{12}^2}$ 为 λ_1、λ_2 的平均值。因为对比度最差时，M_1 的位置对称地分布在对比度最佳位置的两侧，所以相邻对比度最差的 M_1 移动距离 $\Delta d\,[=2(d_2 - d_1)]$ 与 $\Delta\lambda\,(=\lambda_1 - \lambda_2)$ 的关系为

$$\Delta\lambda = \frac{\overline{\lambda_{12}^2}}{2\Delta d} \tag{4-23-9}$$

实验器材

迈克尔逊干涉仪，钠灯，干涉滤光片（546.1 nm），毛玻璃屏，叉丝，白炽灯。

实验内容及步骤

1. 迈克尔逊干涉仪测钠光的波长

(1) 迈克尔逊干涉仪的调节

① 点亮钠灯 S，使之照射毛玻璃屏，形成均匀的扩展光源，在屏上加一叉丝。

② 旋转粗动手轮，使 M_1 和 M_2 至 P_1 镀膜面的距离大致相等，沿 E、P_1 方向观察，将看到叉丝的影子（共有 3 个），其中 2 个对应于动镜 M_1 的反射像（为什么?），另一个对应于 M_2 的反射像。

③ 仔细调节 M_1 和 M_2 背后的三个螺丝，改变 M_1 和 M_2 的相对方位，直至叉丝的双影

(哪两个?为什么?)在水平方向和铅直方向均完全重合。这时可观察到干涉条纹,仔细调节3个螺丝,使干涉条纹成圆形。

④ 细致缓慢地调节 M_2 下方的两个微调拉簧螺丝,使干涉条纹中心仅随观察者的眼睛左右上下的移动而移动,但不发生条纹的"涌出"或"陷入"现象。这时,观察到的干涉环才是严格的等倾干涉。如果眼睛移动时,看到的干涉环有"涌出"或"陷入"现象,要分析一下再调。

(2) 测定钠光波长(D_1、D_2 两波长的平均值)

① 旋转粗动手轮,使 M_1 移动,观察条纹的变化。从条纹的"涌出"或"陷入"判断 d 的变化,并观察 d 的取值与条纹粗细、疏密的关系。

② 当视场中出现清晰的、对比度较好的干涉圆环时,再慢慢地转动微动手轮,可以观察到视场中心条纹向外一个一个地涌出(或者向内陷入中心)。开始记数时,记录 M_1 镜的位置 d_1(两读数转盘读数相加),继续转动微动手轮,数到条纹从中心向外涌出(或陷入)100个时,停止转动微动手轮,再记录 M_1 镜的位置 d_2;继续转动微动手轮,条纹从中心向外涌出(或陷入)每100个时记录 M_1 镜的位置,共测量800～1 000个条纹移动,利用式(4-23-4)算出待测光波的波长 λ 以及平均值并计算不确定度,与公认值比较。

(3) 观察白光的彩色干涉条纹

参照原理部分的分析,思考以下几个问题:

① 在等倾干涉中看到彩色干涉条纹(圆环)的条件是什么?

② 移动 M_1,从看到的现象中,如何判断间距 d 是在增大还是在减小?

③ 向哪个方向移动 M_1 肯定会看到彩色干涉环?

④ 要在等厚干涉中看到彩色条纹,该考虑些什么问题?

先用钠灯看到等倾干涉环,移动 M_1,根据观察的现象认为 M_1 的移动方向正确时,改用白光源继续移动 M_1,直至看到彩色干涉环。

再调等厚干涉的彩色干涉条纹。

注意:由于白光的彩色条纹只有几条,必须耐心细致地慢慢调节微动手轮,如果移动过快,条纹极易一晃而过,难以察觉。

(4) 观察点光源的干涉条纹

自行设计实验步骤,观察点光源照明干涉仪时,干涉条纹的形状、特点、观察条件和变化规律。

2. 测定钠光 D 双线(D_1、D_2)的波长差

(1) 以钠灯为光源调干涉仪观察等倾干涉条纹。

(2) 移动 M_1,使视场中心的视见度最小,记录 M_1 的位置 d_1;沿原方向继续移动 M_1,直至对比度又为最小,记录 M_1 的位置为 d_2,则 $\Delta d=|d_2-d_1|$。由于 λ_1、λ_2 的差很小,对比度最差位置附近较大范围的对比度都很差,即模糊区很宽,因此,确定对比度最差的位置有很大的随机误差。在此可以使用粗调手轮(精度0.001 mm)去测,测出10个模糊区的间距去计算 Δd。这是利用拓展量程去减小单次测量的随机误差。

注意事项

迈克尔逊干涉仪是精密光学仪器，使用时应注意：

1．注意防尘、防潮、防震；不能触摸元件的光学面，不要对着仪器说话、咳嗽等。

2．实验前和实验结束后，所有调节螺丝均应处于放松状态；调节时应先使之处于中间状态，以便有双向调节的余地，调节动作要均匀缓慢。

3．有的干涉仪粗动手轮和微动手轮传动的离合器啮合时，只能使用微动手轮，不能再使用粗动手轮，否则会损坏仪器。

4．旋转微动手轮进行测量时，特别要防止回程误差。

思考题

1．分析扩束激光和钠光产生的圆形干涉条纹的差别。

2．调节钠光的干涉条纹时，如已确使叉丝的双影重合，但条纹并未出现，试分析可能产生的原因。

3．如何判断和检验干涉条纹属于严格的等倾条纹？

4．怎样用实验方法检验干涉条纹的定位区域？

Experiment 4.23　Adjustment and Usage of Michelson Interferometer

Experimental Objectives

1. Understand the characteristics of the Michelson interferometer and learn its adjustment and usage methods.

2. Adjust and observe the interference pattern produced by the Michelson interferometer to deepen the understanding of various interference fringe characteristics.

3. Apply the Michelson interferometer to measure the average wavelength and wavelength difference of the sodium D doublet.

Experimental Instruments

Michelson interferometer, sodium lamp, He-Ne laser, low-pressure mercury lamp, interference filter (546.1nm), ground glass screen, crosshair, incandescent lamp.

Experimental Principle

The most commonly used Michelson interferometer in the laboratory is shown in Fig. 4-23-1 and 4-23-2. M_1 and M_2 are two plane mirrors placed on mutually perpendicular arms, each with three adjusting screws on the back to adjust the orientation of the mirrors. M_2 is fixed and M_1 is controlled by a precision screw, allowing it to move forward and backward along the arm axis, with the movement distance read from the dial. The coarse adjustment wheel in front of the instrument has a minimum graduation value of 10^{-2} mm, and the fine adjustment wheel on the right has a minimum graduation value of 10^{-4} mm, which can be estimated to 10^{-5} mm. The two reading wheels belong to a worm and gear transmission system. At the intersection of the two arm axes, there is a parallel glass plate P_1 at a 45° angle to each arm axis. The second surface of P_2 is coated with a semi-transparent (semi-reflective) film, which divides the incident light into two nearly equal amplitude reflected light 1 and transmitted light 2. Therefore, P_1 is also called a beam splitter. P_2 is also a parallel glass plate placed parallel to P_1, with the same thickness and refractive index as P_1. The purpose of the compensating plate P_2 is to make the light beam 2 pass through the glass plate twice, compensating for the extra optical path traveled by light beam 1 in P_1, so that the interferometer satisfies the equal optical path requirement for different wavelengths of light. The light from the extended source S is divided into two parts after reaching the beam splitter P_1. The reflected light 1 reflects at P_1 and moves toward M_1, while the transmitted light 2 passes through P_1 and moves toward M_2. These two columns of light waves reflect at M_1 and M_2, respectively, and return in the opposite direction of their incident directions, finally reaching point E. These two columns of light waves originate from the same point O on the light source and are coherent light. An observer at E can see the interference pattern.

Due to the reflection of the light returning from M_2 on the second surface of the beam splitter P_1, a virtual image M_2' parallel to M_1 is formed near M_1. Therefore, the interference between the reflections from M_1 and M_2 in the Michelson interferometer is equivalent to the interference from M_1 and M_2', which is produced by an air film of thickness d.

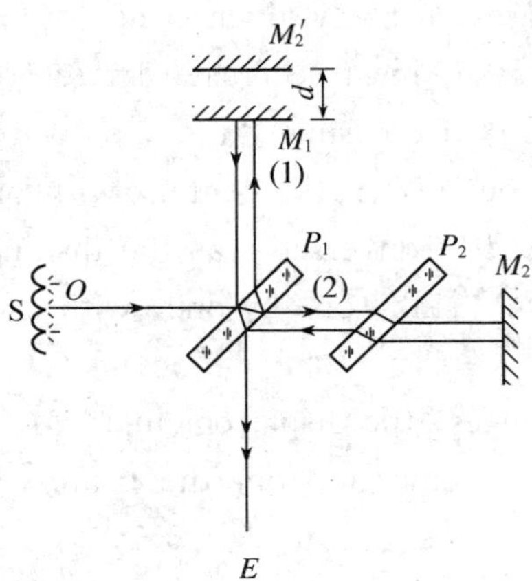

Fig. 4-23-1　Michelson interferometer schematic diagram

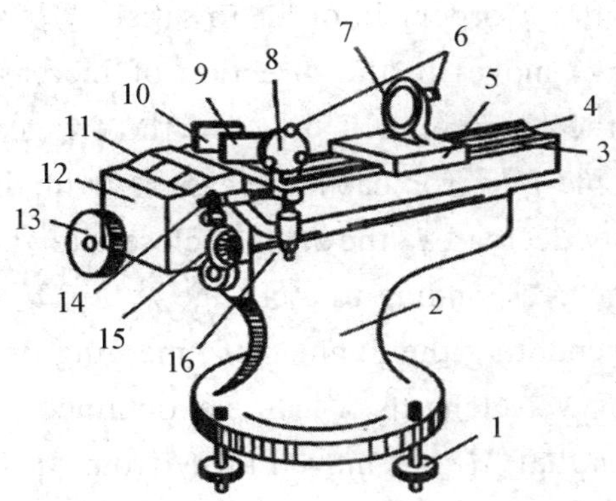

Fig. 4-23-2　Structure

1-Horizontal adjustment screw; 2-Base; 3-Guide rail; 4-Precision screw; 5-Support plate; 6-Mirror adjustment screw; 7-Movable mirror M_1; 8-Fixed mirror M_2; 9-Compensating plate P_2; 10-Beam splitter plate P_1; 11-Reading window; 12-Transmission system cover; 13-Coarse adjustment wheel; 14-Horizontal tension screw; 15-Fine adjustment wheel; 16-Vertical tension screw.

1. Interference Pattern Generated by Illumination from an Extended Light Source

(1) When M_1 and M_2' are strictly parallel, the interference obtained is called equal inclination interference. For an incident light beam with an inclination angle i, the optical path difference Δ introduced by the reflections from M_1 and M_2' is given by:

$$\Delta = 2d\cos i \tag{4-23-1}$$

Here, i is the angle of incidence on M_1, and d is the thickness of the air film. These interference fringes are located at infinity and form a set of concentric bright and dark circles when observed at point E in Fig. 4-23-1 (or with a converging lens placed at E and observed at its focal plane).

The characteristics of these fringes are as follows:

① The order of the fringes is highest at the center. At the center of the interference pattern, the condition for a bright spot to appear, disregarding the phase change between the reflected rays, is given by:

$$\Delta = 2d = k\lambda \tag{4-23-2}$$

where m is an integer. The order of the fringe at the center is determined by:

$$k = \frac{2d}{\lambda} \tag{4-23-3}$$

As the distance d between M_1 and M_2' gradually increases, for any given fringe order, such as the k order, in order to satisfy $2d\cos i_k = k\lambda$, $\cos i_k$ must decrease. Therefore, the fringe moves in the direction of increasing i_k ($\cos i_k$ decreasing), i. e., outward. Then, the observer will perceive the fringes "flowing out" from the center, with one fringe emerging for each increase $\lambda/2$ in the distance d. Conversely, as the distance gradually decreases, the fringes closest to the center will "sink" one by one, with each sinking corresponding to a change $\lambda/2$ in d.

By counting the number of emerging or sinking fringes, the displacement of M_1 in terms of wavelengths λ can be obtained. If N fringes emerge from the center, it indicates that M_1 has moved away from M_2' by:

$$\Delta d = N\frac{\lambda}{2} \tag{4-23-4}$$

Conversely, if N fringes sink, it indicates that M_1 has moved towards M_2' by the same distance. Using Eq. (4-23-4), if the wavelength of light λ is known, the displacement of M_1 can be calculated based on the number of fringe changes. This is the principle of interferometric length measurement. Conversely, if the displacement of M_1 and the number of fringe changes are known, the wavelength of light λ can be calculated.

② The distribution of the interference fringes is wider at the center and narrower at the edges. For adjacent the order of k and the order of $k-1$ interference fringes, we have:

$$2d\cos i_k = k\lambda$$

$$2d\cos i_{k-1} = (k-1)\lambda$$

Subtracting the two equations, and when i is small, we can obtain the angular distance Δi_k between adjacent fringes:

$$\Delta i_k = i_k - i_{k-1} \approx \frac{\lambda}{2di_k} \tag{4-23-5}$$

The above equation shows that for a constant d, the distribution of interference fringes in the field of view is wider at the center (small i_k, large Δi_k) and narrower at the edges (large i_k, small Δi_k). Therefore, when adjusting and measuring, a smaller value of d should be chosen, so that the both distances between M_1 and M_2 from the coating surface of P_1 are approximately the same.

(2) When there is a small angle of inclination α between M_1 and M_2', and the angle of incidence i is also small, the interference obtained is generally called equal thickness interference, and the fringes are located near the surface of the air film. In this case, the optical path difference introduced by the reflections from M_1 and M_2' is still approximately given by:

$$\Delta = 2d\cos i = 2d\left(1-\frac{i^2}{2}\right) \tag{4-23-6}$$

① Near the intersection of the two mirrors, where the thickness d is small, the influence of $d \cdot i^2$ can be neglected, and the coherent optical path difference is mainly determined by the thickness of the film, resulting in a set of evenly spaced straight fringes parallel to the intersection of M_1 and M_2'.

② Far from the intersection of M_1 and M_2', where d is larger, the fringes become curved, and the direction of curvature is away from the intersection of the two mirrors. This is because the term $d \cdot i^2$ in Eq. (4-23-6) can no longer be ignored. Since the interference fringes of the same kth order are the trajectories of equal optical path difference points, to satisfy $2d\left(1-\frac{i^2}{2}\right)=k\lambda$, as i increases, the value of d must be increased accordingly to compensate for the decrease in optical path difference caused by the increase in i. Therefore, the fringes move in the direction of increasing d, forming curved fringes as shown in Fig. 4-23-3. As d increases, the fringes become more severely curved.

(3) The conditions for observing colored interference fringes under white light illumination are as follows:

For equal inclination interference, it can be observed when α approaches zero. For equal thickness interference, it can be observed near the intersection of M_1 and M_2'. This is because at $d=0$, the interference is the same for all wavelengths and does not show color. When d is larger, the interference fringes of different wavelengths overlap each other, resulting in uniform illumination and the disappearance of color. Only when d is close to zero can a small number of colored interference fringes be observed.

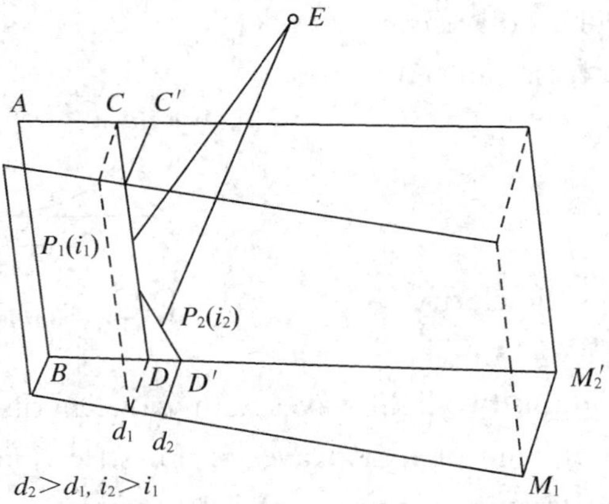

Fig. 4-23-3　Adjusting d

2. Non-local Interference Pattern Generated by Point Source Illumination

The interference phenomenon produced by the reflection of the point source S through M_1 and M_2' is equivalent to the interference generated by two virtual light sources S_1 and S_2, distributed along the axis. Therefore, the spherical waves emitted from S_1 and S_2 are coherent everywhere they meet, resulting in non-local interference. As shown in Fig. 4-23-4, when the laser beam passes through a short focal length beam expander lens, it forms a high-intensity point source S to illuminate the interferometer. By placing the observation screen E at different positions, various interference patterns can be observed.

When the observation screen E is perpendicular to the line connecting S_1 and S_2, circular interference fringes appear on the screen. Similar to equal inclination fringes, at the center of the circular pattern, the optical path difference is maximum, $\Delta=2d$, and the order is highest. When M_1 is moved to increase d, the circular fringes "emerge" one by one from the center. When d is decreased, the circular fringes "sink" one by one towards the center. Each change of one fringe corresponds to a displacement $\lambda/2$ of M_1. Therefore, this setup can be used for length measurement or wavelength measurement.

When M_1 and M_2' are parallel to each other, alternating bright and dark circular interference fringes are obtained. If the light source is strictly monochromatic, even though the fringes continuously emerge or sink as the mirror M_1 is slowly moved, the contrast of the fringes should remain constant.

Set the light strength of the bright fringe as I_1 and the adjacent dark one's as I_2, the contrast can be expressed as

$$V=\frac{I_1-I_2}{I_1+I_2}$$

The contrast describes the clarity of the fringes.

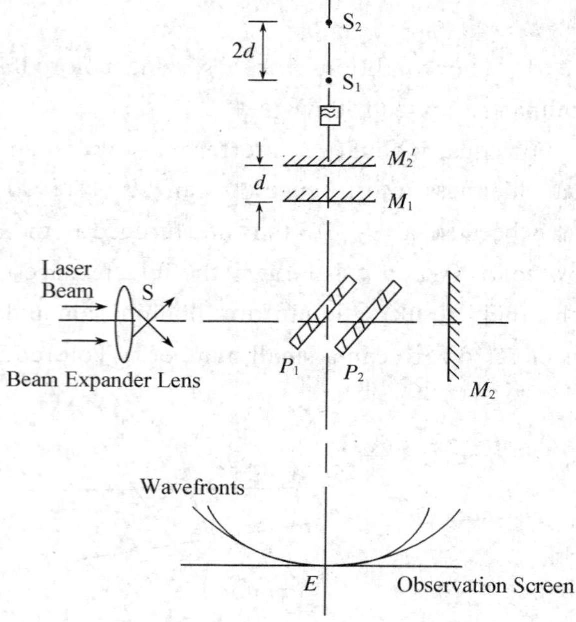

Fig. 4-23-4 Non-local interference

If the light source contains two light waves with wavelengths λ_1 and λ_2 that are close to each other, then each column of light waves is not strictly monochromatic. Taking sodium yellow light as an example, it consists of a doublet with central wavelengths $\lambda_1=589.0$ nm and $\lambda_2=589.6$ nm, with a wavelength difference of 0.6nm. Each spectral line also has a certain width, as shown in Fig. 4-23-5. Due to the small wavelength difference $\Delta\lambda$ between the doublet and the central wavelength, it is referred to as quasi-monochromatic light.

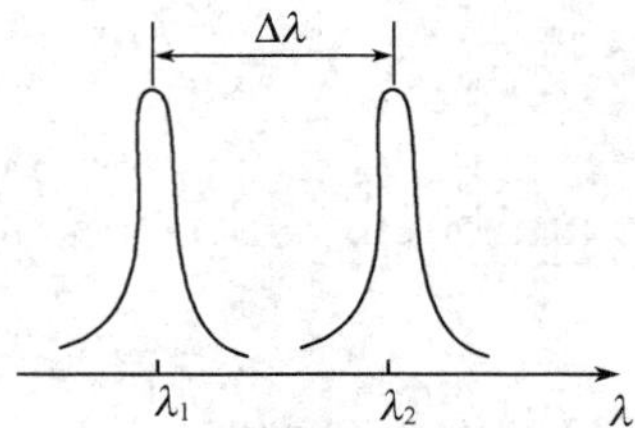

Fig. 4-23-5 Sodium yellow lights

When illuminating the Michelson interferometer with this light source, it will produce two sets of interference patterns. The intensity distribution in the interference field is a non-coherent superposition of the two sets of interference fringes. Due to the slight difference between λ_1 and λ_2, the positions of the bright rings corresponding λ_1 to and the positions of the bright rings corresponding to λ_2 will periodically coincide and shift with the variation of d. Therefore, as d changes, the superimposed interference fringes in the field of view alternately appear "clear" and "blurry or even disappear". Set the value of d as d_1 when it has the best contrast, with both λ_1 and λ_2 as bright fringes, we have:

$$d_1=m\frac{\lambda_1}{2},\ d_1=n\frac{\lambda_2}{2}\quad (m \text{ and } n \text{ are integers})$$

If $\lambda_1>\lambda_2$, when d increases to d_2, if the condition is satisfied:

$$d_2=(m+k)\frac{\lambda_1}{2},\ d_2=(n+k+0.5)\frac{\lambda_2}{2}\quad (k \text{ is an integer})$$

At this point, λ_1 is a bright fringe, and λ_2 is a dark fringe, with the worst contrast (possibly unable to distinguish the fringes). The distance moved by M_1 from the best to the worst contrast is given by:

$$d_2-d_1=k\frac{\lambda_1}{2}=(k+0.5)\frac{\lambda_2}{2}$$

With $d_2-d_1=k\frac{\lambda_1}{2}$ and $k\frac{\lambda_1}{2}=(k+0.5)\frac{\lambda_2}{2}$, by eliminating k we can obtain the wavelength difference as:

$$\lambda_1-\lambda_2=\frac{\lambda_1\lambda_2}{4(d_2-d_1)}\approx\frac{\overline{\lambda_{12}^2}}{4(d_2-d_1)}\tag{4-23-7}$$

where $\overline{\lambda_{12}^2}$ is the average value of λ_1 and λ_2. Since when the contrast is worst, the positions of M_1 are symmetrically distributed on both sides of the positions of the best contrast, the relationship between the displacement of M_1, $\Delta d[=2(d_2-d_1)]$, and $\Delta\lambda(=\lambda_1-\lambda_2)$ for adjacent worst contrast positions is:

$$\Delta\lambda=\frac{\overline{\lambda_{12}^{2}}}{2\Delta d} \tag{4-23-8}$$

Experimental Contents and Steps

1. Adjustment of the Michelson Interferometer

(1) Turn on the sodium lamp S to illuminate the ground glass screen, creating a uniform expanded light source. Add a crosswire on the screen.

(2) Rotate the coarse adjustment wheel to approximately equalize the distances between M_1 and M_2 from the coated surface P_1. Observe direction along the E and P_1, and you will see the shadow of the crosswire (a total of 3 shadows), with 2 corresponding to the reflection of the moving mirror M_1 (why?), and the other corresponding to the reflection of M_2.

(3) Carefully adjust the three screws behind M_1 and M_2 to change the relative orientation of M_1 and M_2, until the double shadows of the crosswire (which two?) completely overlap in both the horizontal and vertical directions. At this point, interference fringes can be observed. Carefully adjust the three screws to make the interference fringes circular.

(4) Carefully and slowly adjust the two fine adjustment screws below M_2, so that the center of the interference fringes only moves with the movement of the observer's eyes left, right, up, and down, without the fringes "emerging" or "sinking". At this point, the observed interference rings are strictly equal inclination interference. If when the eyes move, the observed interference rings exhibit "emerging" or "sinking", analyze the situation and make further adjustments.

2. Measurement of Sodium Light Wavelength (average of D_1 and D_2 wavelengths)

(1) Rotate the coarse adjustment wheel to move M_1, and observe the changes in the fringes. Determine the changes in φ based on the "emerging" or "sinking" of the fringes, and observe the relationship between the value of d and the thickness and density of the fringes.

(2) When clear and well-contrasted interference circular fringes appear in the field of view, slowly rotate the fine adjustment wheel and observe the fringes emerging one by one from the center (or sinking towards the center). Start counting and record the position d_1 of M_1(the sum of the readings of the two rotary scales). Continue rotating the fine adjustment wheel and record the position of M_1 when the fringes emerge (or sink) every 100 fringes. In total, measure the movement of 800 - 1000 fringes, and use Eq. (4-23-4) to calculate the wavelength λ of the measured light and its average value,

as well as the uncertainty, and compare them with the accepted values.

Vibration has a significant impact on the measurement, so attention should be paid. (The three feet of the interferometer should be equipped with soft pads.)

3. Observation of Colored Interference Fringes in White Light

Refer to the analysis in the principle section and consider the following questions:

(1) What are the conditions for observing colored interference fringes (circular rings) in equal inclination interference?

(2) By moving M_1, how can you determine whether the distance d is increasing or decreasing based on the observed phenomena?

(3) In which direction will the colored interference rings definitely be observed when moving M_1?

(4) What should be considered to observe colored fringes in equal thickness interference?

First, use the sodium lamp to observe the equal inclination interference rings, then move M_1, and if you believe the direction of movement of M_1 is correct based on the observed phenomena, switch to a white light source and continue moving M_1 until the colored interference rings are observed.

Then adjust the colored interference fringes in equal thickness interference.

Note: Since there are only a few colored fringes in white light, it is necessary to patiently and carefully adjust the fine adjustment wheel. If the movement is too fast, the fringes may pass quickly and become difficult to perceive.

4. Observation of Interference Fringes from Point Source Illumination

Design your own experimental steps to observe the shape, characteristics, observation conditions, and variation rules of the interference fringes when illuminated by a point source.

5. Measurement of Sodium D Doublet Wavelength Difference

(1) Use a sodium lamp as the light source to observe the equal inclination interference fringes in the interferometer.

(2) Move M_1, so that the visibility of the center of the field of view is minimized. Record the position d_1 of M_1. Continue to move M_1 in the original direction until the contrast becomes minimal again, and record the position as d_2. Then, $\Delta d = |d_2 - d_1|$. Due to the small difference λ_1 and λ_2, the position of the worst contrast has a large random error since a large range of positions near the worst contrast have poor contrast (wide blur zone). In this case, the coarse adjustment wheel (accuracy: 0.001 mm) can be used for measurement. Measure the distances between 10 blur zones and calculate Δd. This uses an extended range to reduce the random error of a single measurement.

Precautions

The Michelson interferometer is a precision optical instrument, and the following should be noted when using it:

1. Pay attention to dust prevention, moisture prevention, and shock prevention. Do not touch the optical surfaces of the components, and avoid speaking or coughing towards the instrument.

2. Before and after the experiment, all adjustment screws should be in a relaxed state. When adjusting, first put them in the middle position to allow for adjustment in both directions. Adjustments should be made evenly and slowly.

3. For some interferometers, when the clutch of the coarse adjustment wheel and the fine adjustment wheel is engaged, only the fine adjustment wheel should be used, and the coarse adjustment wheel should not be used again, as this may damage the instrument.

4. When rotating the fine adjustment wheel for measurement, special attention should be paid to avoid backlash errors.

Questions

1. Analyze the differences between the circular interference fringes produced by expanded laser beams and sodium light.

2. When adjusting the interference fringes of sodium light, it is possible that the double shadows of the crosswire are aligned, but the fringes do not appear. Analyze the possible reasons for this.

3. How can one determine and verify that the interference fringes belong to strictly equal inclination fringes?

4. How can experimental methods be used to determine the positioning region of the interference fringes?

4.24 磁阻效应实验

材料的电阻会因为外加磁场而增加或减少,则电阻的变化称为磁阻。磁阻效应是 1857 年由英国物理学家威廉·汤姆森发现的,它在金属里可以忽略,在半导体中则可能由小到中等。从一般磁阻开始,磁阻发展经历了巨磁阻(GMR)、庞磁阻(CMR)、穿隧磁阻(TMR)、直冲磁阻(BMR)和异常磁阻(EMR)。2007 年诺贝尔物理学奖授予来自法国国家科学研究中心的物理学家艾尔伯·费尔和来自德国尤利希研究中心的物理学家皮特·克鲁伯格,以表彰他们发现巨磁阻效应的贡献。目前，磁阻效应广泛用于磁传感、磁力计、电子罗盘、位置和角度传感器、车辆探测、GPS导航、仪器仪表、磁存储(磁卡、硬盘)等领域。

实验目的

1. 了解磁阻效应和霍尔效应的关系和区别。
2. 测量锑化铟传感器的电阻和磁感应强度的关系。
3. 作锑化铟传感器的电阻变化与磁感应强度的关系曲线。

实验原理

一定条件下，导电材料的电阻值 R 随磁感应强度 B 变化规律称为磁阻效应，如图4-24-1所示，当半导体处于磁场中时，半导体的载流子将受洛仑兹力的作用发生偏转，在两端产生积聚电荷并产生霍尔电场。如果霍尔电场作用和某一速度的载流子的洛仑兹力作用刚好抵消，那么小于或大于该速度的载流子将发生偏转，因而沿外加电场方向运动的载流子数量将减小，电阻增大，表现出横向磁阻效应。磁阻效应与样品的形状有关，不同几何形状的样品，在同样大小的磁场作用下，其电阻不同，该效应称为几何磁阻效应。

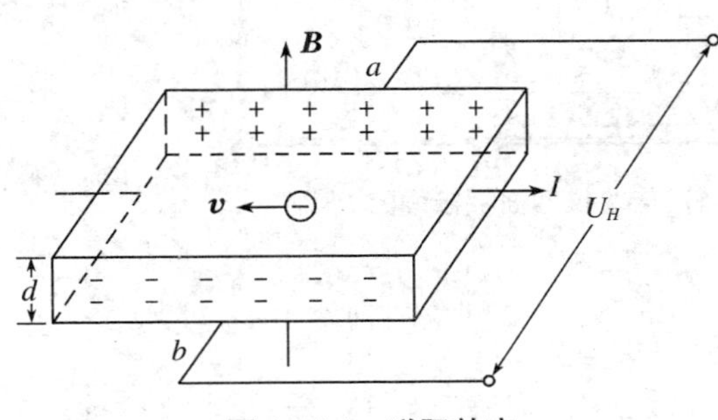

图 4-24-1 磁阻效应

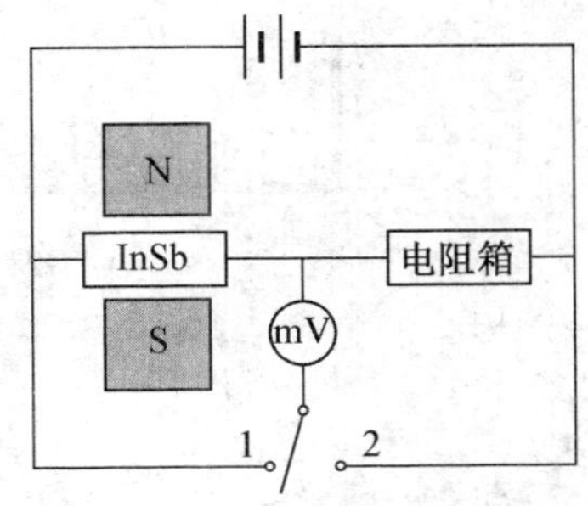

图 4-24-2 磁阻效应实验装置图

通常以电阻率的相对改变量来表示磁阻的大小，即用 $\Delta\rho/\rho(0)$ 表示，其中 $\rho(0)$ 为零磁场时的电阻率。设磁电阻在磁感应强度为 B 的磁场中电阻率为 $\rho(B)$，则 $\Delta\rho=\rho(B)-\rho(0)$。由于磁阻传感器电阻的相对变化率 $\Delta R/R(0)$ 正比于 $\Delta\rho/\rho(0)$，其中 $\Delta R=R(B)-R(0)$，因此，也可以用磁阻传感器的电阻相对改变量 $\Delta R/R(0)$ 来表示磁阻效应的大小。测量磁电阻值 R 与磁感应强度 B 的关系实验装置如图4-24-2所示：

实验证明：当金属或半导体处于较弱磁场中时，一般磁阻传感器电阻相对变化率 $\Delta R/R(0)$ 正比于磁感应强度 B 的二次方，而在强磁场中 $\Delta R/R(0)$ 与磁感应强度 B 呈线性函数关系。磁阻传感器的上述特性在物理学和电子学方面有着重要的应用。

实验器材

FD-MR-II 型磁阻效应实验仪，电阻箱，导线若干。仪器面板图接线如图 4-24-3 所示。

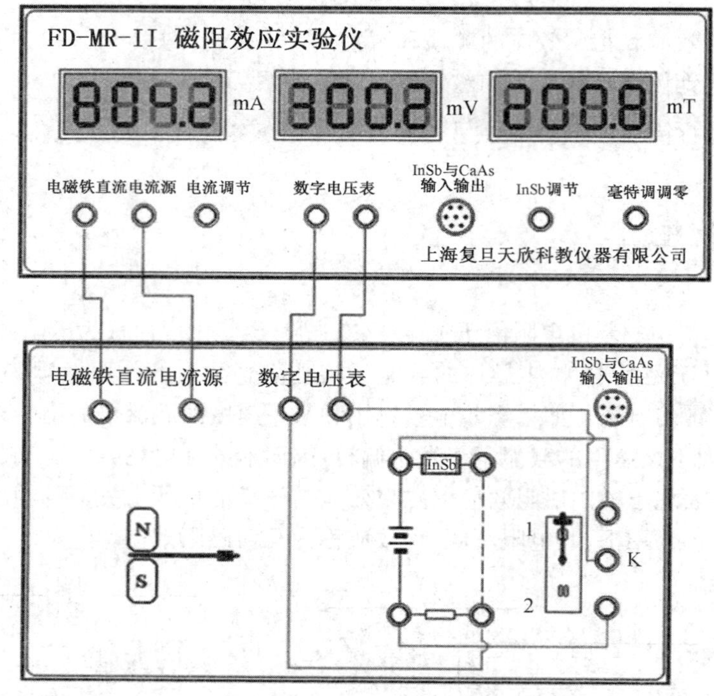

图 4-24-3　磁阻效应实验仪器面板接线图

实验内容及步骤

1. 按图 4-24-3 连接导线，连接时注意要自备电阻箱和注意电压正负极。

2. 调节电磁铁直流电流源，使电流 I_M 为零，调节毫特计调零旋钮，使示数为零。

3. 将单刀双掷开关拨向 2 端，调节电阻箱示数使取样电阻为 200 Ω，调节 InSb 传感器的电流，使得数字电压表示数为 50 mV，则取样电流为 50 mV/200 Ω=0.25 mA。

4. 将单刀双掷开关拨向 1 端，记录此时数字毫伏表的示数，则在磁场为 0 时 InSb 传感器的零场电阻为该示数除以取样电流。

5. 调节电磁铁直流电流源，分别记录数字毫伏表读数，记录在表 4-24-1 中。

表 4-24-1　取样电阻 $R=200\ \Omega$，电压 $U=50\ \text{mV}$，取样电流 $I=0.25\ \text{mA}$

电磁铁电流	InSb 电压值	$\Delta R/R(0)\sim B$ 的对应关系		
I_M/mA	U_R/mV	B/mT	R/Ω	$\Delta R/R(0)$
		0.0		
		10.0		
		20.0		
		30.0		
		40.0		
		50.0		
		60.0		
		70.0		
		100.0		
		150.0		
		200.0		
		250.0		
		300.0		
		350.0		
		400.0		
		450.0		
		500.0		

数据记录与处理

1. 对表 4-24-1 数值进行计算，分别求出 InSb 传感器的电阻 $R(0)$ 和 $\Delta R/R(0)$。
2. 对表 4-24-1 数据在 $B<0.06$ T 时对 $\Delta R/R(0)$ 作出曲线拟合，并写出表达式。
3. 对表 4-24-1 数据在 $B>0.12$ T 时对 $\Delta R/R(0)$ 作直线拟合，并写出表达式。

注意事项

1. 实验时注意不可在实验仪器附近放置磁性物品，将传感器固定在磁铁间隙，不可弯折，InSb 传感器工作电流小于 3 mA。

2. 操作时，应先设置取样电阻阻值，再打开仪器电源。在操作过程中不能调节电阻箱阻值，否则容易损坏仪器。

思考题

1. 什么叫磁阻效应?霍尔传感器为何有磁阻效应?
2. InSb 磁阻传感器在弱磁场和强磁场时的电阻值与磁感应强度关系有什么不同?

Experiment 4.24 Magnetoresistive Effect

Magnetoresistance is the tendency of a material to change its electrical resistance in an external magnetic field. The effect was discovered in 1856 by the British physicist William Thomson. Magnetoresistance is pretty weak in normal metals while it can be significant in semiconductors. Since the first magnetoresistive effect was discovered, giant magnetoresistance (GMR), colossal magnetoresistance (CMR), tunnel magnetoresistance (TMR), bilinear magnetoresistance (BMR), and extraordinary magnetoresistance (EMR) have been observed. In 2007, Albert Fert and Peter Grünberg were awarded the Nobel Physics Prize for the discovery of giant magnetoresistance. Nowadays, people wildly apply the magnetoresistive effect to the development of magnetic sensing, magnetometers, electronic compasses, position and angle sensors, vehicle detection, GPS navigation, instrumentation, magnetic storage (magnetic cards, hard drives), and other fields.

Experimental Objectives

1. To measure the relationship between the electrical resistance of an indiumantimonide sensor and the magnetic field strength.

2. To plot the relationship curve between the resistance variation of the indiumantimonide sensor and the magnetic field strength.

3. To perform curve fitting for the nonlinear and linear regions using curves and straight lines respectively.

Experimental Principle

Under certain conditions, the variation of the electrical resistance R of a conductive material with the magnetic field strength B is known as the magnetoresistive effect. As shown in Fig. 4-24-1, when a semiconductor is placed in a magnetic field, the charge carriers of the semiconductor will be deflected by the Lorentz force, resulting in accumulated charges and the generation of a Hall electric field at both ends. If the effect of the Hall electric field is precisely balanced with the Lorentz force acting on charge carriers at a certain velocity, charge carriers below or above that velocity will be deflected. Consequently, the number of charge carriers moving in the direction of the applied electric field will decrease, leading to an increase in electrical resistance, the so-called transverse magnetoresistive effect.

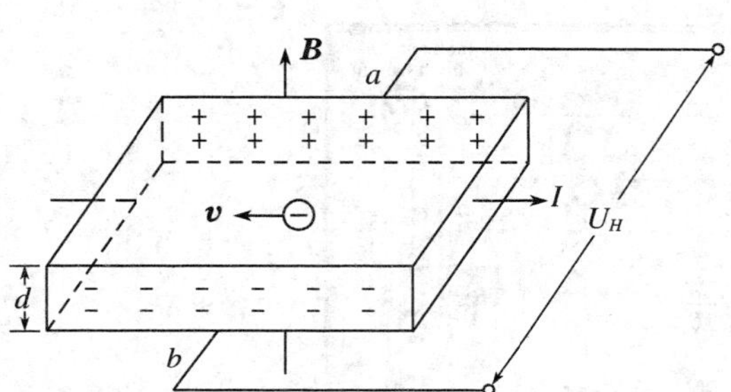

Fig. 4-24-1 Magnetoresistive effect

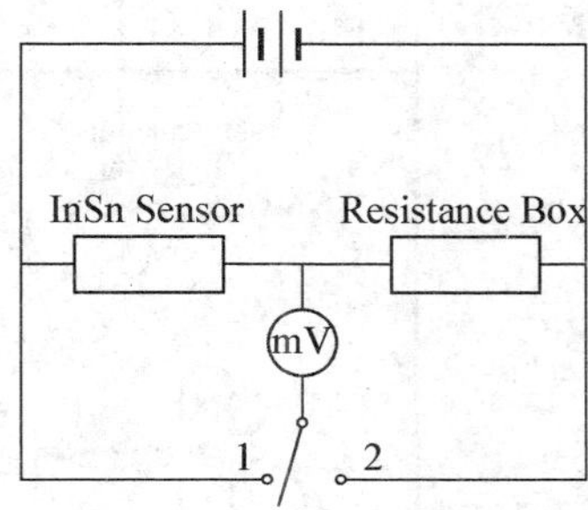

Fig. 4-24-2 Magnetoresistive effect experimental setup

The magnitude of the magnetoresistance is typically expressed as the relative change of resistivity, denoted as $\Delta\rho/\rho(0)$, where $\rho(0)$ represents the resistivity without a magnetic field. Assume the resistivity of the magnetoresistive material in a magnetic field with a strength B is $\rho(B)$, then we have $\Delta\rho = \rho(B) - \rho(0)$. As the relative change in electrical resistance $\Delta R/R(0)$ of the magnetoresistive sensor is proportional to $\Delta\rho/\rho(0)$, where $\Delta R = R(B)-R(0)$, the magnitude of the magnetoresistance can also be expressed using the relative change in electrical resistance of the magnetoresistive sensor, $\Delta R/R(0)$. The experimental setup for measuring the relationship between the magnetoresistance value R and the magnetic field strength is shown in Fig. 4-24-2:

Experimental results should show that when a metal or semiconductor is subjected to a relatively weak magnetic field, the relative change in resistance of a magnetoresistive sensor is generally proportional to the square of the magnetic field strength B. However, in a strong magnetic field, the relationship between the relative change in electrical resistance and the magnetic field strength B is linear.

Experimental Instruments

1. FD-MR-II Magnetoresistance Effect Experiment Apparatus
2. Resistance Box

3. Several wires

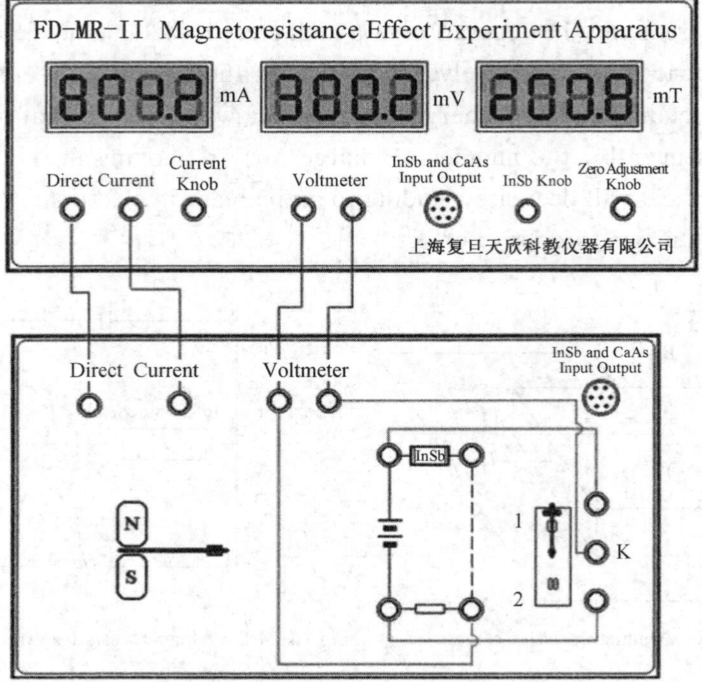

Fig. 4-24-3 The wiring diagram for the instrument panel

Experimental Contents and Steps

1. Connect the wires according to the wiring diagram shown in Fig. 4-24-3. Make sure to use a resistance box and use the correct polarity of the voltage.

2. Adjust the DC current source of the electromagnet to set the current I_M to zero. Use the milliammeter zero adjustment knob to ensure that the reading is zero.

3. Set the single-pole, double-throw switch to position 2. Adjust the resistance box to set the sampling resistance to 200 Ω. Adjust the current through the InSb sensor until the digital voltmeter reading is 50 mV. Therefore, the sampling current is 0.25 mA (50 mV/200).

4. Set the single-pole double-throw switch to position 1. Record the voltage. This reading represents the zero-field resistance of the InSb sensor when no magnetic field is applied. Divide this reading by the sampling current to obtain the zero-field resistance value.

5. Adjust the DC current source of the electromagnet to set the millivoltmeter readings to 10.0, 20.0, 30.0 millivolts, and so on, according to the values in the table.

Record the readings on the digital millivoltmeter for each magnetic field strength.

Table 4-24-1　Sampling resistance $R=200\ \Omega$, voltage $U=50$ mV, current $I=0.25$ mA

Electromagnet Current	InSb Voltage	$\Delta R/R(0)\sim B$		
I_M/mA	U_R/mV	B/mT	R/Ω	$\Delta R/R(0)$
		0.0		
		10.0		
		20.0		
		30.0		
		40.0		
		50.0		
		60.0		
		70.0		
		100.0		
		150.0		
		200.0		
		250.0		
		300.0		
		350.0		
		400.0		
		450.0		
		500.0		

Data Recording and Processing

1. Calculate the R and $\Delta R/R(0)$ for the InSb sensor using the values in Table 4-24-1.

2. Perform a curve fitting between $\Delta R/R(0)$ and B for the data in Table 4-24-1 when $B<0.06$ T and write down the expression.

3. Perform a curve fitting between $\Delta R/R(0)$ and B for the data in Table 4-24-1 when $B>0.12$ T and write down the expression.

Precautions

When performing the experiment, ensure that the sampling resistance value is set correctly before turning on the power of the apparatus. Do not adjust the resistance box during the operation to avoid damaging the apparatus.

Questions

1. What is the magnetoresistive effect? Why do Hall sensors exhibit the magnetoresistive effect?

2. How does the relationship between the electrical resistance of InSb magnetoresistive sensors and magnetic field strength differ in weak and strong magnetic fields?

4.25 电表的改装与校准

电表在电测量中有着广泛的应用,因此,了解电表和使用电表就显得十分重要。电流计(表头)由于构造的原因,一般只能测量较小的电流和电压,如果要用它来测量较大的电流或电压,就必须进行改装,以扩大其量程。万用表的原理就是对微安表头进行多量程改装而来,在电路的测量和故障检测中得到了广泛的应用。

实验目的

1. 测量表头内阻 R_g 及满度电流 I_g。
2. 掌握将 100 μA 表头改成较大量程的电流表和电压表的方法。
3. 学会校准电流表和电压表的方法。

实验原理

一、电流计表头内阻 R_g

常见的磁电式电流计主要由永久磁铁、线圈、游丝和指针所组成。当电流通过线圈时,载流线圈在磁场中产生磁力矩使线圈转动,从而带动指针偏转。线圈偏转角度的大小与通过的电流大小成正比,所以可由指针的偏转直接指示出电流值。

电流计允许通过的最大电流 I_g 称为电流计的量程，电流计的线圈有一定内阻，由 R_g 表示，I_g 和 R_g 是电流计特性的重要参数。

测量内阻 R_g 常用方法有：

1. 半电流法也称中值法

测量原理图见图 4-25-1。当被测电流计接在电路中时，使电流计满偏。再闭合电键 S_2，改变电阻值 R_1 即改变分流程度，当电流计指针指示到中间值，且标准表读数仍保持不变（可通过调电源电压和 R_W 来实现），此时分流电阻值 R_1 就等于电流计的内阻。

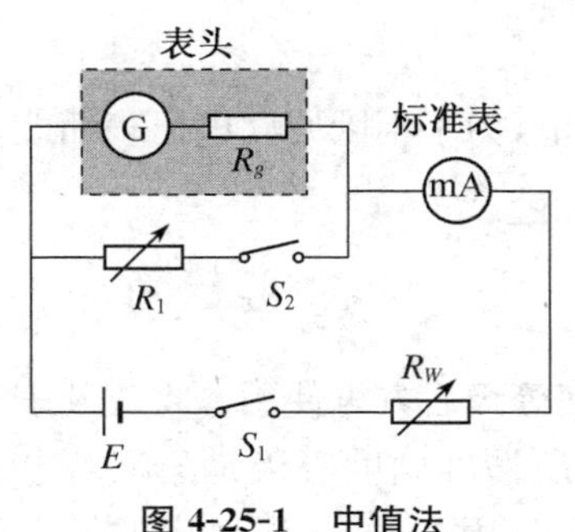

图 4-25-1　中值法

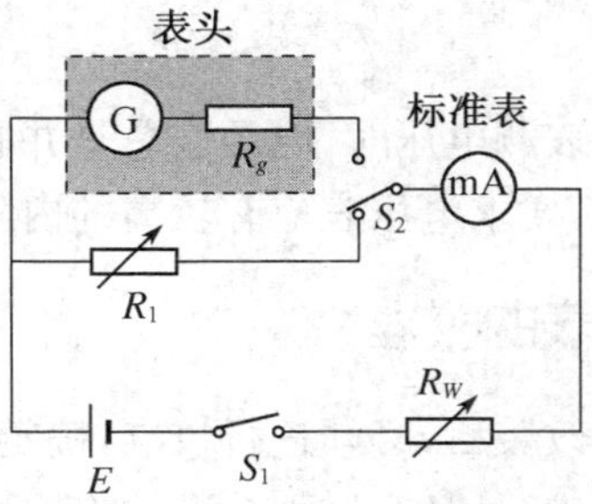

图 4-25-2　替代法

2. 替代法

测量原理图见图 4-25-2。当被测电流计接在电路中时，使电流计满偏。改变电键 S_2 的方向，调节 R_1，当电路中的电压不变，且电路中的电流（标准表读数）亦保持不变时，则电阻箱 R_1 值即为被测电流计内阻。

替代法是一种运用很广的测量方法，具有较高的测量准确度。

二、改装大量程电流表

根据电阻并联规律可知，如果在表头两端并联上一个阻值适当的电阻 R_s，如图 4-25-3 所示，可使表头不能承受的那部分电流从 R_s 上分流通过。这种由表头和并联电阻 R_s 组成的整体就是改装后的电流表。如需将表头量程扩大 n 倍，则不难得出

$$R_s=\frac{R_g}{n-1} \tag{4-25-1}$$

用改装的电流表测量电流时，电流表应串联在被测电路中，所以要求电流表应有较小的内阻。另外，在表头上并联阻值不同的分流电阻，便可制成多量程的电流表。

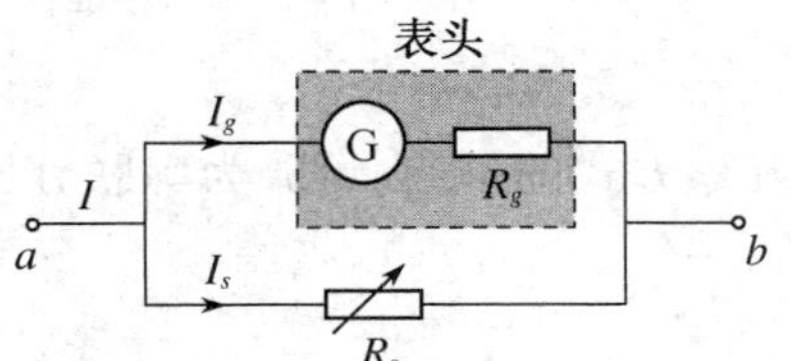

图 4-25-3　改装电流表

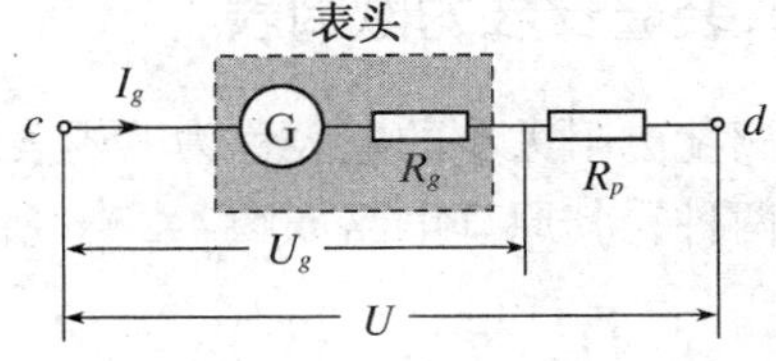

图 4-25-4　改装电压表

三、改装电压表

一般表头能承受的电压很小,不能用来测量较大的电压。为了测量较大的电压,可以给表头串联一个阻值适当的电阻R_P,如图 4-25-4 所示,使表头上不能承受的那部分电压落在电阻 R_P 上。这种由表头和串联电阻 R_P 组成的整体就是电压表,串联的电阻 R_P 叫作扩程电阻。选取不同大小的 R_P,就可以得到不同量程的电压表。由图 4-25-4 可求得扩程电阻值为:

$$R_P = \frac{U}{I_g} - R_g \tag{4-25-2}$$

用电压表测电压时,电压表总是并联在被测电路上,为了不因并联电压表而改变电路中的工作状态,要求电压表应有较高的内阻。

四、电表的校准

电表在改装后,必须进行校准,确定电表的准确度等级后方可使用。将改装表与相应的标准表直接进行比较,这种校准的方法称为比较法。所谓准确度等级是国家对电表规定的质量指标,共有七级:0.1、0.2、0.5、1.0、1.5、2.5 和 5.0,用 S 表示,所对应的最大绝对误差为 $\Delta_{仪}$=量程×S%。

设被校电流表的指示值为 I_x,标准表的读数为 I_s,如果测量一组数据 I_{xi} 和 I_{si},则每个标准点的校正值为 $\Delta I_i = I_{si} - I_{xi}$。如果将它们中绝对值最大的一个作为最大绝对误差,则被校电流表的标称误差为:

$$标称误差 = \frac{最大绝对误差}{量程} \times 100\% \tag{4-25-3}$$

根据标称误差的大小,即可定出电流表的准确度等级(此处没有考虑标准表引起的误差,一般地,标准表的精度至少要比被校表的精度高一个等级)。同理电压表的校准亦如此。

电表的校准结果除用准确度等级表示外,还可以用校准曲线表示,即以被校表的示值为横坐标,以各点的示值误差为纵坐标,画出折线校正曲线,图 4-25-5 为某电流表的校准曲线。

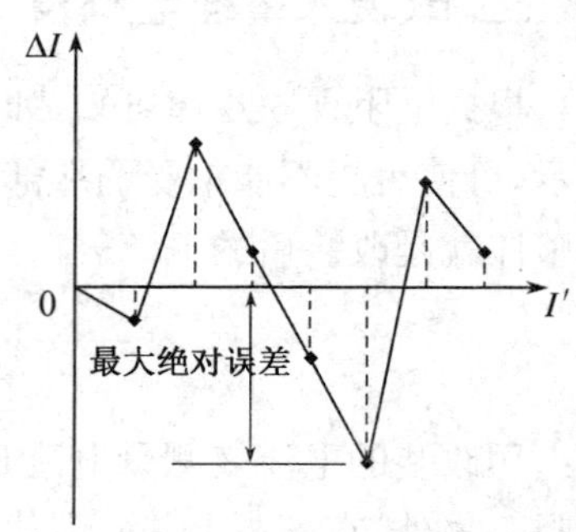

图 4-25-5　校准曲线

五、改装毫安表为欧姆表

用来测量电阻大小的电表称为欧姆表。根据调零方式的不同,可分为串联分压式和并联分流式两种。其原理电路如图 4-25-6 所示。

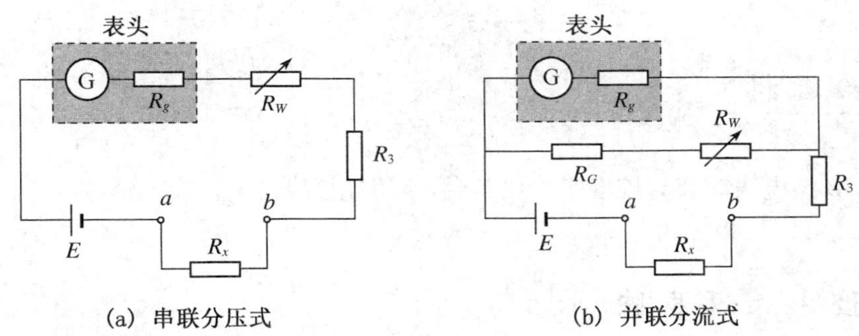

(a) 串联分压式　　　　(b) 并联分流式

图 4-25-6　欧姆表原理图

图中 R_3 为限流电阻，R_w 为调“零”电位器，R_x 为被测电阻，R_g 为等效表头内阻。图 4-25-6(b) 中，R_G 与 R_w 一起组成分流电阻。

欧姆表使用前先要调“零”点，即 a、b 两点短路，(相当于 $R_x=0$)，调节 R_w 的阻值，使表头指针正好偏转到满度。可见，欧姆表的零点是就在表头标度尺的满刻度(即量限)处，与电流表和电压表的零点正好相反。

在图 4-25-6(a)中，当 a、b 端接入被测电阻 R_x 后，电路中的电流为

$$I=\frac{E}{R_g+R_w+R_3+R_x} \tag{4-25-4}$$

对于给定的表头和线路来说，R_g、R_w、R_3 都是常量。由此可见，当电源端电压 E 保持不变时，被测电阻和电流值有一一对应的关系。即接入不同的电阻，表头就会有不同的偏转读数，R_x 越大，电流 I 越小。短路 a、b 两端，即 $R_x=0$ 时，指针满偏。

当 $R_x=R_g+R_w+R_3$ 时

$$I=\frac{E}{R_g+R_w+R_3+R_x}=\frac{1}{2}I_g \tag{4-25-5}$$

这时指针在表头的中间位置，对应的阻值为中值电阻，显然 $R_中=R_g+R_w+R_3$。

当 $R_x=\infty$(相当于 a、b 开路)时，$I=0$，即指针在表头的机械零位。

所以欧姆表的标度尺为反向刻度，且刻度是不均匀的，电阻 R 越大，刻度间隔愈密。如果表头的标度尺预先按已知电阻值刻度，就可以用电流表来直接测量电阻了。

并联分流式欧姆表利用对表头分流来进行调零的，具体参数可自行设计。

欧姆表在使用过程中电池的端电压会有所改变，而表头的内阻 R_g 及限流电阻 R_3 为常量，故要求 R_w 要跟着 E 的变化而改变，以满足调“零”的要求，设计时用可调电源模拟电池变化，范围取 1.3～1.6 V 即可。

实验器材

DH4508B 型电表改装与校准实验仪 1 台,专用连接线若干。

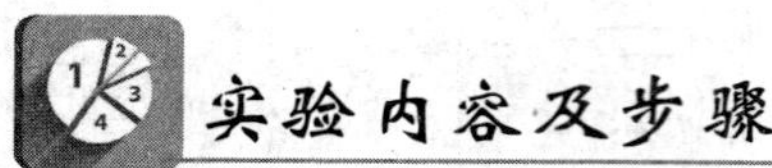

实验内容及步骤

1. 用中值法或替代法测量表头内阻

(1) 中值法

按照图 4-25-7 连线,先不接入电阻箱(断开虚线),先将电源输出电压 E 调至 0 V,接通电路,调节 E 和 R_w 使改装表头满偏,记录标准表的读数 I,此电流即为改装表头的满度电流 I_g。再接入电阻箱(连接虚线),改变电阻箱的数值,使被测表头指针从满度 100 μA 降低到 50 μA 处。注意调节 E 或 R_w,使标准电流表的读数 I 保持不变。记录此时电阻箱阻值为表头内阻 R_g。

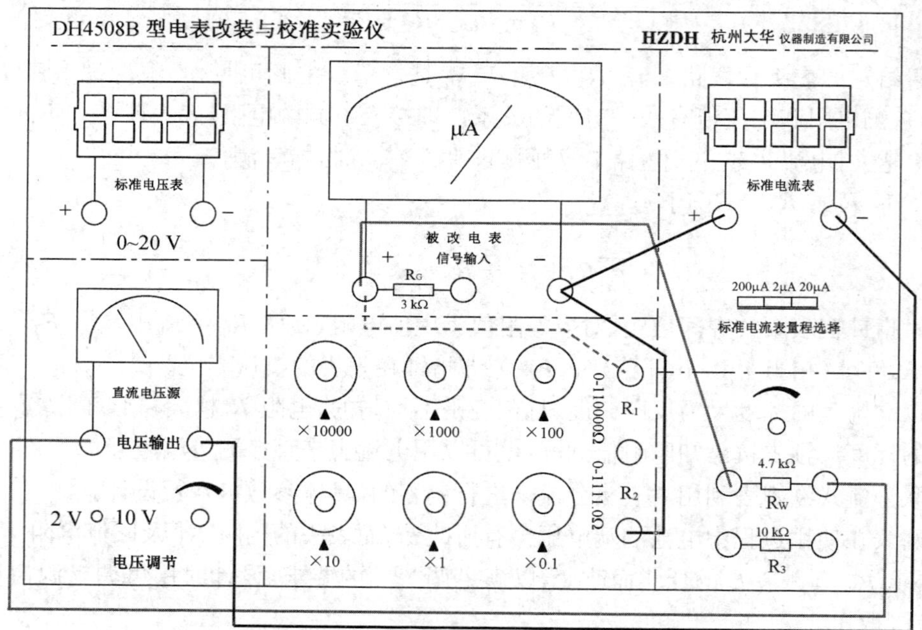

图 4-25-7　中值法测量表头内阻

（2）替代法

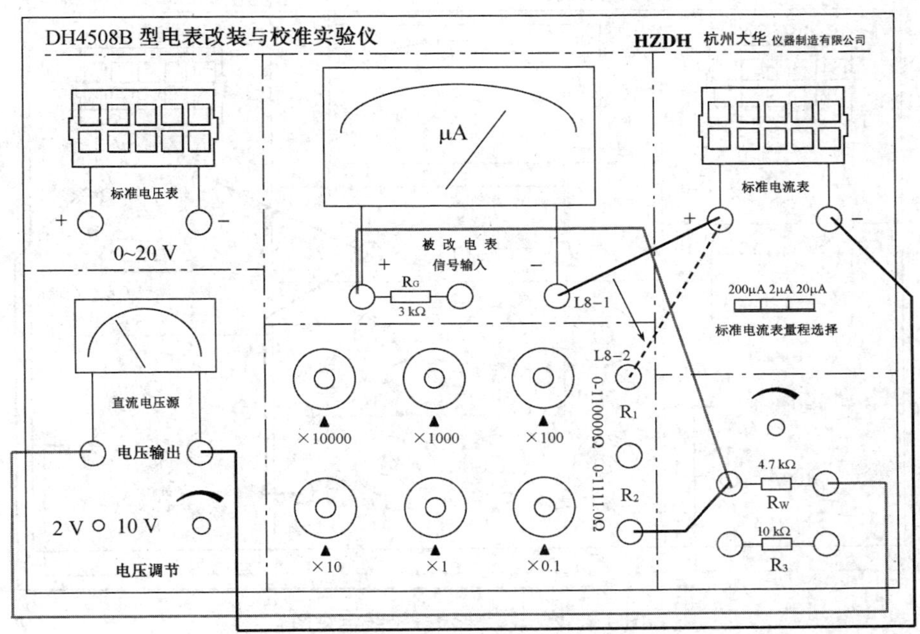

图 4-25-8　替代法测量表头内阻

替代法测量可参考图 4-25-8 接线(先连接 L8-1,断开 L8-2)。先将 E 调至 0 V,接通电路,调节 E 和R_w 使改装表头满偏,记录标准表的读数,此值即为被改装表头的满度电流 I_g;再断开接到改装表头的接线,转接到电阻箱(断开 L8-1,连接 L8-2),调节电阻箱使标准电流表的电流保持刚才记录的数值。这时电阻箱的数值即为被测表头内阻 R_g。

2. 将量程为 100 μA 的表头改装成 1 mA(或自选)量程的电流表

(1) 根据电路参数,估计 E 值大小,并根据公式(4-25-1)计算出分流电阻值 R_s。将电阻箱阻值设置成 R_s 的大小。

(2) 参考图 4-25-9 接线,先将 E 调至 0 V,检查接线正确后,调节 E 和滑动变阻器 R_w,使改装表指到满量程,记录此时标准表读数。注意:R_w 作为限流电阻,阻值不要调至最小值。然后每隔 0.2 mA 逐步减小读数直至零点;再按原间隔逐步增大到满量程,每次记下标准表读数于表 4-25-1。

表 4-25-1　改装电流表测量数据

改装表读数(μA)	标准表读数(mA)			误差 ΔI(mA)
	减小	增大	平均值	
20				
40				
60				
80				
100				

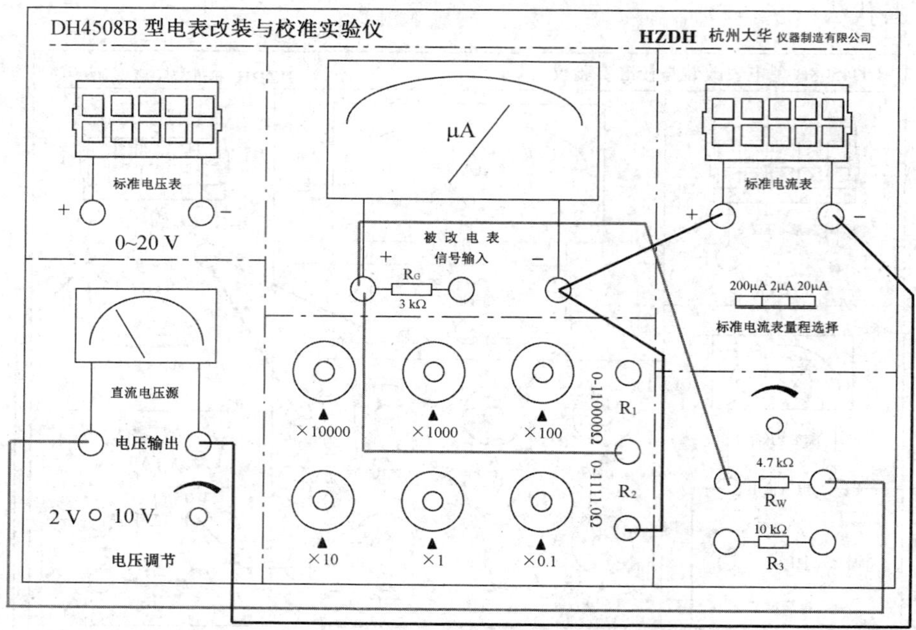

图 4-25-9 改装电流表接线示意图

(3) 在坐标纸上作出改装电流表的校准曲线,并给出电表等级。

(4) 重复以上步骤,将 100 μA 表头改成 10 mA 表头,可按每隔 2 mA 测量一次(可选做)。

3. 将量程为 100 μA 的表头改装成 1.5 V(或自选)量程的电压表

(1) 根据电路参数,估计 E 值大小,并根据公式(4-25-2)计算出扩程电阻 R_P 阻值。

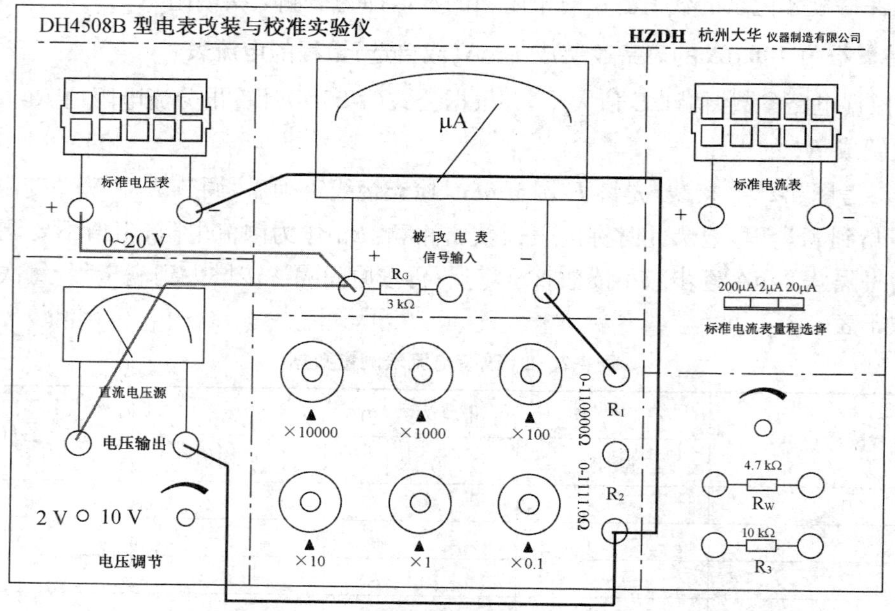

图 4-25-10 改装电压表接线示意图

(2) 按图 4-25-10 连线，先调节电阻箱 R 值至最大值，再调节 E；用标准电压表监测到 1.5 V 时，再调节电阻箱 R 值，使改装表指示为满度。于是 1.5 V 电压表就改装好了。

(3) 调节电源电压，使改装表指针指到满量程(1.5 V)，记下标准表读数。然后每隔 0.3 V 逐步减小改装读数直至零点；再按原间隔逐步增大到满量程，每次记下标准表的相应读数于表 4-25-2。

表 4-25-2 改装电压表测量数据

改装表读数(V)	标准表读数(V)			误差 ΔU(V)
	减小	增大	平均值	
0.3				
0.6				
0.9				
1.2				
1.5				

(4) 在坐标纸上作出改装电压表的校正曲线，并给出电表等级。

(5) 重复以上步骤，将 100 μA 表头改成 10 V 表头，可按每隔 2 V 测量一次(可选做)。

4. 改装欧姆表及标定表面刻度量程

(1) 根据表头参数 I_g 和 R_g 以及电源电压 E，选择 R_w 为 4.7 kΩ，R_3 为 10 kΩ。

(2) 按图 4-25-11 连线。调节电源 E=1.5 V，调节电阻箱阻值为零(即使 R_x=0)，调 R_w 使表头指针正好偏转到满度。如此，欧姆表的调零工作完成。

(3) 测量改装成的欧姆表的中值电阻。如图 4-25-11 所示，调节电阻箱(即 R_x)，使表头指示到正中，这时电阻箱的数值即为中值电阻 $R_{中}$。

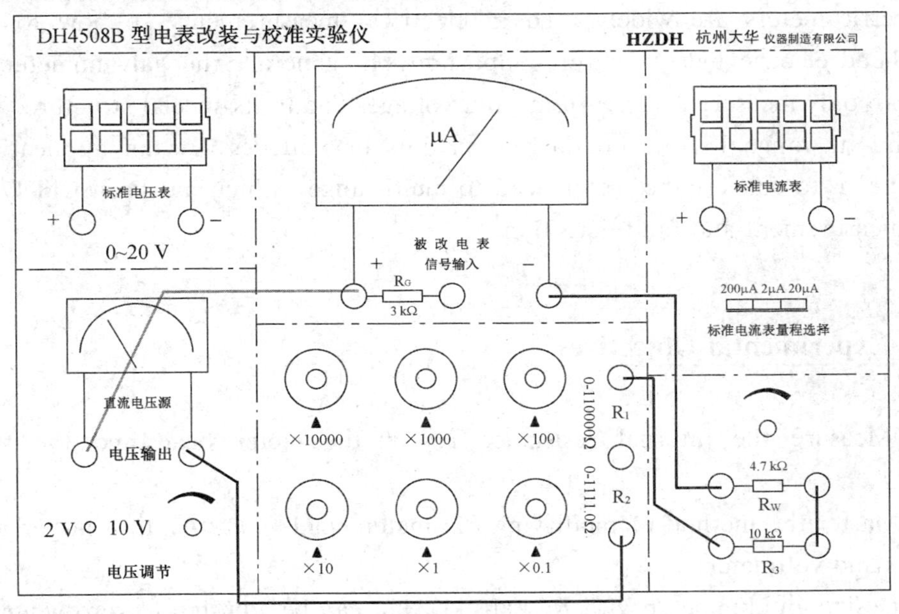

图 4-25-11 改装欧姆表(串联分压式)

(4) 取电阻箱的电阻为一组特定的数值 R_{xi}，读出相应的偏转格数，记录于表 4-25-3。利用所得读数 R_{xi}、div，绘制出改装欧姆表的标度盘。

表 4-25-3　改装欧姆表测量数据

R_{xi} (Ω)	$\frac{1}{5}R_{中}$	$\frac{1}{4}R_{中}$	$\frac{1}{3}R_{中}$	$\frac{1}{2}R_{中}$	$R_{中}$	$2R_{中}$	$3R_{中}$	$4R_{中}$	$5R_{中}$
偏转格数 (div)									

(5) 确定改装欧姆表的电源使用范围。短接 a、b 两测量端，将工作电源放在 0～2 V 一挡，调节 E=1 V 左右，先将 R_w 逆时针调到底，调节 E 直至表头满偏，记录 E_1 值；接着将 R_w 顺时针调到底，再调节 E 直至表头满偏，记录 E_2 值，E_1～ E_2 值就是欧姆表的电源使用范围。

思考题

1. 测量电流计内阻应注意什么？是否还有别的办法来测定电流计内阻？能否用欧姆定律来进行测定？能否用电桥来进行测定？

2. 若要求制作一个线性量程的欧姆表，有什么方法可以实现？

Experiment 4.25　Electric Meters' Modification and Calibration

Electric meters are widely used in electrical measurement, so how to use and comprehend electric meters is very important. In general, the galvanometer (meter head) can only measure small currents and voltages due to its special structure. One has to extend its range to measure larger currents or voltages. As an application, the multimeter is to convert the meter head to multi-range, which has been widely used in circuit measurement and fault detection.

Experimental Objectives

1. Measure the internal resistance R_g of the meter head and the full-scale current I_g.

2. Master the method of modifying the meter head (100 μA) to a larger range of ammeter and voltmeter.

3. Design an ohmmeter with $R_{中}$=10 kΩ that can be adjusted to zero when E is in

the range 1.35～1.6 V.

4. Calibrate the ohmmeter using the resistors, draw the calibration curve, and measure the unknown resistance with the assembled ohmmeter based on the calibration curve.

5. Learn how to calibrate ammeters and voltmeters.

Experimental Principle

Common magnetoelectric meters are mainly composed of a rotating coil wound by a fine enamelled wire in a permanent magnetic field, a spring for generating mechanical torque, an indicating pointer and a permanent magnet. When the current passes through the coil, the current-carrying coil generates a magnetic moment $M_{磁}$ magnetic in the magnetic field, which makes the coil rotate, thereby driving the pointer to deflect. The degree of the coil deflection angle is proportional to the magnitude of the current through, so the current value can be directly indicated by the deflection of the pointer.

1. The maximum current allowed through the meter is called the range of the meter, expressed by I_g. The coil of the meter has a certain internal resistance which is expressed by R_g. I_g and R_g are two important parameters that indicate the characteristic of the meter.

Common methods for measuring internal resistance R_g are:

(1) The half-current method, also called the median method

The measurement principle diagram is shown in Fig. 4-25-1. When the meter being measured is connected in the circuit, the current should be larger enough for the meter to reach its maximum range, and then use the decade resistance box which is connected with the meter in parallel as a shunt resistor. The value of the decade resistance box should be adjusted to change the shunt degree. When the meter pointer indicates the middle value, and the standard meter reading (total current strength) remains unchanged (which can be achieved by adjusting the power supply voltage and the variable resistor R_W), obviously then the shunt resistance value is equal to the internal resistance of the meter.

(2) Replacement method

The measurement principle diagram is shown in Fig. 4-25-2. When the meter being measured is connected to the circuit, one then replace it with a decade resistance box and change the resistance value while keeping the voltage and the current (standard meter reading) in the circuit unchanged, the resistance value of the resistance box is then the internal resistance of the measured meter.

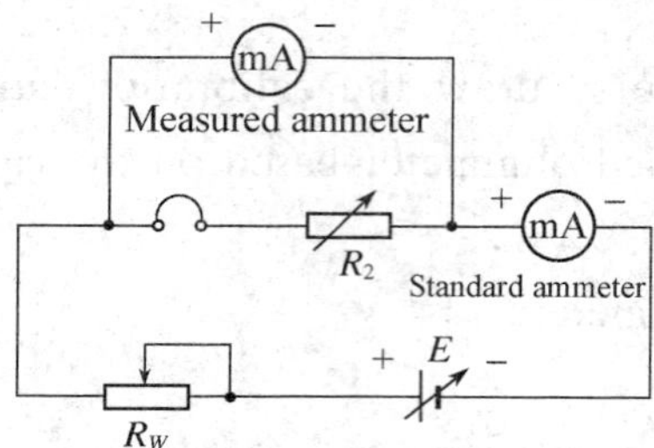

Fig. 4-25-1 The median method

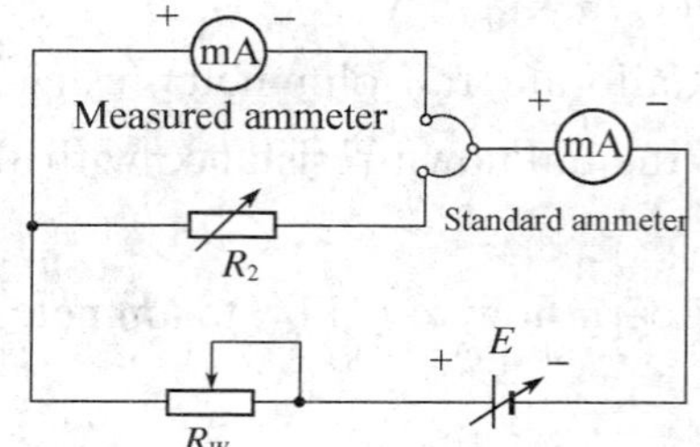

Fig. 4-25-2 Replacement method

Replacement method is a widely used measurement method with high accuracy.

2. Conversion of the meter head into a large range ammeter.

According to the law of parallel resistance, if a resistor R_2 with appropriate resistance value is connected in parallel with the meter head, as shown in Fig. 4-25-3, the part of the current beyond the range of the meter head will be shunt by the resistor R_2. The whole structure composed of the meter head and the parallel resistor R_2 (the part framed by the dotted line in the figure) is the modified ammeter. If one want to expand the range to n times, then

$$R_2 = R_g/(n-1) \tag{4-25-1}$$

When measuring the current with an ammeter, the ammeter should be in series in the circuit being measured, so it is required that the ammeter should have a small internal resistance. In addition, a multi-range ammeter can be made by parallel shunt resistors with different resistance values on the meter head.

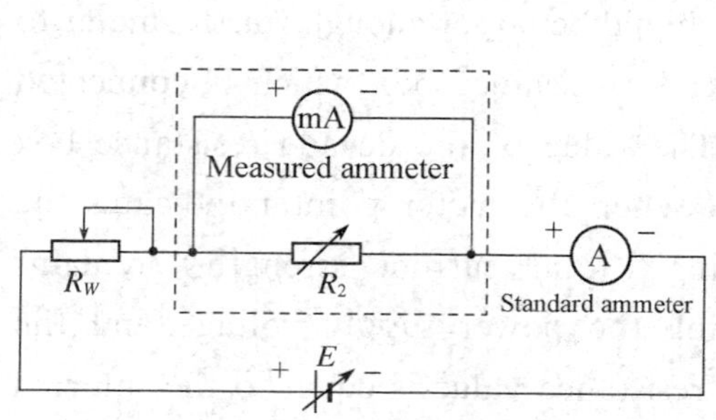

Fig. 4-25-3 Conversion of the meter head into a large range ammeter

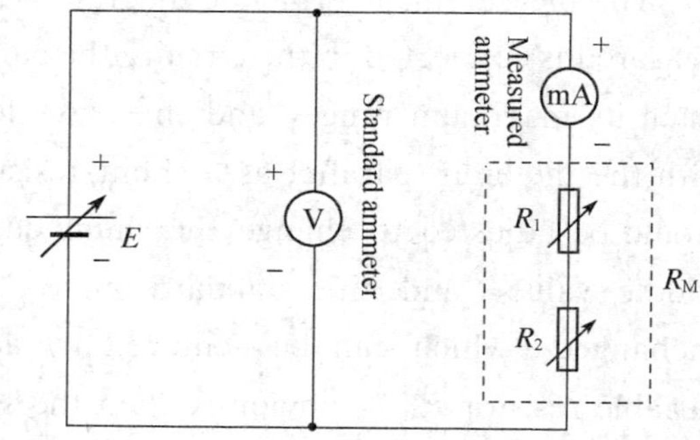

Fig. 4-25-4 Conversion of the meter head into a voltmeter

3. Conversion of the meter head into a voltmeter.

The common meter head can only withstand a very small voltage. In order to measure a large voltage, a resistor R_M with an appropriate resistance value can be placed in series with the meter head, as shown in Fig. 4-25-4, so that the part of the voltage beyond the range of the meter head falls on the resistor R_M. The whole structure composed of the meter head and the series resistor R_M is the modified voltmeter, and the series resistor R_M is called a range-extension resistor. By selecting R_M of different values, voltmeters of different ranges can be obtained. From Fig. 4-25-4, the resistance of R_M is:

$$RM=\frac{U}{I_g}-R_g \tag{4-25-2}$$

When the voltmeter is used to measure voltage, the voltmeter is always in parallel with the circuit to be measured. In order not to change the working state in the circuit due to the parallel voltmeter, it is required that the voltmeter should have a large internal resistance.

4. Conversion of the meter head into an ohmmeter.

A meter used to measure the value of the resistance is called an ohmmeter. According to the different zeroing methods, it can be divided into series voltage-division type and parallel current-division type. The principle circuit is shown in Fig. 4-25-5.

In Fig. 4-25-5, E is the power supply, R_3 is the current-limiting resistor, R_W is the zero-adjustment potentiometer, R_x is the resistor being measured, R_g is the equivalent internal resistance of the microammeter. In Fig. 4-25-5(b), R_G and R_W together form the shunt resistor.

Before using the ohmmeter, the "zero" point should be adjusted. To do so, one can directly short-circuit two points a and b (which is equivalent to $R_X=0$). Then by adjusting the resistance value of R_W the pointer of the meter is just deflected to full degree. It can be seen that the zero point of the ohmmeter is just at the full scale (that is, the limit) of the dial, which is the opposite of the zero point of the ammeter and the voltmeter.

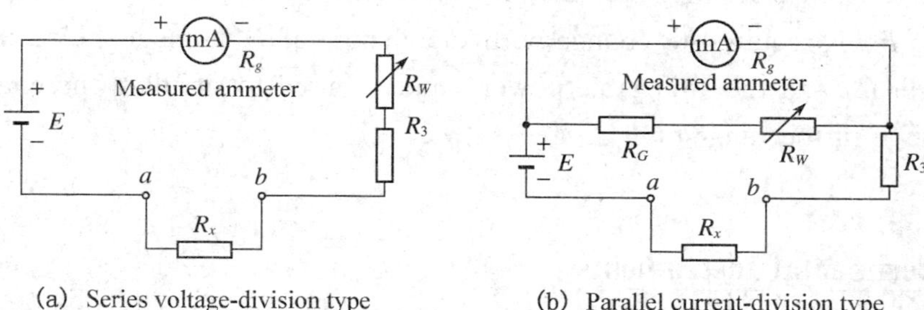

(a) Series voltage-division type　　(b) Parallel current-division type

Fig. 4-25-5　Schematic diagram of the ohmmeter

In Fig. 4-25-5(a), when the resistor R_x is connected to terminals a and b, the current through the circuit is

$$I=\frac{E}{R_g+R_W+R_3+R_X} \tag{4-25-3}$$

For a given meter head and a circuit, R_g, R_W and R_3 are constants. It can be seen that when the terminal voltage E of the power supply remains unchanged, the measured resistance and the current value have a one-to-one correspondence. That is, if different resistors are connected, the meter panel will have different deflection readings. The larger the R_X, the smaller the current I. Short-circuit a and b, that is, $R_X=0$

$$I=\frac{E}{R_g+R_W+R_3}=I_g \tag{4-25-4}$$

The needle of the meter panel now points to the full scale.

when $R_x=R_g+R_W+R_3$

$$I=\frac{E}{R_g+R_W+R_3+R_X}=\frac{1}{2}I_g \tag{4-25-5}$$

At this time, the pointer is in the middle position of the meter panel, and the corresponding resistance value is the median resistance. Obviously, $R_{中}=R_g+R_W+R_3$.

When $R_x=\infty$(i.e., a and b is open circuit), $I=0$, that is, the needle of the meter panel points to the mechanical zero position.

Therefore, the scale of the ohmmeter is the reverse scale, and the scale is not uniform. The greater the resistance R, the closer the scale interval. If the scale of the meter head is calibrated according to the known resistance value in advance, the resistance can be measured directly with an ammeter.

For parallel current-division type ohmmeters the zero-adjustment is implemented by shunting the meter head. The detailed parameters can be set by oneself.

The terminal voltage of the battery will change during the use of the ohmmeter, and the internal resistance R_g and the current-limiting resistance R_3 of the meter head are constant, so R_W is required to change with the change of E to meet the requirement of adjusting "zero", and the adjustable power supply is used to simulate the change of the battery voltage in the range of 1.3~1.6 V.

Experimental Instruments

DH4508B meter modification and calibration test instrument, specific connection wire lines.

Experimental Contents and Steps

1. Using the Median Method or Replacement Method to Measure the Internal Resistance of the Meter Head

See Fig. 4-25-6 for the wiring of the median method. First, one adjusts the voltage of the power supply E to 0 V and connects E, R_W, the meter to be modified and the standard ammeter into the circuit. Then one can adjust E and R_W so that the pointer of the modified meter head reaches its full deflection angle. The reading of the standard ammeter is just the full-scale current of the modified meter, marked as $I_g=$ ________ μA; Second, one then connects the resistance box R into the circuit (i. e., connecting the dashed line). Changing the magnitude of the resistance box R so the needle of the measured meter points from full-scale 100 μA to 50 μA. Meanwhile, one should note to adjust E and R_W so the reading of the standard ammeter remains unchanged. According to Eq. (4-25-5), $R_g=$ ________ Ω.

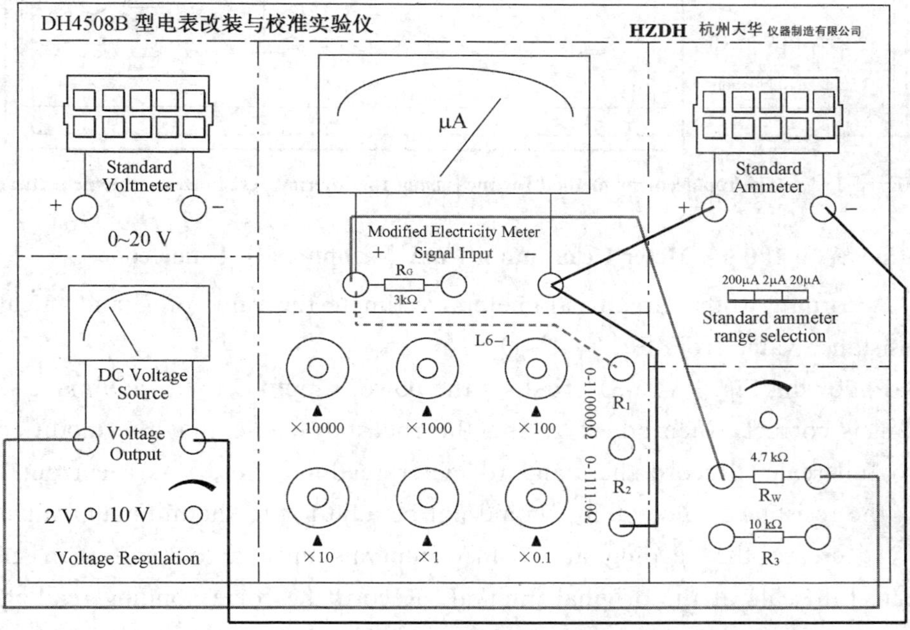

Fig. 4-25-6　The median method for measuring the internal resistance of the meter head

Fig. 4-25-7 shows the wiring of the replacement method. As a first step, connect L7-1 to the standard ammeter, then adjust the power supply E to be 0 V. After connecting E, R_W, the meter to be modified and the standard ammeter into the circuit one then can adjust E and R_W so that the pointer of the modified meter head reaches its full deflection angle. The reading of the standard ammeter is just the full-scale current of the modified meter, marked as $I_g =$ ________ μA. Then disconnecting L7-1 but connecting L7-2 to the standard ammeter, and adjusting R to keep the current of the standard ammeter at the value just recorded. At this time, the value of the resistance box is the internal resistance R_g of the meter head being measured, $R_g =$ ________ Ω.

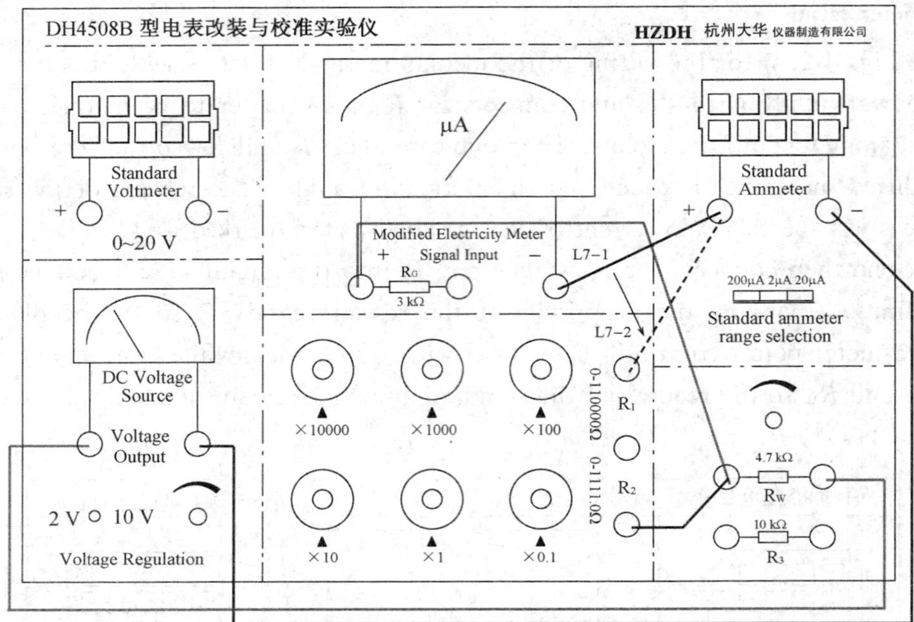

Fig. 4-25-7 The replacement method for measuring the internal resistance of the meter head

2. Convert a 100 μA Meter Head into a 1 mA (or optional) Ammeter

(1) According to the circuit parameters, estimate the value of E and calculate the shunt resistance value from Eq. (4-25-1).

(2) Following Fig. (4-25-8), first set the power supply E to 0V and make sure that the wiring is correct, then adjust E and the rheostat R_W so that the modified meter points to full scale. Record the standard meter reading. Note: As a current limiting resistor, the resistance value of R_W should not be adjusted to the minimum value. Then gradually decrease the reading at 0.2 mA intervals until zero, and then gradually increase to full scale at the original interval. Record the corresponding reading of the standard ammeter in Table 4-25-1 each time.

Table 4-25-1

Reading of modified meter(μA)	Reading of the standard ammeter(mA)			Error ΔI(mA)
	Decreasing	Increasing	Averaged	
20				
40				
60				
80				
100				

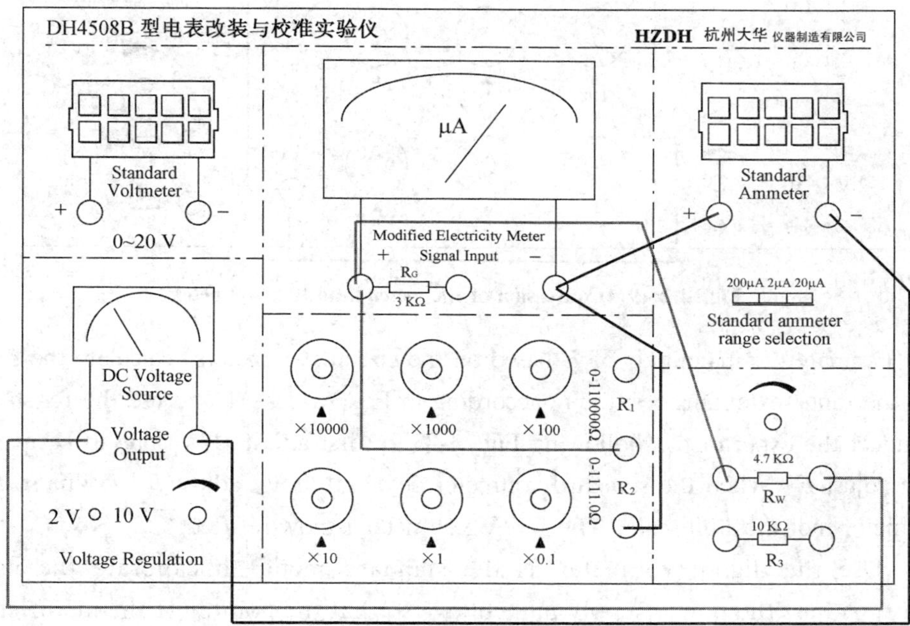

Fig. 4-25-8　Conversion of the meter head to an ammeter

(3) Take the reading of the modified meter as the abscissa, and the average value of the two readings when the standard ammeter is adjusted from large to small and from small to large as the ordinate, draw the correction curve of the ammeter on the coordinate paper, and determine the accuracy level of the modified meter according to the maximum error value of the two meters.

(4) Using 10 mA meter head and repeat the above steps, and choose 2 mA as intervals (optional).

(5) Connect R_G and the meter head in series as a new meter head, repeat the above steps, and compare the differences and similarities of the resistance R_2(optional).

3. Convert a 100 μA Meter to a 1.5 V (or optional) Voltmeter

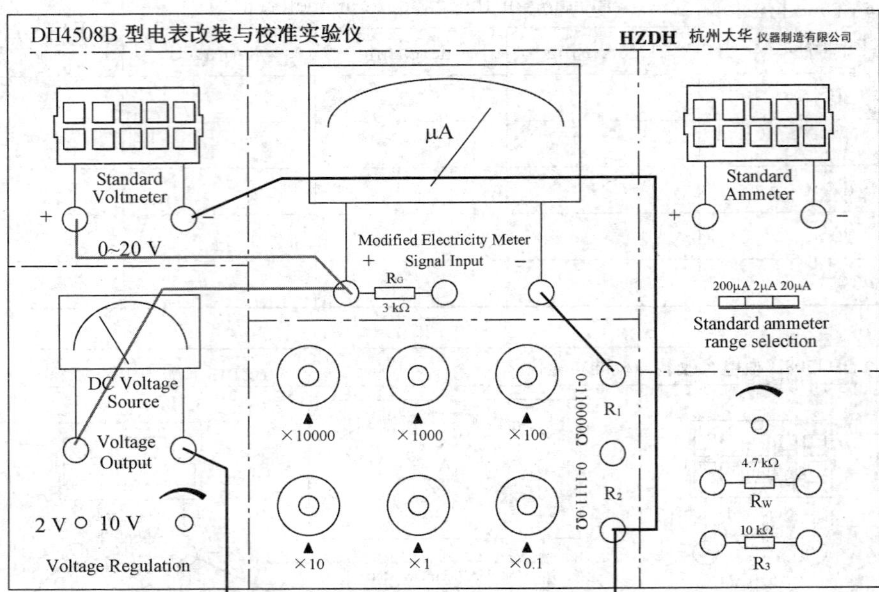

Fig. 4-25-9　Conversion of the meter head to a voltmeter

(1) Estimate the magnitude of E based on the circuit parameters, calculate the resistance value of the range-extending resistor R according to Eq. (4-25-2), and use the resistance box R to conduct the experiment. Following Fig. 4-25-9, first adjust R's value to be maximum, and then adjust E; When the standard voltmeter detects 1.5 V, adjust R's value so that the refitted meter indicates full scale. The 1.5 V voltmeter is now refitted.

(2) Use the digital voltmeter as the standard meter to calibrate the modified voltmeter. Adjust the power supply E's voltage so that the pointer of the modified meter points to the full range (1.5V), and record the reading of the standard voltmeter. Then gradually reduce the modified reading every 0.3V until zero, and then gradually increase it to full scale every 0.3V. Record the corresponding reading of the standard voltmeter in Table 4-25-2 each time:

Table 4-25-2

Reading of modified meter(V)	Reading of standard voltmeter(V)			Indication error U(V)
	Decreasing	Increasing	Average	
0.3				
0.6				
0.9				
1.2				
1.5				

(3) Take the reading of the refitted meter as the abscissa, and the average value of the two readings when the standard voltmeter is adjusted from large to small and from small to large as the ordinate, draw the correction curve of the voltmeter on the coordinate paper, and determine the accuracy level of the refitted meter according to the maximum error value of the two meters.

(4) Using 10 V meter head and repeat the above steps, and choose 2V as intervals (optional).

(5) Connect R_G and the meter head in series as a new meter head, repeat the above steps, and compare the differences and similarities of the resistance R (optional).

4. Conversion of the Meter Head to an Ohmmeter and Calibrating Surface Scales

(1) According to I_g, R_g, and E, set $R_W = 4.7\ \text{k}\Omega$, $R_3 = 10\ \text{k}\Omega$.

(2) Following Fig. 4-25-10, set the power supply $E = 1.5$ V, set the value of the resistance box to be zero (even $R_X = 0$), adjust R_W so the needle of the meter head points exactly to full scale. In this way, the zero adjustment of the ohmmeter is completed.

(3) Measure the median resistance of the modified ohmmeter. As in Fig. 4-25-10, adjust the resistance box R (i.e., R_X) so the needle points to the middle of the panel. At this point, the value of the resistance box is the median resistance, $R_{中} =$ ________ Ω.

(4) Take the resistance of the resistance box as a specific set of values R_{Xi}, and read the corresponding deflection grid number n_{div}. Using the obtained readings R_{Xi} and n_{div} to draw the scale of the modified ohmmeter.

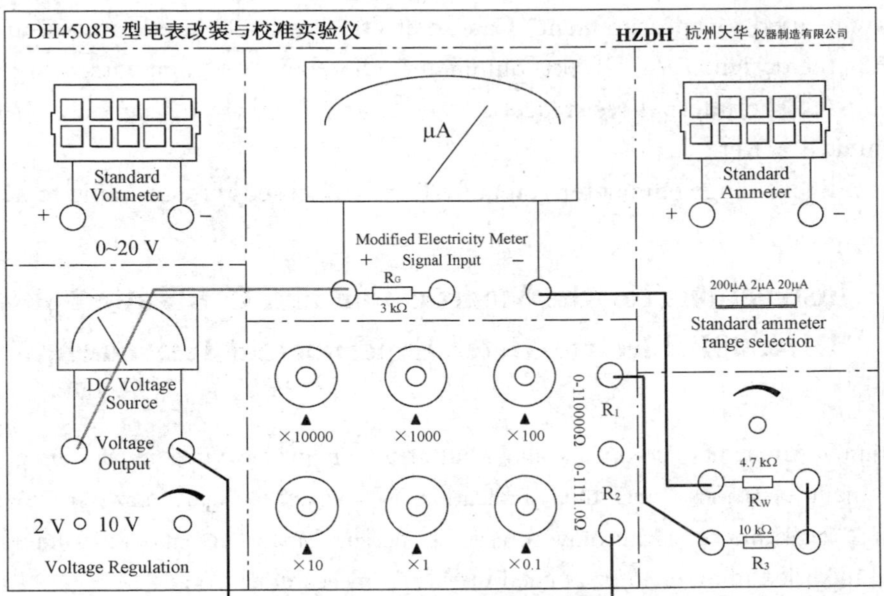

Fig. 4-25-10　Conversion of the meter head to an ohmmeter (Series voltage division type)

Table 4-25-3 $E=$______ V, $R_{中}=$______ Ω

R_{xi} (Ω)	$\frac{1}{5}R_{中}$	$\frac{1}{4}R_{中}$	$\frac{1}{3}R_{中}$	$\frac{1}{2}R_{中}$	$R_{中}$	$2R_{中}$	$3R_{中}$	$4R_{中}$	$5R_{中}$
Number of deflection grids(n_{div})									

(5) Determining the power usage range for modifying the ohmmeter. Short circuit two measuring terminals a and b, place the working power supply in the gear with 0～2 V, set $E=1$ V. First, turn R_W counterclockwise to the end, then adjust E until the meter head is on full scale, and record the value E_1; Next, adjust R_W clockwise to the end, and then adjust E until the meter head is on full scale. Record the value E_2. Then the range $E_1\sim E_2$ is the power usage range of the ohmmeter.

(6) Connect the wires according to Fig. 4-25-5 (b), design a parallel current division typer ohmmeter, then conduct wiring and measurement. Compare it with a series voltage division ohmmeter and see if there are any similarities or differences (optional).

Questions

1. What should be paid attention to when measuring the internal resistance of an ammeter? Is there any other way to measure the internal resistance of the ammeter? Can Ohm's law be used for measurement? Can an electric bridge be used for measurement?

2. For the designed $R_{中}=10$ kΩ ohmmeter, there are two ammeters with measuring range 100 μA. The internal resistance is 2 500 Ω and 1 000 Ω, respectively. Which one do you think is better?

3. If a linear range ohmmeter is required, what methods are available to achieve it?

Instructions for the Modification and Calibration of DH4508B Electric Meter Experimental Instrument

1. Introduction

Pointer Ammeter, Voltmeter and Multimeter are widely used in various electrical measurement occasions, and their indication is realized by galvanometer. Generally speaking, a galvanometer can only be used to measure small currents and voltages, so it must be modified in order to be applied in various measurement fields.

This instrument can complete the experiment of refitting Ammeter, Voltmeter and ohmmeter through wiring, and improve users' proficiency in using electricity meters through experiments.

2. Main Technical Parameters

(1) Pointer type meter to be modified: Range 100 μA, Internal resistance～2 kΩ, Accuracy class 1.5;

(2) Resistance box: Adjustment range 0～111 111.0 Ω, Accuracy class 0.1;

(3) Standard ammeter: Three ranges 200 μA, 2 mA, 20 mA, 4 ½ digital display, Accuracy ±0.1%;

(4) Standard voltmeter: 0～20 V, 4½ digital display, Accuracy ±0.1%;

(5) Adjustable voltage regulator source: Output ranges 0～2 V and 0～10 V, Stability 0.1%/min, Load Regulation 0.1%;

(6) Power supply: AC 220 V±10%, 50 Hz.

3. Instructions

The instrument is equipped with pointer galvanometer, standard Voltmeter and Ammeter, adjustable DC regulated power supply, decade resistance box, special wires and other parts. It can complete various galvanometer modification experiments without other accessories.

The panel of this instrument is shown in Fig. 1:

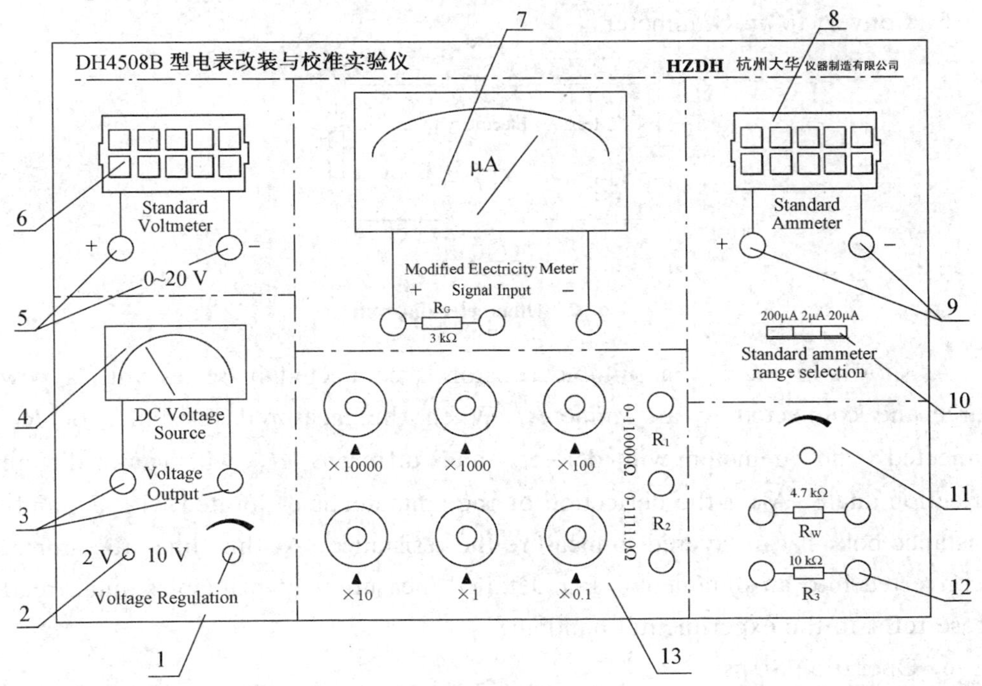

Fig. 1 Panel diagram

1-Regulated power supply potentiometer; 2-Switch of regulated power supply; 3-Output of regulated power supply; 4-Output indication of regulated power supply; 5-Standard voltmeter input terminal; 6-Standard voltmeter; 7-Pointer type ammeter; 8-Standard ammeter; 9-Standard ammeter input terminal; 10-Current range selection switch of standard ammeter; 11-R_W potentiometer (4.7 kΩ); 12. R_3 resistor; 13-Resistance box (R_1 and R_2).

The adjustable DC regulated power supply is divided into two ranges: 2 V and 10 V. Select the required voltage output through the "voltage selection switch", and adjust the voltage required by the "voltage regulation" potentiometer. The indication of pointer voltmeter is also divided into two ranges, 2 V and 10 V.

Standard digital ammeter has three ranges: 200 μA, 2 mA and 20 mA. Different ranges can be selected through the "current range selection switch".

4. Introduction to Principles

(1) Convert to a larger range ammeter

Connecting a suitable resistance in parallel at both ends of the meter head to shunt the current in the measuring circuit, so that when the meter head indicates full degree, the total current of the line is the current value of the range to be modified. At this time, the meter head is modified into an ammeter with a larger range.

(2) Convert to a larger range voltmeter

Connecting a suitable resistance value in series with an ammeter, so that when the ammeter indicates full degree, the voltage on the series circuit is equal to the voltage of the range to be modified, and then the ammeter is modified into a voltmeter.

5. Convert to an Ohmmeter

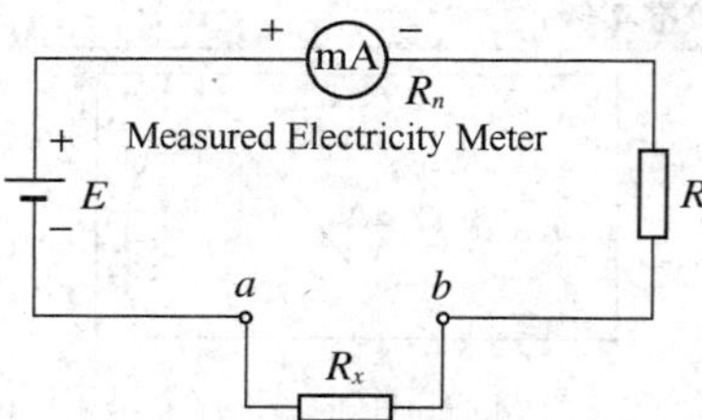

Fig. 2 Ohmmeter diagram

As shown in Fig. 2, a suitable resistor is connected in series with a power source and connected to an ammeter. When the measured resistance of R_X is connected, the ammeter will deflect, and different R_X will cause different deflection angle. After the deflection of the ammeter is calibrated with a standard resistance box, it can be used to measure the resistance. At this time, the ammeter is converted into an ohmmeter. For detailed measurement principles and circuits, please refer to the experimental handout.

6. Operation Steps

(1) Turn on the power switch at the back of the instrument and connect to AC power.

(2) Check whether standard voltmeter and standard ammeter display normally. It is normal for the standard voltmeter to skip words due to high internal resistance during no-load.

(3) Adjust the regulated power supply to output normally.

(4) According to the handout, refit the ammeter and measure the unknown current with the refitted ammeter.

(5) According to the handout, refit the voltmeter and measure the unknown voltage with the modified voltmeter.

(6) Modify series and parallel type ohmmeters according to the handout, and use the modified ohmmeter to measure unknown resistance.

7. Maintenance and Warranty

(1) The instrument should be used correctly according to the experimental requirements.

(2) After use, the power switch should be turned off. If not in use for a long time, the power plug should be unplugged.

(3) The instrument should be stored in an environment free of corrosive substance and kept dry to prevent corrosion.

(4) Under the prescribed usage conditions, the product warranty period is twelve months. If the warranty period is exceeded, the manufacturer will still provide good service.

附　录

我国法定计量单位

附表 1　国际单位制的基本单位

量的名称	单位名称	单位符号
长度	米	m
质量	千克	kg
时间	秒	s
电流	安[培]	A
热力学温度	开[尔文]	K
物质的量	摩[尔]	mol
发光强度	坎[德拉]	cd

附表 2　国际单位制的辅助单位

量的名称	单位名称	单位符号
平面角	弧度	rad
单体角	球面度	sr

附表 3　国际单位制中具有专门名称的导出单位

量的名称	单位名称	单位符号	其他表示实例
频率	赫[兹]	Hz	s^{-1}
力	牛[顿]	N	$kg \cdot m \cdot s^{-2}$
压强;应力	帕[斯卡]	Pa	$N \cdot m^{-2}$
能量;功;热量	焦[尔]	J	$N \cdot m$
功率;辐射能量	瓦[特]	W	$J \cdot s^{-1}$
电荷量	库[仑]	C	$A \cdot s$
电位;电压;电动势	伏[特]	V	$W \cdot A^{-1}$
电容	法[拉]	F	$C \cdot V^{-1}$
电阻	欧[姆]	Ω	$V \cdot A^{-1}$

（续表）

量的名称	单位名称	单位符号	其他表示实例
电导	西[门子]	S	$A \cdot V^{-1}$
磁通量	韦[伯]	Wb	Vs
磁通量密度;磁感应强度	特[斯拉]	T	$Wb \cdot m^{-2}$
电感	亨[利]	H	$Wb \cdot A^{-1}$
摄氏温度	摄氏度	℃	
光通量	流[明]	lm	cd · sr
光照度	勒[克斯]	lx	$lm \cdot m^{-2}$
放射性活度	贝可[勒尔]	Bq	s^{-1}
吸收剂量	戈[瑞]	Gy	$J \cdot kg^{-1}$
剂量当量	希[沃特]	Sv	$J \cdot kg^{-1}$

附表 4 国家选定的非国际单位制单位

量的名称	单位名称	单位符号	换算关系和说明
时间	分 [小]时 天(日)	min h d	1 min=60 s 1 h=60 min=3600 s 1 d=24 h=86 400 s
平面角	[角]秒 [角]分 度	″ ′ °	1″=π/648 000 rad 1′=60″=π/10 800 rad 1°=60′=π/180 rad
旋转速度	转每分	$r \cdot s^{-1}$	$1\ r \cdot s^{-1}=1/60\ s^{-1}$
长度	海里	n mile	1 n mile=1852 m
速度	节	kn	1 kn=1 n mile/h
质量	吨 原子质量单位	t u	$1\ t=10^3\ kg$ $1\ u \approx 1.6605655 \times 10^{-27}\ kg$
体积	升	L;l	$1\ L=1\ dm^3=10^{-3}\ m^3$
能	电子伏	eV	$1\ eV \approx 1.6021892 \times 10^{-19}\ J$
级差	分贝	dB	
线密度	特[克斯]	tex	$1\ tex=1\ g \cdot km^{-1}$

附表 5　用于构成十进制倍数和分数单位的词头

所表示的因数	英文	词头名称	词头符号
10^{24}	yotta	尧[它]	Y
10^{21}	zetta	泽[它]	Z
10^{18}	exa	艾[可萨]	E
10^{15}	peta	拍[它]	P
10^{12}	tera	太[拉]	T
10^{9}	giga	吉[咖]	G
10^{6}	mega	兆	M
10^{3}	kilo	千	k
10^{2}	hecto	百	h
10^{1}	deca	十	da
10^{-1}	deci	分	d
10^{-2}	centi	厘	c
10^{-3}	milli	毫	m
10^{-6}	micro	微	μ
10^{-9}	nano	纳[诺]	n
10^{-12}	pico	皮[可]	p
10^{-15}	femto	飞[姆托]	f
10^{-18}	atto	阿[托]	a
10^{-21}	zepto	仄[普托]	z
10^{-24}	yocto	幺[科托]	y

附表 6　声波在液体中的传播速度

液体	温度 t_0(℃)	速度 v_0(m/s)	温度系数 α(m/(s·K))
苯胺	20	1656	−4.6
丙酮	20	1192	−5.5
苯	20	1326	−5.2
海水	17	1510～1550	/
普通水	25	1497	−2.5
甘油	20	1923	−1.8
煤油	34	1295	/
甲醇	20	1123	−3.3
乙醇	20	1180	−3.6

表中 α 为温度系数，其他温度时的声速可按近似公式 $v(t)=v_0+\alpha(t-t_0)$。

附表 7　某些固体的线胀系数

物质	温度或温度范围/℃	$\alpha/(\times10^{-6}℃^{-1})$
铝	0～100	23.8
铜	0～100	17.1
铁	0～100	12.2
金	0～100	14.3
银	0～100	19.6
钢(0.05%碳)	0～100	12.0
康铜	0～100	15.2
铅	0～100	29.2
锌	0～100	32
铂	0～100	9.1
钨	0～100	4.5
石英玻璃	20～200	0.56
窗玻璃	20～200	9.5
花岗石	20	6～9
瓷器	20～700	3.4～4.1

附表 8　20℃时某些金属的杨氏弹性模量

金属	杨氏模量 E/GPa	金属	杨氏模量 E/GPa
铝	69～70	锌	78
钨	407	镍	203
铁	186～206	铬	235～245
铜	10.127	合金钢	206～216
金	77	碳钢	196～206
银	69～80	康铜	160